SOUND REPORTING

Sound Reporting

THE NPR GUIDE TO BROADCAST, PODCAST AND DIGITAL JOURNALISM

Second Edition

JEROME SOCOLOVSKY

With a Foreword by Michel Martin

THE UNIVERSITY OF CHICAGO PRESS | CHICAGO AND LONDON

The University of Chicago Press, Chicago 60637
The University of Chicago Press, Ltd., London

Published 2024
Printed and bound by CPI Group (UK) Ltd, Croydon, CR0 4YY

33 32 31 30 29 28 27 26 25 24   1 2 3 4 5

ISBN-13: 978-0-226-82466-6 (paper)
ISBN-13: 978-0-226-82467-3 (e-book)
DOI: https://doi.org/10.7208/chicago/9780226824673.001.0001

Library of Congress Control Number: 2024018192

♾ This paper meets the requirements of ANSI/NISO Z39.48-1992 (Permanence of Paper).

To David Gilkey and Zabihullah Tamanna

CONTENTS

FOREWORD

Why do people go into journalism? Some people just like to tell stories. Some people are outraged by unfairness, and they think journalism is a good way to fight it. Some people are nosy and like spilling the tea, or they like attention—and some of us are actually really shy and having a microphone or notebook in hand is one way we have found to be brave.

I have met—or maybe have been—all of these characters in the years I have been in the field. Maybe you can say the same.

I started as a journalist right after college, in one of the classic ways—an internship—and when I say right out of college I mean it. I graduated on June 5, took my driver's license exam on June 6 (long story, don't ask), drove to Washington, D.C., on June 8 (thankfully not by myself) and started my internship on June 10. Since then I have done some of just about everything: local and national news, commercial and public television, breaking news, special coverage, documentary-length features and of course news radio—or audio, as we now call it.

I am telling you all this because there is one through line connecting all these different realms and that is journalism—or to put it another way, work rooted in fact, not fiction. The facts come first: before the elegant turn of phrase, the dramatic picture or the sound bite or audio clip, to say nothing of political spin or clickbait. Those things can be useful, but they need to be used in the service of the facts, not the other way around. Fiction is great, and art can often do things that journalism and other nonfiction forms cannot do—but that isn't what *we* do. All journalists are storytellers, but not all storytellers are journalists.

Which brings me to NPR. This once scrappy little start-up has become a legacy organization with national and international reach. It's grown from one show to many and from one distribution model—produced shows delivered at specific times that you heard on your radio—to shows you can hear whenever you want, however you want: on your phone, your laptop, your smart speaker (and by the time you read this maybe a chip in your head?). Today NPR is also stories you read, digital newsletters and long-form podcasts on all kinds of subjects, from gritty international reports to soul-baring personal narratives to life advice. It's also musical experiences, live events, video shorts, animation and stuff that's in the works that I don't even know how to name because they haven't told me what it is yet.

Common to all of these forms, though, is one principle: try to get the facts as best we know them, and tell them as best we can in the service of truth and our audience. We strive for style, elegance, completeness, fairness, boldness and yeah, sure, attention—because what's the point of doing this if nobody is listening or reading? But, still, the facts come first.

Sometimes we fall short of these big goals. Like all human endeavors, this place is flawed. Like many legacy organizations it can be stodgy, self-righteous, timid and too easily pleased with itself. Like many institutions with roots in the academic world—many of NPR's hundreds of stations are chartered by universities—we sometimes prioritize being "right" in the eyes of the few at the expense of being understood by the many. Conversely, like other groups that are trying to stay current in a world that's changing all the time, we are sometimes too quick to grab on to trendy words and opinions that a lot of people in our audience don't understand, and we don't do enough to help them.

But we still try to get it "right," and the purpose of this latest edition of *Sound Reporting* by NPR reporter, editor and trainer Jerome Socolovsky is to help you—us—get it "right," or as close to "right" as we can, as often as we can. If you take it to heart, you're going to learn something from some of the finest journalists and people you're ever going to meet, my NPR colleagues, who've done a little bit of everything you might ever want to do.

Just take my *Morning Edition* colleagues: Steve Inskeep, who, along with hosting a top-rated news show for two decades, has also managed to turn out a handful of books on everything from the growth of Karachi to how Abraham Lincoln dealt with his critics and adversaries; Leila Fadel, who, despite covering some of the toughest, thorniest and deadliest international conflicts, can always find a human story that will leave you in tears; and A Martínez, who, after a successful career as a sports news talk show host, now covers everything from the ongoing war in Ukraine to where and why men can wear shorts to work.

Here are some questions you might find answers to in this book:

- Most of us know how to talk to people we know and like, but what is it like to interview people you don't know, or don't like very much, or who are going through the worst day of their lives?
- Most of us get up in the morning wanting to do the "right" thing, but what if it isn't always clear what the right thing *is*, or who is telling "the truth"?
- Most of us have some areas of expertise we feel fairly confident about, but what if we're asked to report on something we don't already know? How do we learn, develop sources, expand our knowledge?

You might be getting the idea by now that this book won't just help you be good, or better, at your job. There are clues here about ways to be better at life.

Let's get to work.

MICHEL MARTIN, HOST, *MORNING EDITION*

PREFACE

I sometimes half-joke that I wanted to write this book because my copy of the previous edition was all worn out. Indeed, like many NPR journalists, I read it cover to cover and then kept using it as a reference for many years.

So that part of it was true. But the real reason I wrote this new edition is simple: A lot has happened since longtime NPR senior editor and trainer Jonathan Kern wrote the last one and published it in 2008.

For one, we are bigger. In 2006, NPR produced 10 unique radio programs with a staff of around 700 people. In 2023, we produced more than 30 shows and podcasts, not including special coverage, app-based radio and livestreams, with a staff of around 1,200 people.* In addition to podcasting and streaming audio, we have branched out into other platforms, with a leading news website, long- and short-form videos and a visual newscast as well.

During that time, the media economy has changed. The virtual collapse of the paid newspaper model with the rise of the internet and social media, the closure of thousands of local newspapers and the scaling back of many news organizations' reporting networks, especially overseas, have elevated the role NPR plays in informing the public, together with the more than 1,000 local stations that broadcast a mix of our shows and digital news stories in addition to their own programming.

But one thing hasn't changed. That is the set of core journalistic values that have made NPR a trusted source of news since the inaugural broadcast of *All Things Considered* on May 3, 1971, with reporting on a large protest held that day in Washington, D.C., against the Vietnam War. We

---

* The 2006 figure doesn't include arts programs or podcasts, which mostly consisted of recycled material at the time. In 2023, NPR cut 10% of its staff and canceled four podcasts amid an uncertain economy.

always aim to do our reporting accurately, fairly, thoroughly and with care and respect for both the people we interview and the audiences we serve.

Even so, many of the voices quoted in the previous edition are no longer at NPR. There is a whole new cohort of journalists doing the work. This book is based on more than 80 interviews with producers, reporters, editors, hosts and others working on desks, shows, podcasts and our visuals and digital teams. It was truly inspiring to listen to them talk with passion and fervor about what they do. Whether they were old-timers or some of the more recent hires, they all believed there's something magical and powerful that takes place when we turn on the microphone.

When I first read the previous edition of *Sound Reporting*, I had already been an NPR reporter for six years. I'd learned how to do it partly by the seat of my pants and partly from the advice I received from more experienced editors and colleagues. But when I read the book, I *understood* many of the principles behind what I was already doing. That made me a much better audio journalist.

I wrote this book for anyone interested in how we do audio journalism at NPR, as well as anyone who merely wants a peek behind the curtains. Whether you're a journalism student, already a journalist in public media or elsewhere, or have no plans to become a journalist, this book will give you insight into not only how to report the news but also how our journalism has evolved to include multiple platforms besides audio.

Parts of this book read like a journalistic account of what goes on at NPR, straight from the mouths of the folks in the trenches. Other parts are more like a training manual, with advice on journalistic practices that include interviewing, sourcing, recording and mixing audio and much, much more. I wrote it that way because I believe all of it will be of interest, whether you're reading this as an aspiring or early-career journalist or just out of curiosity. I believe that the best way to become a good journalist is to learn from those who do it well, studying the skills they've developed and the editorial judgments they make plying their craft.

The book opens with a chapter on NPR's values and then roughly follows the journey a story would, from the conception of an idea to the moment the story reaches the audience. I go over how we pitch stories; report, produce, write and edit them; and voice them and tailor them to multiple media formats. The book ends with a look at how shows and podcasts are put together.

You can find audio of the stories discussed in these chapters at npr.org/soundreporting and at the locations cited in the endnotes. A glossary is also included in case you run across terms you're unfamiliar with.

As you read, keep in mind that NPR is but one part of a public media universe that includes member stations and other outlets like PRX and American Public Media. There are now quite a few podcasts that emulate the kind of storytelling we pioneered. Many talented and creative people work at those places, some of which also offer resources for anyone interested in learning how to work in audio storytelling.

What I present is the expertise of people at the editorial heart of the premier public media organization in the U.S. covering daily news 24/7, whose listeners and readers make up some of the biggest audiences in broadcasting. My aim is to put you in a room full of top-notch producers, seasoned correspondents, trusted hosts, rigorous editors and others at NPR, telling you riveting and inspiring stories that give you an inside look at how they do their craft. If you want to learn from their experience, you will—and even if you don't, I think you'll find the book a fascinating read.

CHAPTER 1

What We Stand For

One of the odder moments in NPR history was surely *Fresh Air* host Terry Gross' interview with Kiss bass player Gene Simmons in 2002.[1] The conversation began as might be expected, with Terry introducing the guest and playing a clip of the band's 1975 hit, "Rock and Roll All Nite." But then the conversation went off the rails, as the musician launched into a series of misogynistic rants, mocked NPR as out of touch and bragged about his sexual conquests and wealth.

> TERRY: Well, let's cut to the chase. How much money do you have?
> SIMMONS: Gee, a lot more than NPR.
> TERRY: I know. You're very defensive on money, aren't you?
> SIMMONS: No, I'm not. I'm just trying to show you there's a big world out there, and reading books is wonderful, I've certainly read—well, perhaps as many as you have—but there's a delusional kind of notion that runs rampant in . . .

A few years later, Terry lamented that they "sounded like two first-graders calling each other names"; that was self-deprecating, to say the least, since it was really the musician known for his studded codpiece and fanged platform shoes who dragged the discussion, if you can call it that, into the gutter.

When the interview was first broadcast in February 2002, it made a splash. Terry said it got "more attention than anything we have ever broadcast on the show." Listener emails poured in, *Entertainment Weekly* dubbed Simmons "Crackpot of the Year" and *Rolling Stone* called him one of the biggest disappointments of 2002. Two years later, Terry mentioned it in her book, *All I Did Was Ask*, which included a transcript of the interview, saying, "I guess this proves that controversy sells, and so does a good fight (or even a silly fight)."

Terry is a skillful host, and it's likely the conversation was doomed no matter what she said or did. But whatever you think of the interview and the decision to air it—and re-air it during "Encore Week"—what many at the time couldn't have known was how it foreshadowed a more calculated campaign the following decade to discredit the media.

When the interview aired, vulgar men in public life were nothing new. Neither was making political hay out of bashing journalists or questioning NPR and other major news outlets' legitimacy as purveyors of fact. And this was just one interview, long before Donald Trump's first campaign for president. But by 2016, there were plenty of people willing to believe his characterization of mainstream media as the enemy. And this characterization took root just as the ideologically driven news coverage by some outlets was gaining audiences and as the business model of the free press crumbled: Whereas readers had once been willing to pay for news in print, they were now embracing the credo of the internet age—that information "wants to be free." Between 2005 and 2023, around 2,900 newspapers, or nearly a third of the country's total, stopped publishing.[2]

NPR weathered that new ecosystem relatively well. But the anti-media sentiment didn't make our jobs any easier. NPR reporters faced hostility and shouts of "fake news" at Republican voter rallies. And some progressives were unhappy with what they saw as "both-sidesism" in some of NPR's coverage of the Trump administration.[3] Still, most of the time we were tough, but fair, in our coverage. *Morning Edition* host Steve Inskeep challenged the then former president on his falsehoods surrounding the 2020 election in a Jan. 12, 2022, interview. Our listenership surged during the Trump presidency, as did NPR's website readership. It wasn't the end of the story, of course. Since then, there's been volatility in both audience size and funding streams. But what the tea leaves already made clear in 2002 was the centrality of the struggle for credibility for decades to follow.

In our quest to win and maintain the trust of our audience, NPR upholds a series of core values that underlie the processes and techniques that are discussed in the rest of the book. In this chapter, I focus on those values.

Respect

Tony Cavin spent 20 years at CBS News before NPR hired him as its managing editor for standards and practices in 2021. "As someone coming from commercial broadcasting, I really value the respect we have for our listeners," he says. NPR is selective about calling something "breaking news," and doesn't titillate with voyeuristic crime stories. "We're not covering the trial of that woman in Idaho who killed her children and went to Hawaii with her husband or whatever. We're trying to maintain a level of respect of ourselves and of our audience and of each other."

NPR's journalism is not motivated by the bottom line, because we are not for profit. "We don't have a CEO that has to walk into a quarterly shareholders meeting," says Eric Marrapodi, NPR's vice president for news programming, who previously worked at CNN. "It gives us more runway. And it allows us to take chances on journalism projects that other companies might not because they [those projects] might not have profit at the center."

Many of our member stations across the country periodically suspend regular programming to ask for donations.* Those fundraising drives help the stations pay the fees that are one pillar of NPR's finances. The other two are corporate sponsorship and philanthropic gifts. Keeping those streams flowing is not easy, but at least the risk is diversified. "If one goes down," Eric says, "you still have the other to keep being supported."

Still, audience size matters. We monitor listener numbers and website analytics closely to see what kinds of news stories and programming draw interest. But our efforts to grow audience don't happen at any cost. That basic principle hasn't changed since Terry observed that if she had done the Simmons interview on commercial radio, "the program director would have been in my office the next day encouraging me to fight with my guests all the time to boost our ratings. . . . Thank goodness I'm not in that position."

Credibility

Trust in the news media has eroded, but not only because of Trump's "fake news" charges. Paradoxically, technology is also to blame. Before

---

* The NPR Network is made up of around 250 organizations, including universities, school districts and other state and local authorities and community institutions. Together, they operate more than 1,050 public radio stations in all 50 states, as well as the District of Columbia, Guam and the U.S. Virgin Islands. Some are both radio and TV broadcasters. Many have digital platforms that extend well beyond their broadcast signal range.

the development of laptop-based digital recording software, the only place you could get news and talk shows in high fidelity was on your FM public radio station. "Our technological superiority made us sound more professional than people who weren't serious," says Tony. "If you heard *All Things Considered* or *Morning Edition*, it sounded really good. If you and I sat down with a cassette recorder and recorded a roundtable discussion about why we don't like this, that or the other, it sounded like two guys who were off-mic talking on a cassette with a lot of hiss. Now you and I can sit down with a laptop and do something that will sound close to professional, if not professional."

The upshot is that we've lost that "crutch," as Tony calls it, that gave us an advantage over opinionated voices yelling on the internet, so we must work even harder to win and keep the audience's trust. When he briefs new employees, Tony starts by telling them to imagine that some good friends failed to show up at a party they'd thrown. "And they tell you it was because they were kidnapped by aliens. You would believe them because they have always told you the truth." Meanwhile, friends who have not been straight with you in the past show up late and say it's because of a flat tire. "You wouldn't believe them."

Tony is tasked with upholding the guiding principles in the *NPR Ethics Handbook*.[4] There are 10 of them: accuracy, fairness, completeness, honesty, independence, impartiality, transparency, accountability, excellence and respect. What they boil down to, he says, is winning listeners' trust. "Our most important currency right now is getting people to believe us. And how we do that is by making clear that we will tell them the truth."

Accuracy

Truth begins with accuracy, the first guiding principle. The handbook says it is "at the core of what we do."

> We do our best to ensure that everything we report faithfully depicts reality—from the tiniest detail to the big-picture context that helps put the news into perspective. Facts are incredibly slippery. Studies of press accuracy routinely find mistakes—sometimes many of them—in news media reports. This means that when journalists—even the best ones—think they're getting it right, they're all too often wrong. Errors are inevitable. But our

> best defense against them is constant vigilance. This is why we systematically and rigorously review our facts before we make our reporting public.

Because errors are inevitable, journalists need strategies for catching them. Later in this book I will offer some suggestions to help you identify story elements that are most susceptible to error.

When we do make mistakes, we need to own up. "It helps our credibility more to honestly admit the mistake than to simply not include a correction," says Tony. We don't just delete an error and replace it with the corrected version. We "report" on it by posting a correction explaining the error and setting the record straight. And we don't hide corrections. We put them on all news stories that had errors and, when necessary, at the top of audio transcripts because we've noticed that people don't necessarily read to the end. There's even a dedicated webpage (www.npr.org/corrections) where our mistakes are on display for all to see.

Haste is never an excuse for flawed reporting. Notice that speed is not one of the principles in the handbook. Accountability is. We don't dillydally in our work; we put in the time and effort it takes to confirm a story, even if other news outlets are reporting it.

We learned that the hard way. In 2011, NPR reported, based on a hospital source who wasn't actually involved in the emergency response, that Arizona Rep. Gabrielle Giffords had been shot dead, when in fact she had been critically injured.[5] Other major news outlets also reported the death erroneously, quoting NPR as the source. NPR's executive editor at the time, Dick Meyer, apologized on air the next day.

There aren't a lot of cases where people remember which news organization reported a story first. "But there are cases," Tony says, where you can say, "'Boy, the *Chicago Tribune* really screwed up the 1948 election.'" He's referring to the premature headline on the front page of that paper's predecessor, the *Chicago Daily Tribune*, which famously declared "Dewey Defeats Truman." More than seven decades later, the *Tribune* was still admitting that it had made "the most famous wrong call in electoral history."[6]

NPR's audio corrections policies have changed over time. We used to broadcast corrections on the air, but stopped, reasoning that airtime was valuable and limited, and once a show was broadcast, it disappeared into the ether. But with the rise of the internet, every radio segment we broadcast gained eternal life on the web. Now, when necessary, we put corrections on radio transcript pages, and in the event of major errors we broadcast a correction on air.

Transparency

Credibility also requires transparency, which is at the heart of our decision to grant a source anonymity. It's often a judgment call, Tony says, "because the audience has every right to know who this person is. But we also have an obligation to protect this person who is sharing their story with us."

A source may ask to remain anonymous because they could lose their job or put themselves or their family members in danger if their identity becomes known. Or they may live in a country where freedom of expression is throttled. And there are grades of anonymity. It may be safe, and at the same time strengthen your story, to give a first name, a profession, or describe them in some other way that doesn't give away their identity. "How much can I tell the audience to let them know[that] I didn't just sit here, making up a bunch of quotes? And how much do I have to do so that this person, who is nice enough to share their feelings and their experiences with us and with our audience, doesn't then get penalized for that honesty?"

Impartiality

The handbook also deals with conflicts of interest that may require disclosure, or outright avoidance, as well as potential red flags such as "financial holdings, romantic relationships, family ties, book deals, speaking engagements and others." There's too much in the handbook to cover here, but it's accessible at NPR.org.

One recent change is noteworthy, though. NPR journalists used to be banned from participating in "marches, rallies and public events" and also from weighing in publicly on "controversial" or "polarizing" issues. We revised that policy in 2020 on the heels of the protests that erupted after George Floyd was murdered, and following discussions about the difference between taking sides in a political debate and taking a stand for human rights and against police abuses. The impartiality section now reads this way:

> NPR editorial staff may express support for democratic, civic values that are core to NPR's work, such as, but not limited to: the freedom and

> dignity of human beings, the rights of a free and independent press, the right to thrive in society without facing discrimination on the basis of race, ethnicity, gender, sexual identity, disability, or religion.

Advocating for political issues is still banned. "Don't express personal views on a political issue that you could not write for the air or post on NPR.org," the handbook says, cautioning that "anything you post online reflects both on you and on NPR."

But impartiality does not mean that journalists cannot tap their own experiences to help inform what they cover, nor should those who do that be dismissed as biased. "Those outside the predominant experiences of thought in the newsroom, including people who are facing significant oppression" in society because of their identity, ought to feel free to give input that influences coverage discussions, says Keith Woods, NPR's chief diversity officer. It shouldn't fall on them to carry the responsibility for unearthing or telling these stories; they should just be able to put the ideas in play for the newsroom. "You don't have to go on the air and make personal statements, but you've got to be able to express it in a way that helps everybody else figure out how to make journalism out of it."

Journalists he speaks to typically say that the biggest challenge in their work is being their "authentic self," Keith adds. "They say they feel compelled to check at the door parts of their identity because to talk about them would mean risking being labeled biased. That's especially true when your identity—whether it's race and ethnicity, gender, faith, marital status, etc.—means you are in a significant minority. Then you can't feel comfortable being your full self." Sometimes it's the oft-unspoken requirement of objectivity or neutrality for a journalist that makes them suppress that part of themselves. But newsrooms both in public and commercial media are moving away from that thinking.

"Nobody's telling you that you can't bring your whole self to journalism," says Tony. "But we are saying to you, we're giving you access to this incredible platform. You reach millions of people a week. And in return, you have to give up some of your access to your own personal platform, which probably reaches far fewer people."

NPR's main concern about staff airing their personal views outside the confines of the newsroom is that it could give the appearance of bias. That could make it harder for them and their colleagues to do their work, and get sources who may disagree with the expressed point of view to speak to them. That's why NPR journalists should keep their personal opinions off their social media feeds, and out of their journalism. But does that mean completely suppressing their opinions? I asked international correspondent Daniel Estrin, who has covered stories in the Middle East, Ukraine and elsewhere how he reconciles his own points of view with what he reports. "I often have my own view on a story before I go out to report it. So I'll interview someone who reflects that view. And I'll interview someone with a different point of view. Then I'll ask myself: What surprised me? What challenged my assumptions? It leads to a more complete story that's driven by curiosity and discovery."

Diversity

At NPR, the discussion about staff bringing their lived experiences to their journalism is connected to the network's diversity efforts. Since its founding, NPR has worked to include the voices of people from historically marginalized groups on the air. Back in its early years, however, in many ways it didn't fulfill one of the goals expressed in the newsroom's mission statement: to "look and sound like America." In 1976, five years after its founding, NPR established the Department of Specialized Audience Programming (DSAP) amid concern that its "foundational vision had failed to manifest." Frank Tavares, a longtime NPR employee and advocate for diverse public radio content, and also the "voice" of NPR sponsorships for many years, was a DSAP producer and later director. DSAP hosted workshops and trainings for public radio stations, as well as for journalists from underrepresented groups.

NPR has had women as hosts and in senior positions in the newsroom since almost the beginning: Susan Stamberg, Linda Wertheimer, Nina Totenberg and Cokie Roberts, known as the founding mothers. But having women as journalists didn't correlate with their representation among people interviewed on the air. In 2013, when we started tracking the demographics of people appearing on our newsmagazines, only 28% of on-air sources were women. Even in 2021, the figure was only 40%, even though more than half of the staff were women. (NPR only recently started tracking statistics on transgender, nonbinary and gender-nonconforming staff and on-air voices.) In terms of race, 77% of on-air sources in 2013 were white, though by 2021 that number had dropped to 61%, closer to the proportion of 59% in the general population.[7]

Meanwhile, our staff makeup has also changed. It dropped from 78% to 59% white, while the share of African American employees increased from 9% to 16%, Asian from 9% to 12%, and Latino or Hispanic from 4% to 9%.

Political correspondent Asma Khalid, who came to NPR in 2007 as a producer on *Morning Edition*, says the newsroom has "tremendously diversified." "I realize that sometimes you look around and it feels like change is slow," she tells young producers who may feel frustrated. "But I remember being in meetings where everybody else was white."

To be fair, NPR wasn't the only newsroom in America dominated by white men in the network's early decades. But its chief news executive in the late 1980s, Adam Clayton Powell III, and an NPR president in the 1990s, Delano Lewis, were both Black men. And there have been quite a few hosts of color: Michele Norris was hired in 2002 to host *All Things Considered*; Audie Cornish joined NPR in 2005 and became a host six years later; and Michel Martin joined in 2006 and became a regular host on *Tell Me More* (which was later canceled), *Weekend All Things Considered* and *Morning Edition*. Other nonwhite hosts have included Arun Rath, Ray Suarez, Emil Guillermo, Lee Thornton, Steve Curwood and Farai Chideya.

But NPR did not have a regular Black male host on the daily newsmagazines *Morning Edition* and *All Things Considered* at least through the early 2020s. And there was much criticism of management's efforts to retain journalists of color when Audie Cornish left to join CNN in 2022, a few months after *Weekend Edition Sunday* host Lulu Garcia-Navarro, a Latina, was hired by the *New York Times* to host a podcast, and host Noel King, who is biracial, departed *Morning Edition* and *Up First* for Vox. Even though King was replaced by Lebanese American host Leila Fadel, and Latino host A Martínez had just joined *Morning Edition*, on social media the departures were characterized as a hemorrhaging of talent from marginalized backgrounds, at a time when NPR was trying to attract more Black and Latino listeners.

On the other hand, one of the reasons these journalists were poached by other news organizations is precisely because they had reached the top jobs at NPR. NPR's senior vice president for news at the time, Nancy Barnes, noted the "enormous opportunities for journalists that did not exist a decade ago." While NPR had once been the only game in town for long-form narrative audio, news organizations like CNN and the *Times* were elbowing their way into the audio space, competing for the same talent. But Barnes conceded there were other reasons people were leaving, including "problems in the work environment."

In recent years, there have been efforts to foster broader discussions about race and identity in the entire organization, not just the newsroom. NPR has encouraged staff to join employee resource groups organized around various forms of racial, ethnic, religious, gender and other identities. And in 2021, Whitney Maddox, the director of diversity, equity and inclusion, started biweekly anti-racism discussions to let staff talk about "the ways race and racism affect how people view, engage and invisibilize those around them." Whitney also led efforts to help women of color on staff succeed and remain at NPR.

Whitney says young journalists of color who feel that NPR hasn't resolved all the issues that diversity efforts seek to address shouldn't feel put off. "There is a chance that you don't see yourself or hear yourself in the news that you're hearing from your station, or at NPR in the images that you're seeing being put out on the website. And that's why we need you, because if it's going to change, you have to be in the organization."

Empathy

A lot of media organizations are working on diversity. Many also value the same things we do, such as credibility and impartiality. So what distinguishes NPR from other media? It's a question I asked myself when I set out to write this book, and that I put to the NPR journalists I interviewed. As I mentioned earlier, Tony was impressed by the level of respect we have for the listener. I also found something that goes beyond respect in interview after interview: empathy. In later chapters you'll read about how in covering mental health, Rhitu Chatterjee sometimes drops a reporting idea because "the person is more important" than the story. How Adrian Florido chooses to spend time with people in crisis—"I get emotional with them and that creates a bond"—before he interviews them. And how hosts such as Rachel Martin, formerly of *Weekend Edition Sunday* and *Morning Edition*, show empathy with guests who are distraught: "Just imagine how you would want to be treated if you were in a similar situation."

Rachel, who went on to launch the spirituality series *Enlighten Me* in 2023, says she has always seen public radio as a way "to cultivate empathy in our culture." Listening to a person talk connects you to them in a way that seeing them or reading their words doesn't, she says.

> There's no better way to help another person understand someone's experience than by hearing them in their actual voice explaining what's happened to them in their own words, their own language, their own dialect, with emotion. It connects listeners to a story in a really material way that words on a page just don't. Sure, you can get a lot more detail in print. And TV is so powerful for big events and with those images. But to really understand what a person who's living at the center of a news story is going through, there is nothing like hearing their voice explain it. I think that's a critical tool in getting Americans to understand each other.

Audio has a long history of bringing people together. Our human ancestors were telling stories before they started drawing animals on the walls of caves: It was how they bonded.[8] Developing babies hear their mother's words in the womb, and research suggests they remember some of it.[9]

And it may sound counterintuitive, but listening is a visual experience, which is multiplied when many people listen to one voice. This is how Robert Krulwich, founder of the public radio show *Radiolab*, described it.[10]

> If I say, "Let me take you to a small clearing in the woods where we will find a little, chubby rabbit nuzzling some grass," you can't help it—my words force your mind to create a rabbit of your own (or rather, my rabbit painted by your brain). And those rabbits multiply. If I'm on the radio telling my rabbit story to 10,000 listeners, there will be 10,000 slightly different rabbits in 10,000 slightly (or very) different woods simultaneously.

"That's the genius of radio," says John Burnett, who was himself one of the geniuses of radio during his 37-year reporting career at NPR. "We engage the mind to create the picture. We give them the sound and the words, and our listeners' brains create the pictures."

Nowadays NPR journalists also write stories for the web, shoot photos and record videos. And that's a good thing. It would be a pity to not reach the multitudes of people who prefer to read or watch at least some of their news. By 2023, NPR.org was getting on average about 100 million story views monthly. About two years into an initiative to post news stories in short-video format on Instagram and TikTok, NPR had racked up nearly 2 billion views. But this book is called *Sound Reporting*, and its purpose is to help you learn how to *tell* the news. Other than chapter 13, which covers multiplatform reporting, the book is primarily about audio journalism.

Doing audio journalism means you're at a disadvantage when compared to other media—there are no pictures* or punctuation, and it's not simple to go back if the listener misses something—so you have to make up for it with clear, conversational and descriptive writing. But there are also profound advantages.

INTIMACY. The best public radio programs sound as if a good friend is sitting next to you telling you the story. No matter how big the audience, an adept host always imagines talking to a single person rather than to listeners as a group. With the even more conversational style of podcasts and earbuds that plug the host's voice into your brain, the intimacy is that much more potent. It's no wonder that podcasts can be a cure for social isolation.[11]

PERSONALITY. You get to know people in the story in ways that you can't in other platforms. Newspaper stories often have great quotes, and TV can show vivid images. But people convey what they feel both through their words and through the sound of their voices, and it comes across more clearly when we are not distracted by what they look like. In a radio interview, we often can hear for ourselves when a politician is dismissive, or a protester is angry, or a Nobel Prize winner is thrilled and exhausted; we don't need a reporter to characterize them for us. And public radio usually allows people to speak at some length; an interview in a newsmagazine show might run as long as eight minutes. We don't confine a person's insights and emotions to a 10-second sound bite.

SENSORY EXPERIENCE. Seasoning the story with high-quality ambient sound envelops listeners in the scene. They are smack in the middle of the crowd of sports fans who are jumping up and down as their team sails to victory. Or they are tiptoeing through forest brush, surrounded by the sound of flies humming and exotic birdsong.

VERSATILITY. One of radio's biggest advantages used to be its portability. To listen, all you needed was a transistor radio and a couple of batteries, and you could get your news anywhere. As a reporter, all you needed was a microphone in your hand and recorder slung over your shoulder, and you could get to places and see things that TV crews couldn't, with their glaring lights and reporters doing live hits with their backs to the action. Audio has lost some of that advantage. Anyone can record crystal-clear video on a smartphone. But unlike newspaper stories or news videos, audio stories don't need to be looked at. That means you can listen to the

* Aside from the one that may show up on a podcast episode page.

latest episode of *Code Switch* while driving to work or stream *All Things Considered* while cooking dinner. That is nothing less than a superpower in this multitasking world.

Speaking of multitasking, for decades, millions of people became NPR fans by listening in their cars during their commute. And sometimes, even when they arrived at their destination, the story that was playing would so captivate them that they would stay in their car and listen until it was through—a "driveway moment." The term is thought to have come from a listener letter received in the 1980s or 1990s, though that letter is apparently lost to history. We do know that by 2002, NPR was asking listeners to tell us about their driveway moments; so many submissions came in that NPR created a series of books based on them and published a selection on our website. Here's one sent in 2007 by a KUER listener identified as Joelle:[12]

> Kevin Kling's January 10th *All Things Considered* story about prayer and squirrel monkeys was incredible. Unfortunately, I wasn't in the driveway yet when I started to sob. I had to pull over, so I guess my experience was actually a "side-of-the-road moment." His story was incredibly compelling to me. Thank you.

NPR archivist Julie Rogers sent me a document with a long list of listener notes like this that were read on air. Interestingly, there were no submissions about Terry Gross' widely heard interview with Gene Simmons. But there was a submission about another *Fresh Air* episode from a listener named Lucy, who wrote that it left her "stuck in a rental car in the parking lot of a Wegmans for about 45 minutes listening to the entire show."[13]

The guest that day was comedian Al Franken; it was before he became a U.S. senator. He was telling Terry about the time he was at a racquetball club in Midtown Manhattan, warming up on the court, and his match partner was running late.

> FRANKEN: And I hear this knock on the door and it's Gene Simmons. I didn't know, actually, it was Gene Simmons till later. But he knocks on the door and he said, "I play with you?"
>
> And I said, "Well, I'm waiting for someone. Sure. Why don't you come in."
>
> He says, "I'll kick your ass."

So I said, "Oh, OK. Well, fine." I said, "Well, look, you want to warm up?"

He goes, "No, I'll kick your ass."

And then I said, "Well, go ahead. You serve."

Simmons scored the first two points, but Franken got his game back and crushed him. By then, Franken's partner had arrived, but Simmons insisted on playing again, with the same bluster: "I kick your ass."

FRANKEN: And I said, "Well, no. Look, my friend's here. I'm going to play with him."

And he goes (sounding like a chicken), "Bock, bock, bock, bock, bock, bock, bock. Bock, bock, bock, bock, bock, bock!" And he's taunting me.

And I say, "OK, we'll play this game for 500 bucks." And he just turns around and leaves.

And my partner says, "That was Gene Simmons."

And that's my Gene Simmons story. He was the most awful person I've ever met.

Franken said to Terry that he told the story so that she would "stop beating [herself] up" about the Simmons interview. Think about that. Franken cared about Terry—he was consoling her. And she was gracious. Though the story was really about her, she got out of the way and let him relive the entire memory in such a vivid way that it helped keep at least one listener in her car, in a Wegmans parking lot.

That is what NPR strives for. Not a conversation conducted to see who will win an argument. Not a story done to further a political or activist aim. But a search for the truth, carried out with respect and empathy and in accordance with a code of values whose ultimate purpose is winning credibility in the ears and eyes of the public. And we know we've succeeded when we tell it in a way that keeps people listening a little longer.

CHAPTER 2

Pitching

What is an NPR story? That's a fair question. What makes a news story we do different from one produced by a commercial outlet? We aspire to make it a transformative experience, inviting discovery or inspiring wonder. Whether it's a 50-second newscast spot, a four-minute show segment, a 30-minute podcast episode, or a 500-word digital piece, an NPR story should change the way people see or understand some aspect of the world around them. Correspondent Jasmine Garsd's podcast, *La Última Copa/The Last Cup*, portrayed Argentine player Lionel Messi as an immigrant, which was transformative for many in the audience who solely or primarily knew of him as a soccer star.[1] Across audio, images and text, education correspondent Elissa Nadworny opened our eyes to the toll of the Ukraine war by telling the story of one class of kindergartners.[2]

Our audiences across all platforms come to us because they want to hear something that they didn't know before or that challenges an existing notion or debunks a myth that they may have heard elsewhere. Part of our mission is to "create a more informed public."[3] What people don't want is to be told something they already know, or to be preached at or lectured to. The last thing we want is for them to change the station or click away during a story (when the storytelling failed to captivate listeners, it used to be called "radio death" because we knew they would tune out). So, as you go out looking for stories to tell, continually ask yourself, is this something that will fascinate and challenge my audience, or will it bore them?

Finding and Pitching Stories

Every NPR story is born as an idea in someone's head.

It could be a host or a lead editor thinking about booking a guest to talk about one of the day's top stories. Or a reporter with a hunch that some big question will make for a riveting audio narrative. Or an intern who reads an article on a news site and thinks there's an angle worth exploring further. Or a freelancer who has a story idea that a staff reporter doesn't have the bandwidth or expertise to do. (We also receive story ideas from our audience, but there are so many pitches internally that those outside ideas are usually pursued only if an NPR producer or reporter picks them up.)

One of NPR's greatest strengths is that anyone can pitch a story that ends up on a nationally broadcast newsmagazine, a website with millions of readers or a podcast downloaded by millions of listeners, or perhaps all three. Imagine for a moment if it didn't happen that way. If instead, managers assigned only the stories they deemed important. We would miss a lot. Not because those individuals don't have excellent story ideas or deep professional experience that is of great value in determining coverage. But because even the best editorial leaders can only observe so much of the world from inside their newsroom.

That's why at NPR we value a diverse staff with a multiplicity of perspectives and original ideas. We want them to pitch relentlessly, and we expect editors to keep an open mind when considering pitches on topics that are unfamiliar to them. The result is a collection of shows and podcasts that bring the world in all its variability to the listener's ear.

But let's back up a bit, to that story born in someone's head. Say it's your head. How do you go from an idea to an actual pitch? It's not easy. A pitch involves much more than saying "I think *x* is a good story." Yes, a pitch should be brief, but it should also contain the right combination of facts and outlining of reporting possibilities to be convincing in its promise of something that your audience will find interesting.

Not only that, but pitches vary widely in composition, structure and length, depending on what kind of story is being floated and for what platform. A single sentence might do for a newscast spot ("Do you want a wrap on the president's reaction to the Supreme Court decision?") or for a same-day segment on a topic whose news value is not in question ("Nathan Rott can do a five-minute two-way on the wildfires in California"). In that case, the idea is to communicate the gist and format of a proposed story without getting bogged down in the details.

A pitch needs to be much more developed for a show segment or a podcast episode, to demonstrate that the story, as we journalists say, has legs. Even then, it still needs to be focused. Skillful journalists edit their pitches down, mindful of their editor's prior knowledge and attention span. Here's a pitch I got during the COVID pandemic from WKAR reporter Michelle Jokisch Polo, who made the national significance of the proposed story immediately clear:

> Undocumented immigrants have been deeply affected by the coronavirus pandemic. With online learning for the school year getting under way in Michigan, many working parents cannot help their children with remote learning or afford supplemental tutoring services. While public school systems here have not taken steps to support these families, a grassroots learning center for undocumented and low-income families in the Ann Arbor-Ypsilanti area has mobilized learning pods to help them. As part of this story I will talk with coordinators of a learning pod, undocumented families sending their kids to learn there and have a scene with students.

The pitch ended with an execution plan that gave a sense of the sounds listeners will hear in the story. The more focused the pitch, the better. You'll notice that this pitch wasn't meant to answer every question the editor might have, but it offered enough information to draw interest. The story aired on *All Things Considered* in November 2020.[4]

In this chapter, you'll learn about where to find story ideas, the ingredients of a strong pitch, the art of pitch presentation and how to deal with rejection, since it's inevitable that not all pitches will succeed. But first, I'd like to talk about the ritual that is the pitch meeting.

How Stories Get Pitched

NPR's daily newsmagazines—*All Things Considered, Morning Edition* and *Here and Now* (which is produced by WBUR)—hold daily pitch meetings where staff throw out story ideas. The weekend shows, podcasts and editorial desks also hold the meetings regularly, though not always on a daily basis. If pitches are the lifeblood of NPR's journalism, then the meetings are the heart, pumping out story ideas.

"Everybody from the hosts and managers to the brand-new intern brings their ideas and pitches them. And we kick it around and decide," says co-host Mary Louise Kelly, describing a typical *All Things Considered* pitch meeting. Show staff take turns explaining why their story would interest listeners, and how they envision it being executed—as a conversation with a guest, a piece or narrative by a reporter or host, a two-way or roundtable with reporters or another format. On NPR's editorial desks—national, international, business, culture, music and science—editors present pitches they've collected from reporters and freelancers.

The smorgasbord of stories proposed is impressive. Here are my notes on the pitches fielded at an *All Things Considered* meeting in early 2022, when the COVID pandemic was spreading quickly and Russian president Vladimir Putin was massing forces on the border with Ukraine:

- Efforts to deescalate the Russia-Ukraine crisis
- Who is at fault for the surge in inflation?
- British prime minister apologized in Parliament for holding a party in spite of COVID restrictions
- Conversation with airline CEO over vaccinating employees for COVID
- Is everyone going to get COVID? Propose putting the question to the acting FDA commissioner
- Jailed Iranian poet and filmmaker Baktash Abtin died after contracting COVID in Tehran prison
- Quebec is proposing to tax the unvaccinated
- Shall we mark the unusual date of Feb. 2, 2022 (2/2/22)?
- Controversy in New York over Mayor Eric Adams wanting to hire his brother to paid public office
- Price of meat is going up
- Is there a way to follow up on *Morning Edition* host Steve Inskeep's interview with former president Trump?
- Go to grocery store, gas station, get consumers' views on inflation
- Obituary of a 101-year-old veteran from Saskatchewan, who walked 100 miles to enlist in the Canadian Armed Forces in World War II
- Is the U.S. planning to ease the embargo on Venezuela because of the oil price shock resulting from the Russia-Ukraine crisis and the bans on importing Russian energy sources?
- "Crime tourism"—professional burglars from South America are coming to the U.S. on a tourist visa and targeting high-end homes

Producers, editors and hosts prepare for the meetings by keeping up on what's happening in the world and thinking hard about trends and unanswered questions. "What do I really want to know?" Mary Louise says she asks herself every morning when she wakes up. "What are the most important questions on the stories moving today, and who could answer those? And that's one of the things I love about this job. You can be endlessly curious."

The meeting's leader is usually a senior producer, who fields the pitches, opens the floor for feedback and may also offer opinions. Everyone keeps the commentary brief so that the session doesn't go over half an hour, since *All Things Considered* goes to air about six hours after the meeting's end.

As is typical, only a few of the pitches in the list above made it onto that day's show. Some were held and developed. Others were referred to reporters. And of course, a bunch were rejected or sent back for more work.

All pitches—both those accepted and rejected—ensure that every show covers a variety of topics—so that someone who tunes in for just a half hour (the average commute time in the U.S.) will feel informed and enlightened.

As a supervising producer for *All Things Considered*, Ciera Crawford led many pitch meetings. With two hours of programming to fill every day and a tight turnaround time, she would read ahead of time on the top stories, and glance at coverage during the meeting. It was an intense process, with fact-checking taking place even in the midst of a pitch. While she always sought fresh and original takes on the top stories, she also wanted the reprieves from heavy news that really give the show its character.

"A puppy being reunited with their family. Or a weird science element by Nell Greenfieldboyce, a superspot about a black hole, that most people might not hear about and that you might not see in other publications. That's what makes NPR, NPR," Ciera says. Mary Louise also appreciates those stories: Amid all the domestic and international hard news that needs to be covered, "you also are really desperately trying to carve out room for some humanity, some joy."

Not all pitching is fast paced. When they can, reporters and producers have conversations with editors beforehand about a given idea, or even broadly about the kinds of stories they should be looking for. And when time permits, they should be doing the research and reporting to inform their pitch. NPR journalists need to invest the time to find stories, and let their curiosity lead the way.

Looking for Story Ideas

John Burnett was in Mexico reporting on labor problems at a Korean-owned maquiladora. He was running late for an interview, hurrying across the Zócalo, Mexico City's main square, when he heard an unusual sound coming from a side street. "It was a cross between a whistle, a violin and a kazoo," he recalls. John followed his ears, being a good radio reporter, and came upon a busker who was making music by blowing on an ivy leaf, the way John remembered making sounds with a blade of grass as a child. What's more, the street musician had only one arm, and as John listened to more of his music he decided the guy was a genius. "He would have been a concert violinist if he and his family weren't dirt-poor campesinos."

John rescheduled his interview and spent the next few hours talking with the musician, Carlos Garcia, and doing the story that trusting his own curiosity had led him to, and which he knew would delight U.S. public radio listeners.[5]

When I was reporting from Spain in the 2000s, I also followed my curiosity, while gauging my audience's interest and awareness of what was happening in the country, in looking for stories to pitch to NPR's Europe editor. Many Americans were familiar with the running of the bulls in Pamplona, but did they know how Spanish children are inculcated in the tradition? That led me to a story about an *encierro chico*, a running of the bulls for kids.[6] For another story, I responded to the interest in the U.S. in Spanish food by doing a feature on Jamon Iberico, or Iberian ham, which was about to get import approval by the U.S. Department of Agriculture.[7]

Other times I took stereotypes and debunked them, such as in a story about how lengthy lunch breaks for the siesta were having the unintended effect of forcing people to work late into the evenings.[8] Or I looked for stories that presented alternatives to the American way of life, such as an audio postcard on a summer night in a Spanish village, where the children ran riot in the square and the grownups looked after them, regardless of whom they belonged to.[9] For that story, and the one on bull runs, I didn't even have to leave the village I lived in. Which goes to show that you often don't have to go far for the best stories.

That's not to say I didn't read through a stack of newspapers every morning to stay up on what other reporters were saying, especially the local press. As NPR reporters will tell you, they read voraciously. They scour all kinds of news sites, social media platforms, niche publications, scientific journals and industry and government reports.

When a reporter comes across an interesting subject in another publication, they don't just pitch a story that will match it. Instead, they try to move the story forward—asking themselves what questions are raised by the existing reporting: How can I break new ground with a new story on the subject? Is there an underreported aspect of this story worth pursuing?

Alexi Horowitz-Ghazi, host and producer of *Planet Money*, says that he tries to uncover stories inside of other stories by looking for "the little off-the-main-thread of the news story. Or some minor character that was quoted, but there's a little line of color that suggests that they have some weird, crazy back story." Once, Alexi remembers reading that the initial recipe for synthetic marijuana, which had been responsible for untold overdoses and emergency room visits around the country, had actually been ripped off the internet from research published by a chemistry professor whose aim was to develop a new class of medications. And he wondered, "What if we could tell that guy's story?"

Well, Alexi did, and the story ended up being a revealing look at how a well-intentioned professor inadvertently unleashed a revolution in the illicit drug trade.[10]

Another place Alexi looks for ideas is social media feeds. An Instagram post by an indigenous rights activist tipped him off to a dispute between native lobstermen and commercial fishermen and led to a reporting trip to the waters off Nova Scotia, one of the world's most lucrative lobster grounds.[11]

While he's reporting a story, he also reads widely "to cultivate a garden where new little vines will spin off into other stories." When he did an episode on manatee conservation in Florida, for example, he read a couple of books that were sweeping histories of endangered species and mass extinctions.

Kernels of stories may be buried in other outlets' coverage of big news, waiting for you to uncover them. When President Trump moved the U.S. embassy from Tel Aviv to Jerusalem, correspondent Daniel Estrin was trying "to find ground-level stories to tell." He was watching an interview with the American ambassador on a Christian TV channel when, about halfway through, the envoy talked about church groups that prayed regularly at the dedication plaque. Like a good audio journalist, Daniel is always on the lookout not just for compelling stories, but those that have aural potential. "This is a sound opportunity," Daniel recalls thinking about the prayer sessions. "It's the thing that makes the embassy a real place." Once on the scene, he spent hours outside talking to American Christian pilgrims ecstatic over the embassy's move to Jerusalem and to Palestinians from the West Bank seeking visas to go live in the U.S. The result was a powerful story about the ramifications of the U.S. decision.[12]

Still, if you only pursue stories derived from other journalists' reporting, you may find yourself starting out a step behind. That's why it's imperative for NPR reporters to spend a lot of effort cultivating their own sources.

Shannon Bond, a correspondent on the democracy and disinformation team, checks in with sources regularly to find out what they are working on and to get a sense of what's happening in their fields. Previously, when she was covering tech for the business desk, there was one researcher she called regularly to talk about social media and disinformation. "She was one of the people I would go to for gut checks. 'Hey, we're interested in exploring this thing, is there a there there?'"

At one point, the researcher tweeted about noticing suspicious LinkedIn profiles. Shannon emailed her to ask about it, and the researcher said she'd keep Shannon posted. A month later, Shannon had forgotten about their conversation when she received another email from the researcher with an update, and it made Shannon think, "There might be a story here."[13] She called the researcher up and got enough information to pitch a story to her editor, who approved it, and it became a scoop on marketing companies planting fake, AI-generated LinkedIn profiles.

Jasmine Garsd, a national desk correspondent who previously worked at *Alt.Latino* and *Marketplace*, recommends broadening your interests while looking for stories. "I always tell incoming Latino reporters to be curious about everything," says Jasmine, who was born in Argentina, "because there's only so much cultural identity reporting there's room for on a broadcast." But, she adds, you can also get at some of these topics in other areas of coverage such as economics and science.

At the same time, Jasmine has sometimes hesitated to pitch stories on topics that she's "kind of obsessed with" and has realized later that she was wrong to hold back. For example, daytime gossip TV in Argentina is something she grew up watching with her grandmother. After coming to the U.S., she noticed that a TV show host back in Argentina named Jorge Rial, who was known for stoking fights between scantily clad dancers on his daytime gossip show *Intrusos*, suddenly started holding nuanced conversations about feminism and the growing movement there to legalize abortion.

"It didn't even occur to me to pitch it," she says, until one day when she mentioned it to Jess Jiang, then a producer for *Rough Translation*, who told her to put together an episode on it.[14] Jasmine also did a version for *This American Life* and heard back from many listeners of other cultural backgrounds who told her that they had similar experiences growing up. "I think this happens especially with Latinos and people of color, where you think that what you should pitch is the mainstream and you think this other thing that personally affects your world and your community is too niche and this imaginary larger audience—which I think is often code for a white audience—isn't going to care."

Pitching according to your interests can help you secure a foothold in your career. When I joined *Morning Edition* in 2018, seeing colleagues pitch so thoughtfully about American politics, entertainment and culture was intimidating. I felt out of the loop, having lived abroad for two decades. So, I played to my strengths. I'd been an international correspondent and a religion reporter, so after a blockbuster grand jury report on clerical sex abuse in Pennsylvania, I pitched a conversation with Marie Collin, an Irish abuse survivor and the only one to serve on Pope Francis' advisory council on the protection of children.[15] I knew her from my time as editor-in-chief of Religion News Service.

Alexi had a similar experience when he started out as an intern on *All Things Considered*. He noticed that no one seemed to be paying attention to new documentaries. So, he pitched a conversation with the director of a film on the vibrant pop music bands in Cambodia before the rise of the Khmer Rouge, and the idea delighted then-host Robert Siegel. Alexi realized that pitching documentaries was what he could "do right now" while he worked to broaden his areas of expertise. Many show producers at NPR advise those starting out to find a beat, something they feel is not covered well by others, and then follow their curiosity into other topics.

A Recipe for a Strong Pitch

In my "Intro to Pitching" training session at NPR, I ask people to think of their favorite book, movie or TV series and then say why it would be worth our while to read it or watch it. The point is not to help us decide what to do that weekend, but to identify the qualities that make for gripping storytelling.

"The skills that I learned in film school—in terms of story structure, character arc and how to develop and tell a story—I use them all the time in nonfiction journalistic settings," says Luis Trelles, a senior editor in NPR's Enterprise Storytelling Unit. He used to produce public TV documentaries in Puerto Rico and later joined the Spanish-language *Radio Ambulante* early in the life of the podcast.

Even though journalism deals only in facts, the storytelling devices we use are often inspired by fiction. In my training session I ask the participants, mostly NPR staffers working on shows, to identify the elements of good narratives found in fictional formats and how some of those elements might be used in news stories to bring the facts to life. Usually, we come up with a list that looks like a recipe for a strong pitch:

AN INTERESTING PERSON. Interview someone the audience can identify with or whose motivation can be laid out clearly, and make sure the person is a "good talker." That doesn't mean they have to have a particular accent, or even speak English as a first language. But they should express themselves well in response to your questions. *Planet Money* senior supervising editor Jess Jiang says that her team looks for a central character who is willing to be thoughtful and reexamine their role in something that may have happened a while ago. "What we do requires reflection and really understanding the moment. We need someone who's willing to answer our questions" even when those questions seem off-topic. "'Why do you care what I ate that morning?'"

ACTION. Something should happen. In journalism, it's best when we hear it happen during our story—but action can also come in the form of someone recalling a thing that happened to them or that they witnessed.

CONFLICT. Include obstacles, opposing forces, or another side of the story that creates tension.

SURPRISE. Find unexpected twists or developments. Or what Luis calls the "light-switch element," a phrase he picked up while working with the *Invisibilia* team, that turns listeners on to something they didn't know or realize before.

RELEVANCE. Make sure the story speaks to the audience. If it's a news segment, it should be about or related to a major change or event that's happening now that affects people's lives. If it's long form, it should resonate with something on listeners' minds.

ORIGINALITY. Tell a story that hasn't been told elsewhere. For news, it shouldn't have been reported elsewhere, at least not in the same way, and especially not on our air. Few things will make a journalist look more unprofessional than pitching a story to NPR that we've already done.

GOOD SOUND. Remember how frustrating it was when streaming video would buffer over slow connections? Or, for those of a certain age, when the picture on the TV was grainy or skipped? In the same way, capturing high-quality sound is crucial to great audio stories. In the pitch, there should be a description of the sounds that will immerse the listener in the story.

Identifying elements in the story is half the game of pitching. The other half is presenting it in a way that makes your case for the story to someone who may have very little time to consider it.

Presenting the Pitch

At the *All Things Considered* pitch meeting, the show's hosts like to hear what it was about the story that intrigued the producer or editor pitching it. After all, hosts have to get listeners excited. "A pitch is a lot like [a host's] intro," says *All Things Considered* host Mary Louise Kelly. "The pitch is you selling the story, the same way the host sells it on the show."

In fact, good pitches often sound like intros: They are a few sentences long and are clear about the story's relevance. Here's how former executive producer Cara Tallo describes the qualities of a good pitch for *All Things Considered*:

- The pitch is original, timely, actionable, succinct.
- It moves past existing coverage to highlight a new voice or angle.
- It answers the following questions:
  - What's the story in a sentence or two?
  - Why does it matter (and why does it matter *today*)?
  - What kind of guest should we book?
  - What are the top two to three questions you'd want to answer in a conversation?

Senior newscast producer Carol Anne Clark Kelly says she often wants pitches that show how a national trend is affecting people in the community. "We can talk about policies and numbers and acreage, but it's the human impact—how people are reacting and what people are going to do next—that makes something an interesting story."

A common misconception among newer journalists is that the reporting begins after a pitch is greenlighted. Nothing could be further from the truth. "You don't need to have it exactly down pat in the pitch," Luis says, "but you need to be able to signal, 'Hey, your audience will be interested in this.'"

FOUR QUESTIONS TO ASK YOURSELF

Here's what Luis looks for in pitches for the long-form narrative podcasts he's worked on as an editor, which include *Invisibilia, Rough Translation* and *La Última Copa/The Last Cup*:

- What's the story?
- What is exciting about it?
- What's the light-switch element?
- Why are you the right person to tell this story?

As you'll recall, the light-switch element is something listeners are turned on to. While planning the 2021 season of *Invisibilia*, the podcast's producers felt that some kinds of relationships, like family ties or romantic connections, had been studied to death. But in friendship, the rules and conventions were not very well understood. So, host Yowei Shaw pitched a story about ghosting. The COVID pandemic appeared to be receding, and people were wondering how they might reconnect with friends, if at all. *Invisibilia* ended up doing a whole season about friendship.

Newsy relevance, the notion that a story needs to be pegged to the news of the moment, is also important. But Luis says that's not always necessary in podcasting. If its newsworthiness isn't obvious, "it needs to be a hell of a story, like, 'Oh my God, this character!' or, 'This twist!' You need to be riveted."

As for explaining why you are the right person to tell the story, Luis says you should be able to demonstrate that you have "access to the characters and the world" that the story is about, or that you have "a special, particular understanding of it." Luis was an editor of the first season of the WNYC podcast *La Brega: Stories of the Puerto Rican Experience*, and he says Alana Casanova-Burgess was the right person to host it because she's a Puerto Rican producer and storyteller who has worked extensively in both New York and San Juan.

In other words, representation matters. Don't think you can "go to a community and say, 'I know you better than you know yourself,'" he warns. But he adds, "Does that mean that if you're not in a certain community, you don't get to tell that story? No, because that cuts against all that I believe in: the freedom and power to tell stories."

Again, we're talking here mainly about long-form storytelling of the kind found in podcasting and radio documentaries. In more traditional radio-style news segments, being a journalist is the thing that gives you access. Even if you are not from the community, you can still be a good reporter and tell the story. But it's worth keeping Luis' dictum in mind: "Don't come across sounding like you get the full experience if you haven't lived it."

KEEP IT SHORT

It should be clear by now that a lot of legwork goes into the research phase. Once you've got the elements down, write out your pitch, even if you are going to present it orally. And make it succinct. When international correspondent Daniel Estrin has an idea for a story, he tries to write it as a kind of thesis statement, in under 200 characters. That's the limit for NPR's story summaries, which need to be written early on in the reporting process anyway, because they get sent to member stations as part of the rundown, the list of items on a show, in the network's Direct Access Communications System, or DACS.

Writing the story summary, or DACS line, in the pitching phase helps Daniel identify the focus of the story. "If you can't write the DACS line, then you don't know what your story is," he says. "The DACS line can change. But I just think that's the process of thinking through your story."

Succinct pitches are music to the ears of editors, who are often juggling various tasks that include working on stories with other reporters and coordinating with shows. Luis suggests starting your pitch with the most exciting element, even if you will build up to it later in the actual story.

A Pitching Checklist

ALL PITCHES

TOPIC/RELEVANCE/NEWSINESS. What is the story and why are we doing it now? Why will it engage our audience? How will it be original? What understanding will our listeners gain that they didn't already have?

NEWS PIECES

VOICE(S). Who will we hear from? What role does what they've experienced and what they have to say play in the story? What role does their race, ethnicity, gender, age, economic status, geographical location or other relevant background play?

TENSION. What is the conflict in the story? What unexpected twists or surprises will there be? In other words, what happens in the middle of the story?

EXECUTION. How will you do the story? What sounds and scenes do you expect to have?

ON-AIR INTERVIEWS

VOICE. Who is your guest? How will they offer a fresh perspective, or move the story forward? What role do they play in the story—are they at the center of it, involved in it or affected by it or can they offer an expert view? What role does their race, ethnicity, gender, age, economic status, geographical location or other relevant background play?

EXECUTION. How will the interview be done—remotely or in studio? Is this prerecorded or live?

LONG-FORM NARRATIVES

VOICE. Who will be the focus of the story? Why are they compelling? What role does what they've experienced and what they have to say play in the story? What role does their race, ethnicity, gender, age, economic status, geographical location or other relevant background play?

TENSION. What are or were the obstacles or opposition they face(d)? What unexpected twists or surprises will there be? In other words, how will you keep listeners hooked throughout the story?

ACCESS. Why are you the right person to do this story?

Bouncing Back from Rejection

NPR greenlights plenty of pitches, far more than some other news organizations where rejection is common. But even if you have all the elements of a great pitch, it could still be turned down for reasons that have little to do with its merits. A major breaking story might be taking up lots of real estate on the show, or that particular show or podcast might not be the right fit.

Or your pitch may indeed fall short. Cara says that at *All Things Considered*, the pitches that are most often rejected don't "move past existing coverage in a new and meaningful way (especially if it was on *Morning Edition* that very morning!)." Other times they "can't answer the questions, Why does this matter? And why does this matter now?" Or they are "unfocused, unresearched" or a "premise" rather than a "story."

At the *All Things Considered* meeting I described earlier, a producer pitched a story about "meat inflation." *All Things Considered* was already covering inflation from several angles, so it was rejected. Ciera says the feeling was that the rising price of meat wouldn't sustain a whole conversation. "I can probably write a short intro, right? Maybe one question: 'Why is the price of meat rising?' What's the second or third question?"

Luis says the pitches he rejects tend to be poorly framed, in that they don't tell him "what makes the story exciting or relevant, thought provoking or delightful." Perhaps there is a story there, "but you don't know why it's important or who the main character is or what is the main turn."

Rejection is hard, and we've all been through it. Jasmine Garsd says that when she was starting out, she might pitch seven or eight stories before getting one or two accepted. "I had to really learn to take a no in stride," she says.

That's something for editors and supervisors to keep in mind as well, she says, and not shoot down or ridicule pitches. "I have had the experience of being on a team and someone being very mocking," and it made her reluctant to pitch again. "It's important to have a team that doesn't embarrass you or joke with you when you pitch something that's silly or it's not viable."

A rejected pitch is a learning opportunity, both for you and for the team you're pitching to. Try to find out why your pitch was rejected. Figuring out what you could have done better may not revive a fatally flawed pitch, but it will help you pitch more effectively next time. And many editors will give you that feedback, especially if you're willing to wait until they have a moment to provide it.

Remember, rejected pitches are not wasted effort. Having a clear notion of a show or podcast's identity and storytelling objectives is as much about the stories it does, as it is about the ones it chooses not to do.

More Dos and Don'ts

PITCH A STORY, NOT A TOPIC. When I was looking to break into radio, I lived in Amsterdam and pitched a story to an international broadcaster on the city's cannabis cafes—my story would have me going inside one of them and talking to the clientele. I even got a recorder, did the interviews and wrote up the story—and no one took it. The reason? It was a topic, not a story. There was no conflict or tension or really anything at all exciting. Later, I did figure out how to make drug policy stories interesting—I reported how European countries were starting to emulate the Dutch "harm reduction" approach—in defiance of the U.S. war on drugs. I also did a story on an addict whose family had emigrated to California when he was a child, but he moved back to Holland so he could shoot up without fear of the law.

FOLLOW BUSINESS REPORTER ALINA SELYUKH'S RULE OF THUMB. Find out whether that story you are excited about will also interest other people. Tell your editor about your pitch, but instead of saying "I want to do a story about . . . ," start with "Did you know that . . . ?"

IF YOUR IDEA WAS PROMPTED BY SOMEONE ELSE'S JOURNALISM, FIND A WAY TO ADVANCE THE STORY. Don't pitch by saying "I read this interesting thing in the *New York Times*." Instead, find an unanswered question or a novel way to approach the story.

IF YOUR PITCH WAS INSPIRED BY A SCIENTIFIC STUDY OR GOVERNMENT REPORT, GO TO THE SOURCE. Don't count on other reporters to tell you what's interesting or newsworthy; read it yourself to make sure, and see if there's a novel angle to pursue.

FIND OUT IF A COLLEAGUE HAS DONE A SIMILAR STORY. One of the most embarrassing blunders is pitching a news outlet something they've already done. Do a search before pitching. If they've already covered the story, ask yourself whether enough has happened since the last time to warrant an update. Think about how the new version would be different from the old one.

TAKE TIMING INTO ACCOUNT. Get to know the workflow of the people you're pitching to. Are there hours of the day, or days in the week, when they are slammed? Are there times when they really need content? When I was an overseas correspondent, I used the time difference to my advantage and pitched newscast spots in what were the wee hours of the morning in Washington, when my stateside colleagues were asleep. Needless today, my "batting average" was high with those spots.

MENTION YOUR TRACK RECORD. I began this chapter by saying that anybody at NPR, or a freelancer, can pitch a story. And that's true. A strong pitch will likely succeed, even if it's from an intern who has never pitched before. They don't have to know how to do the story, since many pitches are handed off to producers, reporters and hosts. But if you're proposing to report the story, your track record in getting stories from pitch to air will be a factor in the editor's decision. If you do have a track record of successful pitches, make sure to point that out. At the bottom of your pitch, add a few links to other work you've done. It also can't hurt to point out other editors you've worked with.

CHAPTER 3

Producing

Liz Baker had seen the cable TV show *Mad Men*. It was about the advertising industry in the 1960s when the office secretary was doing all the work behind the scenes, anticipating her boss's every need. And when something went wrong, she took the blame.

"You need to be that," a mentor told Liz when she got an internship at NPR after college and showed interest in becoming a producer.

The advice didn't go over well. "Eww! I hate that so much!!" she remembers thinking.

That was in 2010, and fortunately, a lot has changed at NPR since then. "It's not like this servant job anymore," says Liz, who is living proof. She became a full-fledged producer in 2015 and has partnered with the network's top reporters and hosts, including Michel Martin, Leila Fadel, Nathan Rott, Adrian Florido and Debbie Elliott.

"All the reporters I work with treat you like a fellow journalist," even though your responsibilities are different. And she's been on the air herself doing occasional spots and two-ways and reporting on breaking stories as a producer and reporter. She says: "There's a lot more opportunity for producers if you want to get your name out there."

That's certainly the case on NPR podcasts. The network now names producers and others involved in the production of podcasts at the end of episodes; sometimes they *are* the lead voice, telling the stories they reported. On the radio shows, segment producers often get credit on digital versions published on NPR.org. And many of the faces on NPR's social video feeds are producers.

But the work a producer does is mostly behind the curtain, in support of reporters, editors, hosts and others. They are the driving force that turns a story idea into a broadcast segment.

Working on Other People's Pieces

"You're figuring out how to get that story on the air, whatever you have to do," says Gisele Grayson, who spent more than a decade as a producer on the science and national desks, working with reporters and on stories ranging from the tsunami in Indonesia to black lung in West Virginia and winning a national award in health care reporting for a story she produced on osteopenia.

A producer might call potential interviewees and pre-interview them for a reporter or host, arrange travel, set up in-person or remote interviews, record them, identify sound, cut tape, mix pieces or do all of the above.

The work is tough and can be frustrating, but it's excellent training for a position of leadership in the newsroom, says Kenya Young, one of many public radio leaders who previously worked as a producer. "Day after day, hour after hour," she would lay up other people's pieces, working with the sounds they filed, fading them in and out at exactly the right time to make the story "sing." And that, she says, was key to her rise in the ranks to become executive producer of *Weekend All Things Considered* and then *Morning Edition* from 2015 to 2021. "I wholeheartedly believe that I am the news leader that I am," says Kenya, who went on to become senior vice president of New York Public Radio, "because I sat down for a very long time working on other people's stuff." Many hosts and reporters at NPR began as producers or production assistants. Melissa Block, former special correspondent and longtime host of *All Things Considered*, says that going out on stories with a reporter was the perfect preparation. "When I switched to being a reporter, it was a lot of the same skills," she says.

Producers vary in their expectations of the job.

"I don't really want to be on the air," says education desk producer Lauren Migaki. One of the things she likes most about producing is how much she's appreciated. "If you're a reporter you're very often alone and working kind of isolated. When they send the producer, everyone's pretty happy because that means they have an extra pair of hands."

What does a typical producer's job look like? "It's everything from generating a story idea, [planning] a trip, figuring out where to go, booking all the logistics of how you get there and where you stay and who you're going to talk to, and then making sure that the host is engaged and knows the story and is having the conversations that you want them to have," says Melissa, recalling her assignments as a producer. "It's all the work in advance, but it's also the work when you come back, going through the tape and figuring out how to structure a piece and what sound works and what doesn't and how to make the radio come alive."

The logistical part of the job can involve anything from reserving hotel rooms for a trip to validating parking for guests coming into the studio. It can also be very technical. Producers need to know how to operate recorders and handle microphones in a way that optimizes the acoustics of an interview, or so that ambient sound is clear and dynamic. They must know how to use audio editing software to swiftly put together complex mixes.

Having a Good Ear

But the job has an equally important intellectual component, which begins with brainstorming coverage. "Producers have a huge responsibility and a craving to pitch," says Gisele, who later became editor of NPR's daily science podcast *Short Wave*. "We really depend on them heavily for ideas and smart pitches."

Even for an episode that a host or editor may have pitched, a producer often brings the idea to fruition, thinking through whom to interview and questions for the host to ask.

Finally, there's a technical-slash-creative part of the job in which the producer is thinking about the sounds needed to tell the story, then gathering and placing them in the story in a way that makes for an interesting and satisfying listen.

That's why one of a producer's most important qualities is having a good ear.

"As a producer, I'm always thinking about how we make this listenable," says Brent Baughman, who helped create a number of NPR podcasts, including *It's Been a Minute*, *TED Radio Hour* and the *NPR Politics Podcast*.

How does a producer determine listenability? It's partly the experience a producer builds up through practice in the art of storytelling, and in knowing the audience. And partly gut instinct. "I have the attention span of a two-year-old," says *Morning Edition* producer Barry Gordemer, who moonlights as a puppeteer. That short attention span of his works to his advantage. "I'm supposed to be the one who's the most interested, because I'm the producer," he says. "And if I'm bored, our listeners are probably going to be bored as well."

Barry says one of the most helpful things a producer can do is recognize when a piece, interview or episode is not working, even though everyone may have done what they were supposed to do. "Yes, we've got the facts right. Yes, we wrote it well. Yes, we recorded it well. Yes, we mixed it well. But there's something here that's just not working."

And if the producer doesn't understand an answer given in an interview or if a part of a story doesn't make sense, the producer should speak up and say, "I'm confused," Barry says. "It takes a lot of personal confidence to be able to do that, because when you're a young producer, you're trying to prove how smart you are."

As producers gain experience and branch out, more of their work involves editorial decisions, especially on podcasts with fewer staff members. Gisele says *Short Wave* producers have a lot of leeway and influence in how an episode will sound, because they do the first pass after the host has talked to the guest.

"You might chat through what you heard in an interview, but it's the producer who sits there with the tape and makes the first cut," she says. After listening to a half-hour interview, she might think, "Good luck getting that to 15 minutes!"

Barry says he doesn't see "a lot of gap" between the roles of editor and producer. "The producer might have a little more expertise on the audio side, the editor might have a little more expertise on the writing. We are at our best when those lines are so blurred, where you can't remember whose idea was what."

So, while NPR counts on the producer to be the audio expert, it's really a role that requires much more: a gravity-defying ability to work fast, adapt to unpredictable situations and make people sound good.

Producer Roles and Levels

Print journalists—whether rookie reporters or managing editors of news sites—work with words. What's important is how well they know how to use those words.

In the same way, audio producers work with tape, recording, cutting and mixing it.[a] Their job can seem purely technical. But to see it that way is to miss what they really do: use tape to tell a story. And what's important is how well they know how to do it.

At NPR, producers' roles are classified in two ways. The first is by seniority level, starting with production assistant, assistant producer and associate producer, then producer I, producer II and producer III up to senior producer.[b]

Barry Gordemer, the senior producer of *Morning Edition*, says that he got his start as a production assistant at *All Things Considered* back in the reel-to-reel age, when his mentor was "a young man named Ira Glass," the celebrated host and producer of *This American Life*. Barry says that for producers, "each level of promotion up is a statement more about your editorial abilities than it is your production abilities."

So, while a production assistant may clean up tape, removing ums and ahs and other distracting noises, and make other cuts as instructed, a full-fledged producer may take out sentences or sentence fragments from a conversation, which requires editorial judgment. And while many senior producers edit and mix audio, they also decide which production techniques and stylistic approaches make for the best journalism. For example, they may be the ones to choose the format for a given story, whether it be an interview with the host, a reported piece, a non-narrated piece, a tape and copy or something else.

Experienced producers also pay attention to the editorial implications of production decisions, such as why an interview recorded in an office has the "ambience of a jungle," Barry says. "Production is not just about making audio pretty. Production has editorial impact."

The other way NPR producers' job descriptions vary is by where they work and what they do. These are some of the most common roles.

DESK PRODUCER

Assigned to one of NPR's editorial desks—national, international, Washington, business, science, culture, education or investigations—the desk producer works with reporters assigned to stories, going through interviews and pulling clips and offering other production assistance. The desk producer also provides editorial support, brainstorming ideas for how to report a story and sometimes taking on some of the legwork as well, such as attending press conferences, recording interviews and researching and gathering information for the story. When a reporter gets sent out on a story that requires production assistance, frequently the desk producer will come along, acting as the field producer. A desk producer may also pitch stories that either they or a reporter ends up doing.

FIELD PRODUCER

The field producer goes out and gathers news with a host or reporter. Generally, it's a temporary role focused on a specific assignment. The field producer plans the trip, books and records interviews, brainstorms and supports editorial coverage, makes sure team members are safe and fed and communicates with the shows. On major stories, NPR may send several reporters as well as hosts supported by their own production teams. In such a case, the field producer becomes a coordinating producer to serve as a kind of air traffic control, working with other producers and reporters to make sure efforts are not duplicated. They act as the point person for the team's communications with the shows and other people in the newsroom and handle a lot of the logistics that a field producer does. On overseas assignments, the producer might also hire and oversee fixers and translators to assist a reporter.

SHOW PRODUCER

Producers of a live show such as *Morning Edition*, *All Things Considered* and *Weekend Edition* are its intellectual foot soldiers. They are relentless in coming up with story ideas, especially for host interviews, and then book guests, record conversations, cut interviews and work with an editor to get the story on air. They mix reporter pieces, plan feature segments and report and write tape and copies. A show producer also acts as board builder (the person who plans and moves around segments on a given show) or as director in the control room during the show.

LINE PRODUCER

The line producer heads the show's production staff, leading pitch meetings and making decisions, in coordination with other show leadership, on proposed story ideas. They are in charge of the daily show, making assignments, rearranging the board for news and making sure that stories within segments are paired well together and that each hour of the show has a varied selection of stories that reflects the day's news. The line producer also works with hosts directly, making real-time decisions if segments need to be busted or moved at the last minute. And crucially, they are the last set of ears on each story before it airs.

UPDATE PRODUCER

The update producer takes over from the line producer after a show's first two-hour feed. They are responsible for incorporating any major news that breaks during the subsequent feeds. Depending on the show and the amount of staffing, they may need to fill other production roles, such as booking guests and directing newly booked conversations.

PODCAST PRODUCER

Producers are the intellectual troops of podcasts in the same way as for shows, and it is often their ideas that get turned into episodes. Even when an idea is not theirs, they may be the ones figuring out how to turn it into a 30-minute episode. Podcasts tend to have smaller staffs and fewer resources, so their producer may have more editorial responsibility as well as more creative freedom. Podcast producers, especially those working on long-form narrative or interview-based productions, are also less likely than radio producers to do field reporting.

a. We don't handle physical tape anymore at NPR, but we still use "tape" to refer to digital audio that we gather, mix and broadcast, so I'm using it that way throughout the book.

b. At the management level, each show or podcast has an executive producer who oversees the operation.

Producer as "Travel Agent, Mom, Journalist"

As a producer on *Weekend All Things Considered* and then on the national desk, Liz Baker covered wildfires, hurricanes, elections, protests, you name it. She describes the job of producer as "half travel agent, half mom, half journalist." Her math is not wrong: producing can seem like doing several jobs at the same time because it involves such a dizzying array of responsibilities.

"A producer is kind of everything," Liz says. "But first and foremost, you're a journalist."

Producers often propose what angles to pursue, whom to interview and even how to structure a story. At NPR, the producer may have pitched the idea for the story, even if it's passed on to a reporter to do. "You're just as much a reporter as the reporter you're working with," Liz says. "I think a lot of producers are never really told that, and then they're playing catch-up."

Producers are often called on to be the extra eye on things. Liz says that might involve monitoring social media, weather reports or other outlets' coverage or going out to talk to people.

"Sometimes it's even going to the press conference and not being afraid, just because you're a producer, to stand up and say, 'Liz Baker, NPR. Governor, why is the death toll wrong?' "

There's a common image of the producer as the person who holds the microphone while someone else conducts the interview. "That's the most visible part of the job," says Lauren, "but it's actually the smallest."

Lauren, who teamed up frequently with education correspondent Elissa Nadworny and did several stints as a coordinating producer on the ground during the war in Ukraine, says the producer is also a second set of ears. During interviews, she's listening to whether the reporter has touched on all the essential points and isn't forgetting an important question, or isn't missing an unforeseen one raised by something the interviewee just said.

"You're thinking about what questions need to be answered," Lauren says, "because we're not going to get a chance to come back and re-interview this grandmother again."

After an interview, the producer will get together with the reporter and the editor to talk about which cuts to pull. It's often after that conversation that a considerable amount of editorial discretion is given to the producer. "You're making a lot of decisions about the rhythm of things," says Lauren.

Having been an extra set of ears can be helpful when the reporter sits down to do the story. "I'm often writing it side by side with them," Lauren says, "or if not, I'm sitting in on the edit making sure everything is true to what actually happened as we both saw it."

Producers are sometimes surprised at how much writing is part of the job, says Brent.

"If you're producing a piece for a segment on the *TED Radio Hour* or an intro to the *NPR Politics Podcast*, someone has to write that," he says. "A person whose voice listeners hear—that person didn't always write those words. Someone had to write them and often that's a producer."

He always advises young producers to "read a lot and listen a lot and think about the language."

I would add one more thing: read chapter 8 in this book, on writing conversationally.

Producer as Logistics Expert

On assignment, producers often become a kind of quasi travel agent. But it would be a serious understatement to say that they arrange travel the way the rest of us do. Yes, they go online to book hotel rooms and rent a car. But this kind of travel planning requires journalistic chops.

When Liz went with reporter Brakkton Booker to cover Hurricane Florence, they needed to set up a base. The path of the hurricane as it approached the Carolinas kept changing, and no hotel booking site she knew of had data on flood risk factors. So, she did what can only be described as reporting, studying historical hurricane paths, elevation levels on topographical surveys and 100-year flood data.

"Looking at the maps I was able to figure out, OK, there's a river, and we need to be [on] the south side of the river, otherwise the river will flood and we'll get stuck on the north side." She also called around to hotels to see which ones had generators. It paid off. The hotel she selected not only had a generator, but it also remained clear of the flooding, and "all the other news crews ended up there the next day."

Lauren went with the first NPR reporters sent to cover the war in Ukraine in early 2022 and did several stints as a coordinating producer. When she arrived the first time, Russia had just invaded, and the team had evacuated from Kyiv and was camping out in their interpreter's living room in Lviv. With the help of fellow NPR producer Julian Hayda, Lauren looked for accommodations at a local university dormitory. "He and I went to check the place out," Lauren says, "making sure they had an adequate bomb shelter [and] robust internet, that it wasn't located near any potential military targets and that our team could do basic things like make coffee and grab food somewhere close."

On moving day, they loaded all their equipment—recording and studio-connection gear, satellite phones, bulletproof vests, helmets, etc.—into several vehicles and headed over to the university. There, they built a studio by piling pillows on a blanket-covered desk. And Lauren made sure everyone kept quiet while reporters recorded their stories with blankets over their heads.

They needed a car to get around, but renting was not a reliable option, so Lauren and Julian were told to buy one—not an easy task in a war zone. The DMVs weren't open, for one thing, "and we had to make sure that we weren't doing something illegal by not finishing the registration," Lauren says. And because there was a run on fuel, she bought an extra tank of diesel but had to do some research to figure out its shelf life. It was a similar story when she went to Puerto Rico with Mandalit Del Barco to cover the aftermath of Hurricane Maria. Lauren brought a solar panel to power their laptops and satellite phones. "You're really thinking about basic necessities and managing your reporters and making sure they have what they need to do their job."

For breaking news stories, planning has to be done quickly. But that doesn't mean a producer should be impulsive. Liz has some hard-won advice to offer:

> The number one thing I tell people is to try to slow down. Because so many of the assignments that get tossed at them—it's like, "Hurricane! Coming! We need you to go now! It's going to hit tomorrow!"
>
> Sometimes you can't slow down, because you have to leave or your flight is not going to be able to land. But if you can, slow down a little bit. "OK, I'm not going to panic and throw everything in my bag and run to the airport in two hours and try to get the next flight. I'm going to take a flight at the end of the day, or first thing tomorrow morning, so that I can settle myself and do some research and figure stuff out."
>
> Because once you're on the go, it's way harder to find the time to do all that. And a lot of times that's what the reporter is doing anyway. They're going on the next flight, and what they really need is for the producer to figure out all the extra things, like where to stay and what is our car situation and what are some of the places we could go to and who should we follow on Twitter.
>
> Breaking news is urgent, but it doesn't need to be panicky.

Liz recommends thinking through everything that might happen on a trip. "If this happens, what do I need? If that happens, what do I need? I'm going to [cover] a hurricane, so obviously I need all the rain gear, and I also need waterproofing stuff. But then after the hurricane there's going to be press conferences, so I need a mic stand and a long cable."

She plans for food, water, internet connectivity and security, and maintains what she calls "the most paranoid packing list of all time."

It includes recording gear, cables, headphones, batteries, chargers and other basic necessities such as press ID, a Sharpie and maps downloaded to her phone. But it's really much more than a packing list. It's actually 12 alternate checklists, each for a different situation she could get sent to—including fire, earthquake, civil unrest, power outage or extreme cold (see appendix 2).

The lists represent all the "things I learned the hard way that I feel nobody else should," such as bringing a battery-powered fan to cover a hurricane, or a neti pot to a wildfire. "If you're going to be in a fire zone where you're inhaling all this smoke, it's really nice to be able to clean the soot out of your sinuses!" she says.

Keeping the Team Safe and Fed

Liz wasn't kidding when she said part of the job is being a parent. Often the top priority is making sure everyone is safe and fed. She and Eric Westervelt were in California reporting on forest fires, doing a live two-way with *Weekend Edition Sunday* host Lulu Garcia-Navarro from inside their rental car, when a tree branch fell on a power line above. She quickly and quietly started the car and backed out of the danger while Eric continued talking.

Gisele Grayson went with reporter Joanne Silberner to cover the 2004 tsunami and earthquake in Banda Aceh, Indonesia, and a major aftershock hit the city. Gisele went into Joanne's room and found her asleep.

"Joanne would not wake up. And I said, '*Wake up!!!!*' And I'm dragging her and shouting, 'Get out, get out!'" Finally, she did get up. "But I would have carried her out!"

When Liz was in New Orleans covering Hurricane Ida in 2021, her team ran low on food because restaurants and grocery stores were shut down. So, she went out in search of sustenance for her reporters, because they "don't have time to walk around for an hour and stand in line to get a hot dog from the guy who's cooking up his entire freezer down the street."

NPR producers are trained to work in hostile environments, which includes administering first aid. Before going out, Liz finds out any medical conditions her team may have. She says reporters sometimes neglect their most basic needs while chasing a story. They may forget to hydrate or they may risk frostbite to get another interview at a wintertime protest. That's why, Liz says, it's helpful to have a producer telling them, "We have a lot of tape already," or "It's pretty hot today, let's make sure we're drinking water!"

Field Producing: On the Road With a Host or Reporter

NPR often sends journalists out to do on-the-scene reporting. Back when the equipment was heavy and bulky (reel-to-reel tape recorders once weighed around 20 pounds!), a three-person team would be sent: an engineer, who did all the recording; a producer, who handled the logistics and then mixed the story later in the newsroom; and a reporter or host. Digital recorders, however, are small and weigh just a few ounces, so now it's usually just a two-person team going out, with the producer handling both recording and the mixing. On big stories, a field producer might be working with several reporters, whereas on a feature assignment, it could be just the producer with one reporter.

An internal guide to field producing that Liz and her colleagues Marisa Peñaloza and Walter Watson wrote for the national desk describes the role like this:

> There may be days when you never leave the hotel, and days when you run around recording pressers, gathering vox, and chasing down leads all on your own. You might work closely with one reporter who needs extra help on breaking news files, and barely see another reporter who is working on a feature. Trust your instincts, and ask the desk editor for advice if you are unsure how to prioritize.

With a varied and unpredictable schedule, field producers may find themselves putting in 20-hour days, thinking or talking about their stories during every waking moment with a reporter or host. Not all reporting trips are that intense, but even under the best of circumstances the job still demands working closely with another journalist—often for days at a time, and sometimes for weeks.

"The first few times you go out and record anything, you're going to be scared out of your mind that you're going to mess it up," says Lauren, recalling her first field-producing assignments.

You may be afraid that you're not miking correctly or that you haven't set the levels properly. Or that you've missed a sound opportunity. But that's a good sign. "When you lose that fear," she says, "that's when you kind of get a little sloppy."

COLLABORATING WITH HOSTS AND REPORTERS

The success of field-producing trips depends on collegiality and collaboration. A reporter usually wants the producer's input on questions to ask during the interview, not just before meeting a source, but also while talking to them.

"In most interviews," Lauren says, "almost every reporter I've worked with generally turns to me and says, 'Do you have anything to ask?' "

So, while making sure the recording levels are good, listen closely to the interview. Education correspondent Elissa Nadworny often defers to Lauren when she needs to pause and think. "I know that she needs a moment to regroup when she does that. And so, I often have questions at the ready so she can take a minute to tick through her mental list of what she needs to hit."

Barry Gordemer tries to maintain eye contact with the reporter while also scanning the surroundings for "little things on the periphery" that might be relevant to the topic of the interview.

Once, he was out with *Morning Edition* host Bob Edwards, who was interviewing country singer Glen Campbell at his home in Phoenix, when Barry noticed a photo of Campbell and Johnny Cash on a wall full of gold records.

"I pointed the picture out to Bob, and Bob looked up and said, 'Oh, there's a picture of you and Johnny Cash, two sons of sharecroppers.' I didn't even know that Bob knew that!" he recalls. "Bob just seamlessly wove it right into the interview."

The producer-reporter collaboration can boost speed and focus on a deadline assignment.

"You're getting things done way faster and more efficiently," says Lauren, who recalls covering the aftermath of Hurricane Maria in 2017 with Mandalit Del Barco.

"Mandalit would be writing her radio piece while I would be pulling cuts, and she would file the piece and start writing web copy while I would mix the radio story." When a reporter is on breaking news and being asked to file for newscast, the newsmagazines and digital, she says, "one of those things is going to fall to the wayside if you don't have a little help."

Also in 2017, Jane Greenhalgh traveled to Borneo with reporter Michaeleen Doucleff for a story on the causes of pandemics (a prescient piece of reporting, three years before the COVID-19 outbreak). It was not a deadline assignment, but the reporting was more effective because the two were working together, both going into a bat cave with all their recording gear.

"I'd be doing the recording so that the reporter can really concentrate on asking the questions," says Jane, "and not having to worry, 'Are all my levels OK and not overmodulated?' Especially when you're recording in a bat cave, where it's dark and there are cockroaches all over the place."

When hosts travel, they rely on producers to free them up for their other duties.

"The host is inevitably pulled in 700 different directions at once," says Lauren. "Sometimes they're literally hosting the show. So, you have to kind of be in the background doing the things they don't have time for."

They may need you to go to a press conference or gather vox while they're writing. Or they may need you to find a place to launder a shirt.

"You don't want that person to feel weird about going to their interview tomorrow because their shirt is seven days old," says Liz. "You have to be really humble and able to ask what people need and how you can be helpful and—this is something I have a lot of trouble with—don't feel bad if the thing that is most helpful is something that seems easy to you or stupid."

That doesn't mean catering to every whim. *Morning Edition* host David Greene loved listening to country music on long drives. But on one trip across Montana, Lauren made him turn it off so that she could "plug in the laptop and be pulling the audio cuts while he was driving."

Still, producers should roll with the different working styles and rhythms that hosts and reporters have, says Liz.

"Some people really like to be left alone to write," she says. "Other people like to talk about their entire script as they're writing it. Some people like it when you write a draft. Other people just want your help with an outline."

Asked what she looks for in a producer, *All Things Considered* host Mary Louise Kelly echoes the point Brent Baughman made earlier: "beautiful, clear writing."

She also values honest feedback. Almost every interview an NPR host does will also have a producer and editor assigned to it. "We often will book 15 minutes on the calendar just to talk through what's our plan," Mary Louise says. "And often, even when I thought I had a plan, listening to what other people are curious about, I'll realize, 'Oh, I don't think everyone knows that.' Or, 'There's this whole aspect of the story I knew nothing about that this other person is really interested in.' Or, 'I have 47,000 questions I would like to put to the secretary of state. Let me kick it around with my team a little bit, because I'm not going to get to put 47,000 questions to him. What do we think are the top three that we really need to ask?'"

GETTING THE STORY OUT ON TIME "WITH MINIMAL AMOUNTS OF PAIN"

Unless the piece is very rich in ambi, or natural sound, most field producers on fast-moving news stories will send all the audio elements to be mixed by a show producer.

"An early heads-up that something has to be crashy, or that something's going to have a lot of ambi, is really, really, really important. Because if the show producer doesn't know that, they're going to plan their day differently. And that could really be the difference between a nice mix and a shitty mix."

While gathering tape, Liz tries to share her plans because she knows the show has to commit resources to bringing it across the finish line. If she's out with a correspondent like John Burnett, doing interviews for a same-day turnaround for *All Things Considered*, she'll make sure to send a message early in the day that goes something like this:

> Hey, I'm with John Burnett. We know we have a piece on *All Things Considered*. We need two more interviews that we're doing by 11 a.m. and then we're going to come back to the hotel. We're hoping to edit by 2 p.m., and I'll file as soon as possible. I'll send you the acts before we edit. Some of the acts might change, but most of them will be the same, so you can get them fixed and line them up on your board. And I'll put a tease in the collection.

One last piece of advice Liz gives is to ask for help. Producers take on a lot of responsibility, especially on reporting trips, and may feel pressured to stay up 24 hours a day. But that's not healthy. There are people in the newsroom who can take a lot of the pressure off: producers and engineers who can record televised press conferences and pull cuts or fix audio problems. "There's no heroism," Baker says, "in taking it all on yourself."

Producing is about working with the resources you have to meet your deadline, says Christina Cala, who produced NPR's award-winning coverage of the Trump administration's family separation policy and efforts to limit asylum protections. "For a long time, I was a perfectionist about everything. And you just can't be," she says. "Success is getting the thing out on time with minimal amounts of pain."

How Liz Baker Covered the 2023 Monterey Park Shooting

On Sunday, Jan. 22, 2023, at 6 a.m., national desk producer Liz Baker got a call from Washington, D.C. "Can you go to Monterey Park?" an editor asked. There had been a shooting overnight during celebrations for the Lunar New Year at a dance hall in the predominantly Asian American town in California's San Gabriel Valley. Liz, who lives in Los Angeles, tossed a couple changes of outfits in her kit, which she always keeps at the ready—with a recorder, extra cables, memory cards, an external laptop battery and assorted necessities like a Sharpie and AA batteries—for moments such as this one.

A correspondent who lived near the site, Sergio Olmos, had been there since the early hours of the morning and was filing newscast spots and updates for *Weekend Edition Sunday*. Liz planned to meet up with L.A.-based culture reporter Mandalit Del Barco to cover the shooting for *Weekend All Things Considered* and for the later newscast feeds. Mandalit set aside reporting she was doing on the upcoming Oscar nominations to join the breaking news coverage.

Liz and Mandalit had only a few hours until *Weekend All Things Considered* was scheduled to air at 2 p.m. PT. They talked over the phone and decided that the gist of their reporting would be, "What do we know now?" They aimed to get one or two clips from a press conference and one or two pieces of community

4:00 Norinko handgun from cargo van registered to suspect also clothing he wore during crimes
4:29 · Wrestled away semi-auto assault weapon
5:27 arrest in 1990 for unlawful posession of fire arm
6:06 · was only 1 person Brandon Tsai who disarmed suspect
7:03 a lot of details are coming forward
* 7:19 we still don't have a motive
Solis: 2 survivors will be released

* 1:56 we do not have a motive yet. we want to know as much as you
2:31 jealousy - hearing those things too but not confirmed
3:05 · Weiss - dispatch recieved 911@10:22-10:23, officers on scene within 3 mins
5:45 we believe on victim was shot outside
Question about timing of public notice
* 7:21 Questions: was it planned, what drove a madman
7:40 this is disturbing. how someone can reason to do this
8:06 · Hemet PD put out info on poisonings

Liz Baker's notes from a press conference given by Los Angeles County sheriff Robert Luna on Jan. 23, 2023, after a mass shooting two days earlier in Monterey Park, California. As she often does while listening and taking notes, Liz put a star next to each of the salient quotes; two of the three with a star ended up in reporter Adrian Florido's two-way with Ari Shapiro on *All Things Considered*. The numbers in the left margin refer to the recorder's clock, not the time of day.

reaction. Liz knew that they wouldn't make it to Monterey Park in time for a press conference by the Los Angeles County sheriff, so she asked the operations desk in Washington, D.C., to record it and pulled over on her way out to Monterey Park to take notes as she listened to it on a local radio station.

Once Liz arrived in the area, she looked for a place where members of the local Asian American community might gather. "I started taking the side streets and came across a Taoist temple that, because it was Lunar New Year, was holding festivities outside," Liz says. "There were lion dancers, there was a drum. There were a ton of people."

"I wish I had taken Mandarin when my brain was plastic enough to become fluent in it," Liz says, reflecting on not speaking the language. Still, she did what good reporters do when they don't have access to an interpreter: find someone on hand who can translate. Seeing lots of multigenerational families, she asked English-speaking members of the younger generation to put her questions to their parents.

Authorities were searching for the killer, so Liz made sure to get answers that didn't presume the shooter was still at large. That was wise, because the gunman died in a standoff with police before *Weekend All Things Considered* went on air.

From the temple Liz went to city hall, where she met Mandalit, and they waited for a follow-up press conference. It started late, and showtime was approaching. Liz connected her recorder to get the audio feed. "So I was rolling on the press conference and listening and writing time stamps," Liz says, with Mandalit standing beside her. When the reporter heard something she wanted to include in the story, she would alert Liz, who would mark a star next to the time stamp. "And then once it wrapped up, it was like, OK, here are two things that we starred," Liz says. The news conference ended just 10 minutes before the top of the show, but Liz had already isolated the cuts and sent them to the newsroom. She took out her phone and opened an app that allows a high-quality live connection to the studio; once the show started, Liz recorded Mandalit's side of her live two-way with the host, Michel Martin, just in case the connection failed.

After Sunday's reporting, Mandalit went back to her Oscars story, and a couple other reporters got assigned to cover the Monterey Park shooting. Liz covered press conferences and got vox for their stories and also filed her own spots. One press conference needing coverage the next day was being held by California governor Gavin Newsom, but it was scheduled to begin just before Adrian Florido was going on *All Things Considered* with an update on the shooting investigation. Liz told him, "I'll go, and if [Newsom] ends up being late, because he's always late, you can go on air just fine and I'll get the tape and fill you in on what he says." As it turned out, the governor didn't show. He was called away because another shooting had just happened hundreds of miles away, in Half Moon Bay.[a] The press conference in Monterey

Park ended up featuring the same officials who had briefed the press earlier, and although they said nothing new, Liz took notes as before and starred quotes for Adrian's two-way.[b] "Producers sometimes go on these wild goose chases, because your time is not as valuable as the reporter's," Liz says. Still, that doesn't make her job any less important, because her reporting was essential to NPR's coverage of the story.

a. NPR correspondent Eric Westervelt was sent to cover that shooting, which was not related to the Monterey Park killings.

b. Ari Shapiro, "The Latest on the Monterey Park Shooting," interview of Adrian Florido, *All Things Considered*, NPR, Jan. 23, 2023, https://www.npr.org/2023/01/23/1150844954/the-latest-on-the-monterey-park-shooting.

CHAPTER 4

Sound Gathering

In the late 1990s, NPR ran an ad campaign proclaiming, "NPR Takes You There." It was the idea that radio can transport listeners to places they've never been and introduce them to people doing incredible things, and that with their ears the listener can experience it as though they're there right next to the reporter.

"We're an aural medium, and sound is our currency," says Midwest bureau chief Cheryl Corley, who spent more than 28 years as a reporter and correspondent* covering some of the country's most important news stories, including Hurricane Katrina, the Trayvon Martin shooting and the presidential campaigns of Barack Obama. "You want to make sure that you have things that are going to be interesting to the ear."

The journalism we do involves many of the skills practiced by reporters for other media—finding sources, talking to people, digging through documents, getting to the scene of the action, observing carefully. But there's one skill that's unique to our medium: listening, or "reporting with your ears."

That doesn't just mean hearing what someone is saying, but also how they are saying it—the texture, the cadence, the intonation, the emotional quality. And in deciding which words to isolate for an actuality—an excerpt of an interview or press conference that is the audio equivalent of a quote—the *how* may actually matter more than the *what.*

Listening in audio journalism means finding the sounds that tell a story: the whine of an air raid siren in a war zone, the echoes in a building abandoned because of a chemical spill, the roar of a trading pit in Chicago.

* At some news outlets, reporters work out of the main newsroom while correspondents are based in the field. At NPR, the distinction is not about location but rather experience, with *correspondent* signifying a more senior ranking than *reporter.*

They help form the images in listeners' minds that take the place of what they might see in a TV news report or in a photo on a news site.

Sound plays several roles in storytelling. It is immersive, because it draws listeners into the auditory experience of being there. It can be descriptive, sometimes demonstrating a key point or effectively conveying elements of the narrative. And it can serve as a signpost, letting listeners know where they are in the narrative, usually by signaling the start of a new chapter or scene.

Sound also makes for efficient storytelling. Hearing someone slam a car door shut, jangle the keys on the keychain and start the motor might take a few seconds, but certainly less time than it takes to hear someone read, "She slammed the car door shut, jangled the keys on the keychain and started up the motor." The right sound is worth a multitude of words.

This chapter is about gathering actual sounds for use in news stories. Later, in chapter 12, when I talk about sound design in podcasting and long-form features, I'll consider when it's permissible to use sound effects to help tell a story. In news stories, the sound we record is always real, whether we record it with our own microphone or obtain it from another source.

Eyder Peralta on the Intimacy of Radio

I fell in love with audio because of the intimacy. I've always loved film, and if you watch Pedro Almodóvar's films, they're all super-intimate. It's not about what you see. It's about what you hear. It's very "radio" because the microphone in his films is very close to the person's mouth. I didn't realize, until much later in my career, that the reason his films feel so intimate is because they're recorded like we record our audio.

The first time I noticed this, we were in Sudan following a military takeover of power. I was trying to get this young doctor who led a huge popular rebellion that ousted the 30-year dictatorship of Omar al-Bashir. He [the doctor] was on the run, and he tells us through some intermediaries, "Be at this parking lot at 9 p.m." And we get to the parking lot and we get into his car, which he parks facing outward so he can haul ass right as soon as something happens. And we do this interview in the car. It's a really small car so I have this shotgun mic right up at his mouth. And then when the story airs, I listen to it and I'm like, Whoa! You got all of that tension of that moment! All of it from his voice. It felt like you were in that cramped car.

So imagine somebody speaking really close to your ear. That's what's happening when you put your microphone right up to someone's mouth. No matter what you're talking about, it creates an intimate environment.

Types of Sound in Stories

At NPR, the catchall term for the sounds and voices we collect is "tape." It harks back to the days when the recordings were on magnetic ribbons wound around reels. Tape signified the audio that had been gathered outside the newsroom and that brought the stories to life, as opposed to the parts of the broadcast that were created in the studio, such as the host's introductions or interviews with guests. This term has stuck, even though all the audio we work with nowadays is on computer files.

For the most part, there are two types of tape in news stories: acts and ambi.

An act, short for "actuality," is a cut from an interview, or from a speech or news conference. It's similar to the notion in TV news of "A-roll," the footage taken by the main camera, as opposed to "B-roll," the secondary shots used in cutaways or as filler. The actualities, or acts, form the backbone of the audio piece, much as A-roll does in TV news.

Ambi, or ambient sound, is also known as natural sound, or nat sound. As its name suggests, it's sound recorded in the real world that captures the feeling or mood of a place. And it's more or less like B-roll in TV, in that it gives the rest of the piece vibrancy and texture.

Ambi can be used in a variety of ways:

AS A POST. The sound is heard by itself ("in the clear") and not mixed with a reporter's tracks. The most effective post tends to be a sound that is sharp and evocative, such as a door closing, a soda can being popped open, a gymnast's feet hitting the mat, a diver's body splashing into the water.

AS A BED. The sound has a more continuous nature—such as the din of a crowd, chatter on the street, a live band playing, a recorded musical release—and is usually heard underneath a track.

AS A POST AND BED. Continuous sound can be used as a post first, then faded down to form a bed under the reporter's tracks.

Another thing about ambi: you record it with your own microphone. There are, of course, sounds featured in stories that are recorded off the internet or obtained from outside sources. These can be audio files or videos from social media, feeds from press conferences or speeches, recordings of other news broadcasts or files downloaded from audio archives. They may be used in a story as either a post or a bed—but remember that, legally, you may need permission to use it, and, ethically, it's important to authenticate the clip and provide proper attribution to the source.

Never assume you have the rights to any audio or video. For example, even though Martin Luther King Jr.'s "I Have a Dream" speech, delivered from the steps of the Lincoln Memorial in 1963, is such an iconic moment in U.S. history and can readily be found online, you still have to pay to use it. That's because the King family owns the rights. Just because a piece of audio is available online, does not mean you have the right to use it. Sometimes you can use copyrighted audio pursuant to the "fair use" doctrine, which protects freedom of expression, but you should make sure you are familiar with the fair use principles and consult with an editor and with a media lawyer, if you have one in your organization, before deciding that your use is fair. If your use would not qualify as fair use, then you would need permission to use the audio, and you may even have to pay for it unless the copyright holders agree to let you use it without a fee. Make sure to factor in time to get in touch with the owners, whether it's an audio library, another news broadcaster, a commercial institution or a private individual, to obtain the rights to use it in your story. Some audio files are old enough that they are no longer protected by copyright; they are considered to be in the public domain and can be used freely. Also, works created by the federal government, like recordings made by NASA or the military, are not copyrightable and may also be used freely.

It's important to respect the rights of copyright holders, because infringement lawsuits can be expensive and time-consuming. No one wants to pay damages for using unauthorized audio.

In cases where it is permissible to record sound off the internet, you still have no control over how it was recorded or whether it even is what it purports to be, so it's important to authenticate the tape. And even when it is authentic, it's rarely the kind of tape that will tell a story as well as the tape you gather with your own gear.

In learning how sound is gathered in the real world, it's helpful to consider the similarities with how visual stories are told. Next time you watch a TV show or a movie, pay attention to the images. Notice the varying sequences of wide, medium and tight shots? TV news is similar. News photographers on assignment will move in toward and out from their subjects, not just by using a zoom lens but also by physically shifting their camera, in order to get each of these types of shots.* In audio, it's not all that different.

So, a photographer's wide shot of a barbecue party might show the entire crowd in the yard, a medium shot could focus on the chefs at the grill, and a tight shot would be a close-up of a burger being flipped. For an audio reporter, the corresponding wide shot of sound would be the hubbub of the crowd, the medium shot might be the chefs' conversation about the food, and a tight shot could be the clank of the spatula.†

It's worth noting that the best videographers recognize the power of sound and also listen for ambi as they are filming. And by the same token, the best audio producers think about sound recording visually.

"I'm very careful about thinking about it like a movie," says Liz Baker. "You want to get the sound in the stairwell versus the sound outside, and the sound of somebody cheering while everyone else is banging the pots."

---

* While videographers can zoom in to get a tight shot, audio journalists really have to move their microphone closer to the subject to get the same effect. That's because even the microphone equivalent of a zoom lens—the directional mic—cannot completely shut out extraneous noise coming from around the subject the way a camera can exclude any unwanted visual elements from its frame.

† Getting a wide shot can sometimes be done by turning a microphone away from the source of a sound. So, if you're standing in the midst of that garden party, just point the microphone toward the periphery.

Recording close to the source will often give you good material for audio posts, while wide shots are better for audio beds or for posts that fade down and become beds.

One story where I relied on posts and beds was one I mentioned in chapter 2, an audio postcard on the annual running of the bulls for children in a village near Madrid.[1] (Real bulls weren't used; grownups pushed wheelbarrows with life-sized stuffed animal heads on the front.) On the scene, I recorded a marching band playing at the starting line, a bottle rocket marking the start of the run and the joyful squeals of the little ones. Here's how the opening scene featured those sounds with my narration:

> (*Post of marching band and crowd noise, fades under tracks.*)
> JEROME: The band warms up the crowd in the street and then—
> (*Bang, posted over music and crowd noise.*)
> JEROME: —a bottle rocket signals the start of the *encierro*, the running of the bulls!
>
> Many young Spaniards dream of one day sprinting a hair's breadth ahead of a thousand-pound charging toro . . .

As I was recording, I knew that I would need a variety of sounds—and different takes on each sound opportunity. When I found a group of kids ambushing a "bull" and whacking it with rolled-up newspapers, I recorded the action from close in and from a little farther out, where some of the grownups were cheering them on. I also recorded a local bullfighting youth group singing their fight song.

It's relatively easy to capture the wide shots. You don't have to be in the thick of the crowd or uncomfortably close to your subject. But they aren't particularly exciting. That's why you have to overcome your inhibitions for the tight shots—getting right up to the person with the old photo album so that you can record their fingers turning the page, or right next to a gardener's trowel turning the soil.

"Get close," says Christina Cala, who did a lot of field recording as a producer for *All Things Considered*, "closer than you think you might need to."

It's normal to feel a little self-conscious when you're holding a microphone and wearing bulky headphones in the middle of a crowd. Christina deals with that by smiling and giving people around her a look that says, "I know this is weird."

"People will kind of forget that you're there," she says. "Even though it doesn't seem like that'll be the case."

It helps to explain what you're doing, especially if you're with one person or a small group and have the time to do it. That way you will have won their trust, or maybe their indifference. After a while, Christina says, they get used to it. "Oh, that's just the weird girl with the microphone."

I subscribe to Christina's school of the smile and the "I know this is weird" look. That's how I was able to overcome the feeling of sticking out like a sore thumb as I ran across the village square of Miraflores de la Sierra with all my recording gear into a horde of tiny would-be matadors and make that audio postcard.

Choosing the Right Microphone

Which kind of microphone should you carry? It depends. Microphones are classified according to the direction or directions in which they are most sensitive, and how they are powered. Here are some you will most often find in newsrooms:

DIRECTIONALITY

OMNIDIRECTIONAL. Captures sound from all directions equally, so it doesn't matter where the source of the sound is relative to the microphone. But it must be held close to the subject if there are competing background sounds. Omnidirectional microphones experience less wind noise and handle *p* pops—distortion created by air blasts coming from the mouth—better than other types of microphones.

CARDIOID. Has an upside-down, heart-shaped pickup pattern. It captures sound mostly from the front, not so well from the sides, and rejects sound at the rear. It can be quite sensitive to plosives. Good for handheld, back-and-forth style interviewing. And it's more susceptible to wind noise.

TYPE OF POWER

DYNAMIC. Driven by sound pressure. It's forgiving with handling noise, like from your knuckles creaking or fingertips moving ever so slightly. It's good for crowded environments where you may need to pull it out fast and start an interview. But it's not very sensitive and doesn't do well with very fine or low-volume sounds.

CONDENSER. Draws voltage (known as "phantom power") from your recorder, which is an extra drain on its batteries, though sometimes these mics have a replaceable battery built in to provide the phantom power needed. The mic is very sensitive, and excellent for getting very low-volume or distant sounds. But you'll want to use it in controlled environments, where there's not a lot of wind and not a lot of noise.

SPECIFIC EXAMPLES

SHOTGUN. Usually a condenser type of mic with a supercardioid pattern, which extends like a long and narrow cone from the microphone's axis, efficiently rejecting sounds outside it. The mic is very sensitive, and can be really helpful if you can't get close to your subject, such as in a media scrum around a politician standing in a capitol corridor. Shotgun mics often are susceptible to handling and wind noise. Bring a shock mount and a windjammer (a long-haired type of mic screen also known as a "dead cat") to cut down on these sounds.

LAVALIER. A small lapel clip-on, often wireless, condenser microphone that can be omni or cardioid. It's usually used on television sets and for people giving speeches to an audience. But some audio reporters and producers favor them when the person they're interviewing is walking or engaged in another activity that makes it difficult to hold a microphone to their mouth.

At NPR, many producers and reporters find that a condenser shotgun and a dynamic omni will get you through most field-recording situations.

Sound Setting Up Scenes

Most reported audio stories are made up of scenes, a term that is used in much the same way as a movie or play. And ambi can really help usher in a scene, as it does with the help of descriptive writing in the opening lines of this story by Ryan Lucas, on Ukrainian civilians setting up checkpoints after the Russian invasion in 2022:[2]

(*Ambi of passing vehicle.*)

RYAN: Bundled up against the cold and wind, the men huddle around a rusted oil drum fashioned into a wood-burning stove. There's smoke billowing from a crooked pipe jutting from the top.

(*Ambi of metal clanging.*)

RYAN: This, for now, is their war, manning a checkpoint on a road outside the city of Lviv in western Ukraine, hundreds of miles from the battles raging to the east. One of the men is Oleh Pokhrovetsky.

(*Pokhrovetsky speaking in Ukrainian.*)

Ambi can also serve as a transition from one scene to the next, as in this piece I did on a Buddhist community trying to limit the time spent on their phones.[3] I started it at a Zen worship space, and then used an ambi post to change scenes.

JEROME: Carlos Moura and Leslie Cohen are among the people taking part in the screen mindfulness workshop. Afterward, Cohen, a tourist from San Diego, says the chance to turn off is what brought her here.

COHEN: We were in, like, Ocean City. And just—you know, the TV was on. The kids were on their screens. And I had a moment of, like, I've got to find a place to meditate as soon as I get to Washington, D.C.

(*Ambi of gong being struck and resonating.*)

JEROME: A meditation session begins at a different Buddhist center a few miles away . . .

While the first decision that often comes up in planning a story is whom to interview, another equally important decision is what sounds are needed for the story.

The second decision is rarely simple. You can schedule interviews, but sound is harder to predict, especially when you're flying halfway around the world, the way Jane Greenhalgh did for her story on bats in the previous chapter.

Before any reporting trip, Jane tries to set up three scenes. In Borneo, she and Michaeleen Doucleff had arranged to go on a bat-hunting expedition with a scientist, visit a lab where the flying mammals were being tested for viruses and enter a cave frequented by tourists.

Inside the Gomantong Caves in Borneo, she and Michaeleen both took out their shotgun mics and recorded the dripping water and hissing cockroaches, and the conversation they had with two Australian tourists, Jenny and Graham Whitaker, who were trying to navigate a path slick with bat dung:[4]

MICHAELEEN: When you walk inside, it's breathtaking.
(*High-pitch trilling sound fades up.*)
JENNY: My goodness!
GRAHAM: Fantastic!
MICHAELEEN: The cave . . . is gorgeous. It looks like a cathedral. Forty, 50 feet high with this light streaming in from the side—it really is spectacular.
JENNY: How amazing!
MICHAELEEN: So we're going deeper into the cave now.
JENNY: Be careful.
GRAHAM: It's slippery.
JENNY: Whoa, almost just slipped!

This scene demonstrates the power of getting people reacting to ambi, says Jane. "Sometimes people think of ambience in scenes as, 'OK, I just need to get some crunching through the forest and I need to make sure I get the sound of the bat squeaking as it gets caught,'" she says. "But a lot of the action is in the voice of the people who are in the scene."

That's why it's best to have your interviews on location where you plan to get your ambi, "as opposed to just going out there doing an interview and then recording a couple of minutes of traffic sound or a couple of minutes of a waterfall."

Again, think about how TV and film tell stories. In addition to showing wide, medium and tight shots, image sequences often show action and reaction. The camera might focus on a gift being opened, then cut to a child's ecstatic face; or it might show a pastor preaching, then the raptured expression of a congregant.

Jane says it's the same with audio—the human reaction is powerful. That's why, when she and Michaeleen were out in the rainforest looking for bats with the scientist, she clipped a lavalier mic on the scientist's collar so that they would get his reaction when he heard a bat. Even if all he said was, "There's one!"

Recording Interviews and Getting Room Tone

A producer should constantly be checking levels when recording ambi and interviews. *All Things Considered* producer Jonaki Mehta tries to keep the needle in the middle of the range, which is usually between −12 and −8 decibels. "I'll usually hold the mic—they say about a fist length away—but if you have a really soft-spoken person, maybe a little closer and a little off to the side so you're not catching the *p* pops." That's the plosive effect of the air coming from their mouth, which happens most frequently with the letter *p*. Jonaki will then have a conversation with the person before the interview starts. Some producers throw an easy question, asking what the person had for breakfast. Jonaki makes small talk about any subject to get a sense of the person's normal volume, and even then she's ready to adjust once the conversation gets rolling. "Inevitably they're much louder or much softer when they're actually doing the interviews," Jonaki says. In extreme cases, you may need to interrupt the conversation, readjust the level and ask the person to repeat an answer. When in doubt, set the level conservatively. "It's better to be on the lower end than the superhigh end because the sound can get distorted and blown out, whereas you can usually boost softer sound."

And get ambi before and after every interview so that you have sound to start the scene with, if you need it. It's easiest when you're talking to people whose work or activity makes for interesting sound opportunities—a potter forming clay, a harpsichordist plucking the strings, a chef preparing a meal. Get the sounds of them in action.

Unfortunately, the ambi generated by the work of many of the people we interview (maybe too many?) is the sound of typing on a keyboard. It does little to enhance the story. So, you need to get creative.

One way is to have your recorder rolling when they greet you, and you should make it clear that you're recording. A good idea is to let them know ahead of time. You can decide later if the hellos work in your story.

Another way is to ask for a tour, especially if you're at their home or workplace. Or get them to show you something. Maybe there's an item related to the story (a relic, a gadget, a payment stub) that they can dig out of their closet or filing cabinet (if they still have a real one) or attic. Or they can show you pictures. Even if the pictures are stored on their phone, it will still get them talking in a way that evokes the images, which often makes for good tape.

Barry Gordemer recalls an interview he did with a woman who had grown a magnificent herb garden in her backyard. "And there was a great moment in the piece where she says, 'Come over here.' And you hear the snap of a twig. And she goes, 'Smell that!' That was such a neat radio moment. It was so three dimensional."

Sound-Gathering Ethics

At NPR, we don't "manufacture" scenes for purposes of news coverage. We never tell people to act something out and pretend it's real. We don't ask an interviewee to pick up the phone and pretend they are talking to their mother, in order to get a bit of scene tape. We don't ask a baker to open an oven to get the sound if they aren't doing it to bake their bread or their pie or whatever. They have to be really doing what they do. By the same token, if you arrive after an event is over, you cannot ask participants to reconvene so that you can record a simulation of the event. Plan your reporting and find out what's going to happen, and when, so you can be there and record the real thing.

You also shouldn't use effects that could be mistaken for real ambi. In news stories, we don't fabricate sounds and pretend they are real. If you interview an environmentalist who is researching whether pesticides are causing abnormalities in frogs, you cannot add the generic sound of frogs croaking when you mix your piece, as if to suggest they are the frogs being studied.*

But you can use sounds that you legitimately generate as part of your reporting. If you're trying to record a scene and your microphone accidentally bumps into someone, that could make for a good moment. For a Thanksgiving segment, the *Morning Edition* team wanted to see if it

---

* In a case that shows how much damage simulated sound effects can do, some keen birdwatchers caught CBS Sports adding canned birdsong to the broadcast of the 2000 PGA Championship golf tournament in Louisville, Kentucky. CBS apologized and promised to subsequently air only real birds' chirps. But for years the deception prompted jokes by late-night comedians and speculation of further trickery.

was possible to cook an elaborate dinner—like the kind superstar chefs are always making on TV—in a small home kitchen.[5] So they invited Christopher Kimball of the PBS program *America's Test Kitchen* into the home of one of the show's producers.

It was clearly a tight fit as Kimball made the meal with hosts Steve Inskeep and Renee Montagne:

RENEE: Chris Kimball usually works in a spacious kitchen—actually, a TV studio—in Boston. This year he came to the nation's capital for a make-ahead Thanksgiving in a kitchen that's probably a lot more like yours.

(*Kerplunk.*)

RENEE: This is the Washington, D.C., home of *Morning Edition*'s senior producer, Madhulika Sikka. She has a charming but snug galley kitchen. It's about as narrow as a spatula and you have to be a bit of a contortionist to avoid bumping into things.

(*Microphone bumps.*)

MADHULIKA: Watch your foot, by your right foot.

(*Microphone bumps again.*)

KIMBALL: So which way should I—can we go this way?

(*Tap.*)

STEVE: When there's not a lot of elbow room or counter space, Chris Kimball says planning is critical.

KIMBALL: For example, make mashed potatoes ahead of time . . .

Notice Renee's description of the kitchen being "about as narrow as a spatula" and that you have to be a bit of "a contortionist" to cook there. That gives the scene a sense of place, along with the pots and pans and the rustling of the equipment.

"Let the seams show," Barry says, adding that even when you're checking your microphone—"Testing, one, two, three," or "Is that sound in the background going to be too loud?"—can end up being great moments.

In fact, it's arguably more ethical at times to leave them in, because you are letting the listener in on the process.

Noisy Environments

A reporting trip is a big investment in a story, and it should be reflected in the tape you bring back. "One of my frustrations with field reporting," says Barry, "is when people come back and they put together a piece that could have been donc right back here in the studio."

So do your best to record your interviewees in an interesting setting. If you're talking to voters about what's important to them, have dinner with them, or go to the site of the factory where they used to work. "I like recording in noisy environments. Yeah, it makes the editing a little bit tougher. But it doesn't sound like you're in a studio."

And in the same way that it is ethically unacceptable to add to a scene ambient sound that wasn't recorded there, you also want to try to preserve the authenticity of your interview subject's aural environment, Lauren Migaki pointed out in an email she sent me after field producing in Ukraine.

> When I was a less experienced producer, my inclination was to always make sure the recording was pristine—no interruptions, making sure the AC unit is turned off, no background noise, etc. I don't do that anymore because the background is a part of the story. It puts us in a place. Nowadays, when I interview a school principal, I don't ask them to silence their cellphones or office phones because the volume of calls and content of calls they receive says something about their job.
>
> When I interview a parent, they sometimes offer to meet in a café because their kids will be loud at home—but that's part of the story. If we wanted quiet, we would have booked them a studio.
>
> Sound offers us many facts, it builds a character, it puts listeners in a place—the cattle rancher who listens to classical music in his truck, the teacher whose ring tone is Baby Shark, the little girl who is trying to pay attention to her online class while her baby brother screams in the background.
>
> Sure, it's not ideal if the war refugee is telling you her story in a loud café while upbeat pop music blares (a thing that just happened to me)—but if you explain to the listener that this café is a place of comfort to the refugee, that it's the place she stops before her son has karate practice, that it's a little slice of normalcy during a turbulent time, then the scene will be that much richer to your listeners.

Of course, if there are ambient sounds that make it hard to understand the person you're interviewing, find a quieter location.

When you do interviews in environments with even the slightest background noise—like the hum of an air conditioner or distant road traffic—you'll need to get "room tone" for use as a bed to ease in and out of actualities.

Getting it is an art. You might have to stand there alone, holding your microphone while doing nothing. Many journalists find that awkward, but you'll get used to it. If it makes you uncomfortable to stand idly, or think it might look weird, take out your phone and pretend you're checking messages. And let the recorder run for at least a minute.

It's very important. Because without it, the background noise will come in and out abruptly at the beginning and end of each actuality or ambi post. You may have heard what it sounds like in some poorly produced podcasts.

Getting room tone often requires human management, because sometimes you have to do it with your interviewee present, and you need them to be still. What you don't want is for them to start a conversation with someone else in the room.

"I always tell people," says Cheryl Corley, " 'This might sound a little crazy, but we both have to be quiet, and I'm going to record us for a minute. And, *I'll let you know* when a minute is up.' "

If the interviewee is still mystified, Cheryl sometimes explains that the recording is necessary for production work. And she looks at her recorder while recording, not at the person, so as not to encourage them to talk.

Lauren Migaki on Doing Field Interviews

BEFORE THE INTERVIEW

I always, always, always test my kit. Doesn't matter how recently I've used it. I make sure that I have fresh batteries in my recorder, that the memory card isn't full, that the microphone and cord work, and that I have a secondary device if my recorder fails.

And I think about where I'm going. You can't just walk into the CIA with the microphone out, but at a school I may be walking in with the mic already rolling and gathering ambient sound. I also think through the situation: could my microphone be construed as a gun? I almost always put the big fuzzy windscreen on, even indoors. Especially when you're working with children, it's much more charming.

I think about what I'll wear because I can't tell you how many times I've been sitting on floors holding the mic up. You have to be able to move and be comfortable and appropriately dressed for your environment.

And I have a million extra sets of batteries.

ON ARRIVAL

I sense the room and think about where the sound bounces. Generally, you want soft surfaces that will absorb the echo. Any hard surfaces, it's going to bounce and create more echo. That's why your bathroom is such a bad place to record. Think of a lush studio with a nice carpet, drapes and things like that.

I've been to a lot of offices where people want to sit you down at a big glass table, and that's terrible for sound. If it's the vice president, you just deal with it. Otherwise, I suggest that we move to the couch or a quieter room.

That said, if my story is about a protest, I don't want it to sound like it's in a studio. You want people to be understood and intelligible but you don't want to take them out of the scene. So it's about finding that balance.

DURING THE INTERVIEW

I always hold the mic. I don't use a mic stand, because people move. Sometimes they think they need to lean into the mic and put their mouth right in front of it. So I'm always adjusting to make sure it's to the side of their mouth so that they're not popping their *p*'s. Some people will slowly lean back over the course of the interview, and you have to kind of chase them with the mic!

While I record, I'm always checking levels and making sure the recorder's clock is rolling. I'm also watching my battery levels. If I have to change batteries, I try to find the right moment to break in and take a pause. And I'm listening for the moments of the interview that need to be in the story.

Studio interviews are more straightforward. I'm checking to make sure that things are recording, but I can take notes and focus more on the content.

AFTER THE INTERVIEW

The thing I always like to say to the interviewee is, "If you like to meditate, now is a good time." And I grab about a minute of ambi. I actually find this to be incredibly hard because people don't feel comfortable with silence in a social situation, and very often they will start fidgeting or whispering to someone. And you have to insist, "No, we actually need to be totally silent."

Also, before packing up, I think, *Are there other sounds I need to get?* I can't tell you how many times we have stayed after a long school day just to gather the sounds of after-school activities that will probably be in the piece for one second, but it's enough to give you a sense of scene and place. You're always adding sounds, even if it's just a straight interview, like the sound of someone saying hello at the front desk.

Generally, I won't have taken notes during the interview because I'm holding one or two mics, so when I leave an interview I almost immediately either write down or record a voice memo with my top five moments.

Then I talk with the reporter about the highlights while calling for a car, and I may call ahead to the next interview if we're running late.

And then, I'm listening to make sure the recording sounds OK and uploading the files to my computer. I don't like to let things sit on my recorder too long because it's easy to forget what files you have on there, and losing a recording can be devastating. So the sooner you can get the file onto a second device, the better.

Recording "Like a Vacuum Cleaner"

In the olden days of public radio, the amount of tape in the recorder forced tough decisions about what to record, and how much. Even when digital recorders came along, compact flash cards had enough memory for about an hour of recording. But now even the tiniest memory cards allow more recording time than you will most probably ever need. "Since you don't have to worry about wasting physical tape, you may as well just record as much as you can," says producer Lauren Migaki. And you won't miss a moment that could have been the star of the piece.

Lauren says she learned that the hard way. Once she was in Crimea with *Morning Edition* host David Greene, and they went to see a soccer game. "And the one time I wasn't recording is when they scored the goal!"

So when you're out in the field, she says, you should be "like a vacuum cleaner gathering up sounds." Many producers say the best moments come when you least expect them: You're waiting to go inside a building for an interview and a crowd of protesters shows up. Or a person suddenly sees a friend and calls out to them. If an interviewee is interrupted by a phone call, don't stop recording unless they ask you to. It might make for a good scene.

If the message you're getting here is that you have to spend long stretches of time recording to get good tape, you're right. If you leave before the best sound opportunity happens, you simply won't have it for your story. So be prepared to wait hours and hours for that train to trundle by, or for the crowd to give that really wild roar.

Organizing and Labeling Tape

A downside to all that "vacuuming" is that you can get buried in tape. When you come back with hours and hours of material, trying to find that spectacular ambi you got of the footsteps on the gravel or the soccer fans cheering can feel like looking for a needle in a haystack.

Fortunately, there are hacks. Here are a few tips from NPR staffers:

SLAP THE MIC. While recording, whenever you hear something that might make good ambi, "slap the mic," says Lauren. That way, it will be easy to find later, when you've loaded the file into the editing software: just look for the peak formed by the slap. That method works when no one is speaking; for interviews, see the tips below.

CARRY A SECOND RECORDER. Try using two recorders, one for getting tape, the other as a marker. Whenever you hear something good, stop and start the second recorder again. The best cuts are likely to be in the last 30 seconds or so of the second recorder's files, and you can use that as a guide to locate them in the main tape, says Brian Mann. This is a good method for interviews since it doesn't disrupt the flow of the conversation and you can just hold the secondary recorder in your hand.

MAKE A NOTE, OPTION 1. Summarize what you heard and when you heard it in a notebook, on your phone or in a message you send yourself. "Really squeaky bat, day 1, around 11 p.m.," says Jane Greenhalgh, giving an example. For it to work, your recorder's clock has to be accurate. Also, you may not have time to do this until the end of the day. But that may be an advantage, because if it still sticks in your mind then, it's probably worth having in the story.

MAKE A NOTE, OPTION 2. When noting the time, use the time on the file, not the time on your watch, or take a photo of the recorder. Lauren used this approach during one of her stints covering the war in Ukraine. "A mother started speaking passionately about her life during the war at 15:37 into the file, so I wrote '15:37' in my notebook," she says.

When you're recording all the time, you don't want to have unwieldy files or risk losing them, so stop the recorder every half hour or so and start a new file. Think of it as similar to saving a document periodically in case your computer crashes.

Or you trip and fall, and your recorder's batteries pop out.

"I just didn't see that they had this raised-cement thing in the backyard," says Christina Cala, recalling a long tearful interview with the son of a detained immigrant. After it, she continued recording, hoping to get the raw material for a natural transition, as they walked outside where the kids were jumping on the trampoline.

"That's when I fell, and we lost the whole interview"—two and a half hours' worth of tape, she says. It was a hard lesson. "I definitely start and stop my files a lot more now."[6]

Many producers periodically transfer their files onto their laptop and rename them. "Always when I'm in the field, I will go through at the end of the day and name the files so that I don't just have scores of files," Jane says. The names she gives the raw files help identify the recording. So for the bat virus story, she named them "day one-bat testing," "day two-florida guy," "day two-australians."

Later, as you're listening to your tape and pulling cuts, you might want to give them names that reflect the content, as well as the larger file they were pulled from: "day two-florida guy-guano" or "day two-australians-buying tickets."

When you're done transferring and labeling, don't delete the files on the recorder. Let them be your backup. And remember to take enough memory cards so you don't have to delete any files while in the field. In the extreme case that you do, find another place to back up your files, such as a USB stick or a folder in the cloud.

Never, ever leave a location without backing up your files.

Cliché Ambi

Certain sounds are so identifiable that we almost expect to hear them in certain stories—a school bell in an education story, for example, or a sportscaster screaming "*G-o-o-o-o-o-o-l!!*" at a soccer match in Spain or Latin America.

Though sometimes dismissed as cliché, there is nothing inherently wrong with this kind of ambi. In fact, it's often the sound you want. Lauren, whose work has included producing stories on teachers strikes, children's mental health and schooling during the COVID-19 pandemic, says a school bell is often the perfect way to start an education story.

"It's cliché for a reason," she says. "It immediately tells your listener, 'Hey we're in school right now.' "

But occasionally good ambi that is easy to get becomes hackneyed. And when it plays off or reinforces a stereotype, it can offend.

"As a Black woman, when you're doing a story about Black people, I dislike the 'Oh, here we are at the church, in the choir scene,' " says Cheryl. She recognizes that Black churches offer rich sound opportunities, and that sometimes it's warranted. "But it doesn't necessarily mean that that's a thing that you always need to get."

When science reporter Ari Daniel was reporting on cholera in Lebanon, he visited a health clinic. Just before his interview, a call to prayer emanated from a nearby mosque and he recorded it. But he and his editor decided not to include it because the story wasn't about religion, and it would have felt like he was using a trope to say, "Here I am, in the Middle East."

So, always ask yourself: is this ambi germane to the story? If it feels tired or clichéd, look for an alternative. When I was gathering tape for a story on the Islamic history of southern Spain, I met a Moroccan-born shopkeeper in Granada who had studied Arabic literature.[7] He mentioned a famous writing by a Syrian poet extolling that heritage, so I asked if he could recite it from memory. He did, and I put that near the top of the piece, after the sound of water trickling in the fountains of the Alhambra, a storied palace built under Muslim rule. I did record a call to prayer, but I put it later in the piece to introduce an interview with an imam.

What If There's No Sound?

Don't let this long discussion of natural sound give you the impression that every single story needs it. Some stories simply don't have sound opportunities, such as a think piece that is all about arguments and ideas. But sometimes there's no ambi because all the reporting is being done remotely, a practice that became more common during the COVID pandemic.

If you have such a story:

- Ask yourself whether it really has to be done remotely. Are you really so far away, or so pressed for time, that you can't do it in person, and while you're on location look for a sound opportunity?
- Be creative. Record people introducing themselves and see if they can give you a virtual tour of where they are. Or maybe they can share a video of something related to the story. Eva Tesfaye started out in public radio in the middle of the COVID pandemic, when most interviews had to be done remotely. For one story, Eva interviewed a young woman who was practicing dancing at home so she'd be ready to party when it was all over. Eva asked her to record her dance steps and send the sound file, and that became a post that signaled a transition in the story.[8]
- Compensate for the lack of ambi with vivid writing (see "Write Pictures" in chapter 8).

Then there are the stories that you'd assume would offer sound opportunities, but don't. The bird that had been chirping for weeks outside someone's window has an off day. Or the laboratory that your source assured you would make for great sound doesn't quite live up to the promise.

At that point, remember: Sound is important, but it's not an end in itself. The end is the images you create in the listener's mind. And you can get there by thinking about your stories visually.

Lauren went to Brazil with reporter Lulu Garcia-Navarro and photographer Kainaz Amaria for a drought story in 2015. "Everywhere we went," Lauren recalls, "she kept saying, 'I'm sorry, I can't make a visual story of this because I can't see the drought.'" They finally found a marina on a reservoir outside Sao Paulo and took gorgeous pictures for the story.[9] Lauren was able to create a powerful scene with visual writing and a poignant actuality:

LAUREN: It happened slowly at first. The reservoir's water level dropped, so the resort extended the boat launch ramp. Then they had to add another extension. Eventually, the water dropped so much the resort gave up, and business dried up too.

(*Resort director speaking in Portuguese.*)

LAUREN: Manager Francisco Carlos Fonseca looks out at the brown pit that used to be the reservoir.

(*Resort director speaking in Portuguese.*)

INTERPRETER: For this coming weekend, there's not one reservation. This business was 98% dependent on the water. Now that the water's gone, the customers are gone as well.

"When you think with your other senses," says Lauren, "it leads to way better audio."

CHAPTER 5

Reporting

It had already been a rough night in international correspondent Ruth Sherlock's household. Her feverish 3-year-old daughter had taken forever to fall asleep, then did so in her parents' bed, forcing Ruth's husband, Paul Wood, to go to another room. Ruth, too, finally fell into a slumber. Then, at about 3:20 a.m., she woke up. The building was swaying. Ruth and Paul grabbed their daughter and infant son and ran out into the freezing darkness. There was no damage in their neighborhood, and it soon became clear that Beirut was mostly spared in the earthquake that measured 7.8 on the Richter scale and sent shock waves across the region on Feb. 6, 2023.

The worst damage was in Turkey and Syria. "My phone started blowing up," Ruth recalls. An activist in Syria whom she had previously followed for updates on the civil war that had been raging for more than a decade started streaming videos that showed "people screaming and shouting names in the darkness as these houses collapsed in complete chaos." Ruth reached a member of a civil defense group who gave her a list of villages where he said the buildings had collapsed, trapping entire families under the rubble.

Reports of even more devastation from the original quake and powerful aftershocks came in from Turkey. Ruth did a two-way for *Morning Edition* and wrote up a digital story with what she knew from her own sources and wire reports and reporting by Beirut-based producer Jawad Rizkallah. After consulting her editor, Ruth looked for flights to Adana, the nearest major city in Turkey. "It just so happened there was a plane in sort of an hour. I threw my clothes into a bag and rushed to the airport." Ruth's international SIM card wasn't working, so she called her Lebanese mobile carrier to see if they could make her SIM card work abroad. They told her it couldn't be done, but she persisted. "I said, 'I'm a journalist. This is a matter of life and death. I can't have zero connectivity from an earthquake zone.' So they agreed."

Why am I telling you about Ruth's amazing feats of packing, rushing to catch a flight and getting her cellphone to work abroad? Because without the tenacity and quick thinking to overcome the obstacles in the way of getting the story, there's no story. Or, as Ruth puts it, "So much of reporting is just pushing, pushing, pushing."

Over the next week and a half, Ruth did quite a bit of pushing to get the story of the earthquake and its aftermath. After getting herself to the scene, she dealt with all kinds of uncertainty—language barriers, logistical puzzles, her own safety—while striving to build a network of reliable sources, establish the accuracy of the information she was getting and tell the whole story in a compelling way on NPR's different platforms. Later in this chapter I'll talk about how Ruth did that with this story, as well as how other NPR reporters cover stories on a variety of beats under circumstances that are often just as trying. But first, I'll look at some of the traits you should develop if you want to be a reporter.

What Makes a Great Reporter

In the first few moments of working on a story, and during much of your reporting, the job is not about topical expertise. You can build that up as you do the legwork, whether it's on a fast-breaking story such as a natural disaster, or a new development in a story that you've already been following. What distinguishes the best reporters is a series of traits that make them adept at responding to new situations and gathering the information needed to tell a story. Here are some of the most important ones:

TENACITY

Planet Money host and producer Alexi Horowitz-Ghazi had lined up a series of interviews before flying to Nova Scotia to talk to Mi'kmaq lobstermen about the threats to their livelihood, for the story he pitched in chapter 2.[1] Many indigenous people in Canada's Atlantic provinces are wary of outside press, and when Alexi got there, his calls went unanswered. Rather than give up, he buttonholed random people on the docks until he found a skipper whom he'd wanted to interview but didn't know how to reach. They did a 90-minute interview and went out in the skipper's boat, and the skipper ended up being the protagonist in the story. Alexi then studied the tide charts to see when more boats would return. He staked out the docks again and found another fisherman whose mother was one of the people who had ghosted him. He got her address, and drove to her house and knocked on the door. Her boyfriend answered, saying she was in the next town two hours away. "And I said, 'All right, can I just hang out until she comes back?'" Eventually she showed up, agreed to an interview and invited him out on her boat, Alexi says, "and we both pretended like it wasn't weird that she had totally disappeared."

Such setbacks don't happen on every story, but it is common for things not to turn out as expected. "Your story is always kind of like your dream of the story of meeting the reality of the world," he says, and you figure out how to make it work.

You have to be tenacious, but sometimes you will be following your leads, and your reporting yields information that turns the story on its head. That's why you also need to be flexible. "Do not go in with any kind of predetermined ideas about what you're going to get," says Cheryl, "and always be open to change." You may need to find a plan B—figuring out what needs to be different to make the story work. How else can we get to the scene? What other person can we interview? Alexi says a light bulb went on in his head when he realized that about journalism. "My job is to not give up. I can do that!"

CURIOSITY

Speaking of not giving up, reporters should be asking questions all the time, even to the point of being annoying, because pursuing the answers will often lead to interesting stories and to a deeper, better understanding of a story line you are pursuing. The questions should reflect a curiosity about how the world works, but also what people are thinking and what makes them tick. And they are the kinds of questions that can lead you to news, uncovered stories or fresh story angles, which should always be your aim. Seeing a Confederate statue on a main square in Texas led John Burnett to a story about a Black fifth-grade teacher who was leading a battle to get it taken away. And hearing an unusual sound on Mexico City's Zócalo led him to the one-armed busker making music with an ivy leaf and the story I mentioned in chapter 2. John said that for him, it was about "not just following the news, but getting the story behind the story and finding oddball topics that no one else is covering."

SKEPTICISM

A reporter should have a built-in "bullshit detector," says Mary Glendinning, deputy chief of NPR's Research, Archives and Data Strategy group, who regularly works with reporters and editors. "If I tell you that the sky is blue, we probably have a shared understanding. But if I tell you something with all certainty that you don't know to be true, stop and think," she says. "Especially if you're starting out, it's OK to say, 'I'm not sure this is true.'" You shouldn't be afraid to ask follow-up questions, which I'll talk more about in chapter 6.

PRECISION

That penchant for truth includes an obsession with accuracy. Good reporters double- and triple-check all facts, including names, ages, dates, spelling, pronunciations, titles, pronouns and numbers in their stories.[2]

Accuracy is the currency of journalism; the trust our listeners and readers place in us depends on it. At NPR we also own up when we make mistakes, with corrections on the bottom of any story or audio transcript where there was one, and a separate page with a running list of corrections.[3]

Time was when a radio news script could contain a misspelled name or word and it didn't matter, so long as it didn't cause a mispronunciation. That's no longer the case, because there's always the possibility, even the probability, that the script or parts of it will end up on the web, where spelling does matter.

So start every interview by asking people to pronounce their name and spell it for you on tape, and any other relevant facts such as their age, title and pronouns. And then ask the interviewee how they know the facts that they are asserting. NPR Training has an accuracy checklist that can help you avoid some of the most common mistakes.

On longer pieces or podcasts, you may be pulling in information from a broad array of sources, and you need to have a plan for keeping track of everything, which may include annotating where your information comes from. Mary cautions that honest mistakes can end up looking like plagiarism.

Whenever you include a fact in your story, make sure it's true, without exception. If you're not sure and want to look it up later so that you can keep writing, that's fine so long as you have a system for calling attention to

THE NPR ACCURACY CHECKLIST WHAT YOU SHOULD DOUBLE- OR TRIPLE-CHECK npr

- [ ] **AGES** (Get the date of birth and do the math.)
- [ ] **DAYS, DATES** (Are you sure it happened then?)
- [ ] **GRAMMAR & SPELLING** (Readers notice such things.)
- [ ] **HISTORICAL "FACTS"** (Don't trust your memory.)
- [ ] **LOCATIONS** (Get them right, say them correctly.)
- [ ] **NUMBERS** (Check all math. Don't say "millions" if it's "billions." Learn percentage vs. percentage point.)
- [ ] **PROPER NAMES** (Get the correct spelling and pronunciation for people, businesses, groups, etc.)
- [ ] **PRONOUNS** (Don't guess or assume.)
- [ ] **PRONUNCIATIONS** (Of names, places, terms, etc. Dictionaries are your friend!)
- [ ] **QUOTES** (Make sure they're accurate and correctly attributed.)
- [ ] **SUPERLATIVES** (Claims to be the first, best, worst, etc. are often wrong; never trust them.)
- [ ] **TITLES** (CEO, president, professor)
- [ ] **WEB ADDRESSES & PHONE NUMBERS** (Never report them without testing them first.)

NPR Training's accuracy checklist.

facts that need checking. Some journalists write "TK," short for "to come"; I put three asterisks together (***) because they're easy to find. Whatever you do, make sure to go back and comb through the text one final time to make sure no inaccuracies slip through. Everyone makes mistakes. Good reporters get rid of them before publication.

WARMTH

"At the end of the day, we're selling vacuum cleaners," says John. Don't assume someone will talk to you just "because you're NPR. We have to make them like us and want to talk to us and open up to us." You have to genuinely like talking to people and charming them, so cultivate your personality to warm them up. "It's an art. And it's a seduction," John says.

A warm and friendly approach can make it easier when you have to push back. If you first establish trust, says Adrian Florido, "people tend to be more willing to consider the challenge and be a little bit more self-critical."

Working on Deadline

As Ruth approached downtown Antakya, the worst-hit area of the earthquake, a stream of ambulances passed in the other direction, sirens wailing. And when she got there, she understood why. This is how she described the scene later that day in her conversation with *All Things Considered* host Juana Summers:

> RUTH: Then we began to see the destruction and, I mean, building after building collapsed. In one area, they were, you know, on either side of the road. There was just debris, and I could smell rubber, dust and clearly the putrid smell of dead bodies as well.

The roads were impassable, so to see the worst damage, Ruth had to walk part of the way. She was traveling with Erin O'Brien, a freelance journalist writing for the *Economist*, and they were running behind schedule because the drive took much longer than usual. Also, when their driver came to pick them up, he hadn't filled up his gas tank as Ruth had asked him to. "It triggered this kind of fear in his mind, and he didn't actually want to go to Antakya," she says. "So we had to talk to him and promise that we weren't going to put him in any danger."

Hours later, when they reached Antakya and walked far enough to get some tape of rescue operations, Ruth had to make a decision, given the unreliable cellphone reception in downtown Antakya. Should they head further into town to see if the devastation there was worse and get more voices, or turn around and get back to the hotel to file? The deadline for getting material into *All Things Considered* was 11 p.m. her time. Would the 20 minutes' walk into town yield anything different from what she had already?

She decided against it. They went back to the car, and on the drive back she started going through her tape and scripting the two-way. "And then I got back to the hotel 10 minutes before taping."

Perfection in reporting is a double-edged sword. We want to report thoroughly and write beautifully and accurately. But no matter how comprehensive and compelling your story is, it's worthless if you miss the filing deadline. Even when you don't have a deadline, there comes a moment when you have to decide, is it worth more reporting to have another scene, or maybe a better actuality? Or do you work with what you have? Here are a few tips for meeting your deadlines.

BACKTIME YOUR REPORTING

Give yourself interim deadlines. Base them on a realistic estimate of how long it takes you to go through each step of the process: getting interviews done, transcribing cuts or full interviews depending on deadline, writing, going through editing, tracking and uploading your files. Will you need to file a newscast spot or two while you're preparing your magazine piece? Are you also writing a digital story or contributing elements for it? Factor in time for your colleagues back at headquarters: the show producer who will mix your audio, the engineer who might have to fix any audio problems, the line producer who auditions your story before it goes live. You also don't want to cause needless stress for show staff who are juggling a gazillion other things.

"If I'm on *All Things Considered*—the show hits at 3 p.m. Central Time—I like to be finished with my interviews by noon," says Cheryl.* If she has advance notice of a story happening, she'll start working on it the night before. "Do as much research as you can and get those phone calls done."

A lot comes down to planning. You can't anticipate every story, but those you can you should approach like triage. "Being able to say to your editor, 'I think we're going to get done at *x* time, that's going to give me this much time until the show. I need somebody back at headquarters who can pull cuts I identify, and somebody who can help pull together the web story,'" says national political correspondent Sarah McCammon, recalling her reporting on the anti-abortion March for Life in January 2023, a few months after the Supreme Court overturned Roe v. Wade. "Because if I sit down in the freezing-cold tent and write my web story, even if I've pre-written some of it, I might not make it for *All Things Considered* in time, or I'm going to only have time to write something really short and not very insightful."

IDENTIFY CUTS AS YOU GO

Many reporters routinely do what Sarah did at the rally: identify possible actualities for their story already while they are doing interviews or following a press conference or speech. Cheryl writes down time codes as she's taking notes. "So I just go back to that and pull the actuality and go from there." That's just one of the hacks mentioned in chapter 4, like slapping your mic, or taking a photo of the recorder display, that will help you organize your tape and work faster.

The thing to avoid, even on nondeadline stories, is having to sift through hours and hours of tape. Even though transcription software has

* *All Things Considered* is a two-hour program that begins broadcasting at 4 p.m. ET and 1 p.m. PT.

made this task easier, a long interview will generate pages and pages of text that you have to comb through to identify potential actualities. "You're a stenographer when you're supposed to be a storyteller," says national desk correspondent Brian Mann, who has put a lot of effort into learning the hacks. "Because I now mark my tape efficiently, I come back and after a four-hour field thing, I have the clips that I'm almost certain to use established in my queue within about 15 minutes. It's not only faster, but it's also just vastly less drudgery."

SHARE YOUR WORK EFFICIENTLY

When the previous edition of this book was written, most reporters had relatively straightforward workflow for filing. For breaking news, they would do a newscast spot or two and then, depending on which show would be up next, work up a piece for either *Morning Edition* or *All Things Considered*, or for one of the weekend shows. Now, reporters can get pulled in many directions from the get-go. Should they post on social media first before writing a spot? Should they also write a digital post?

NPR's breaking news workflow now emphasizes getting the story out to audiences as quickly as possible. The reporter or editor will first send a "reportable" to the entire newsroom. A reportable is an email that consists of a couple of paragraphs on what NPR has confirmed and can be reported on our website, newscast, shows and other platforms. It comes with a coverage plan, detailing who will be filing for which platform. That allows all desks to immediately treat the news as reportable fact. Then the reporter generally files first for newscast, then a piece for NPR's website, and then a two-way or short piece for the daily newsmagazines, while the digital desk posts story updates on social media.

Correspondent Shannon Bond often uses her newscast spot as a draft to build on. "You've got to get the point of the story across really quickly, and that is often a good way to build a first digital version," she says. Many reporters try to feed multiple platforms simultaneously. "When I'm at a political rally or at a speech or press conference, sometimes I'll live tweet it," says Sarah, "and those highlights are also like notes for me, because I'm being very careful to quote the person accurately. Or even if I just can't quote them, I can summarize what they're saying. There's a time queue, so I can go back to my notes and my audio later and pull those quotes. And often whatever stood out to me in the moment starts being a scaffold on which to construct whatever else I'm doing."

Sarah says getting quotes and notes out on social media is also a way to collaborate internally. When a different reporter is working on the digital version of the story, they pull quotes from her live feed.

In Turkey, in the early hours of the story, Ruth's most urgent focus was getting tape. But she needed to be able to share the fruits of her reporting with other NPR journalists, especially those working around the clock on digital stories and scripts for newscast anchors. So she created a folder and uploaded all her tape there—the ambi, the interviews and her own descriptions of what she saw. She also wrote up some notes describing the audio and indicating where to find some of the highlights, and continued uploading material throughout her stay. One editor called the folder a gold mine.

RELY ON YOUR EDITOR

Editors at NPR are in charge of the individual stories reporters do and of the overall coverage, which includes logistics and coordination between teams. On breaking stories, editors can help set some of the priorities on the workflow: They strive to balance the needs of the newsmagazines, newscasts and digital platforms while being mindful of the pressures on the reporter. "I leave it to my editor to worry about the whole world of 'Does this need to be on a show?'" says Shannon. And that allows her to focus on following the news developments.

Essentially, the editor shields the reporter from the demands of the rest of the organization. Since Ruth was the only NPR correspondent near the epicenter in those early days, everybody wanted her reporting. "This is where a good editor really matters," Ruth says. Her editor, Larry Kaplow, fielded all the requests so that nobody reached out to Ruth directly. "My job was to go and gather tape."

The editor can also be a sounding board as the story is developing. In July 2022, when science reporter Ari Daniel was on the ground covering a breaking story on the first case of community transmission of polio in the U.S. since 1979, he did a series of interviews with doctors and health officials, and his editor, Rebecca Davis, told him to check in after each one. "So between interviews in the car, I'd call her and basically just do a dump," Ari says. Rebecca would ask, "What stood out to you? What was the most important thing? What did they say that you remember?" Then, she would prep him with possible questions for the next interview. That made it easier for him to focus on getting tape, and since Rebecca already had a good sense of what Ari had, she spent less time editing later.

WORK IN A BREAK WHEN YOU CAN

On the way to the airport, Ari tried to write the story in his head while driving, but he ended up pulling over. Yet even after leaving the car, he struggled. "I was going up this hill, and I was trying to script the host intro. And I thought, 'This isn't working. I can't concentrate.' So then I went back down, I crossed the street and there was a flat path on a road. And I walked along that. And then I was able to think through the host intro." He recorded it as a voice memo on his phone and transcribed it later.

Reporting is tough; it can be dispiriting. Breakthroughs occasionally come when you're not at computer or when you're doing something that uses just part of your brain, like driving. So go for a jog or a walk or do something that lets you process what's already in your head. You may actually end up ready to file sooner.

IF YOU HAVE TIME, OVERREPORT

Not every story on deadline is a mad scramble. If you plan well, you should find time to do some extra reporting. And that can be a good antidote to parachute journalism, going somewhere with only superficial knowledge of the place.

When she travels to unfamiliar places to cover political stories, Asma Khalid tries to do as much pre-reporting as she can. "There's so many people I talk to that don't actually end up in the tape." It's partly because she thinks the criticism of parachute journalists is sometimes warranted. As a Muslim and a Midwesterner, she's experienced how outsiders come in and "assume that people think a certain way."

Dipping into stories is unavoidable. "The job of any journalist is sometimes to unfortunately have to go into communities that they're not familiar with," Asma says. "I don't think it means that you shouldn't go into other places. I think it's your job to be as understanding, empathetic and eyes-wide-open as to what you're seeing."

Research Vs. Interviews: Which Comes First?

Let's go back to Ruth's story and the moment she left her apartment in a rush and headed to Beirut international airport. As she boarded the plane, she realized that she'd be in a pinch once she landed in Turkey. "I had no driver, no friends, no interpreter, no fixer." So the moment Ruth took her seat, she started developing sources. She talked to the aid worker sitting next to her, and to a man two rows away who had just heard that his wife and brother were among the dead and that his children were missing. When Ruth landed at the airport in Adana, she spoke to the rescue teams who were arriving from around the world with search dogs and equipment. Information she got from these conversations would make up part of the spots she filed later that day. It would also inform the two-way she did the next day on *Morning Edition*. "You just have to be open to what's happening around you," she says.

In Adana, Ruth caught a taxi and, using Google Translate, instructed the driver to take her to buildings in the city that were destroyed. She got back to her hotel at around 3 a.m. local time and started scripting the *Morning Edition* hit. While she was doing that, she decided to go to Antakya, the worst-hit city. So she texted the driver and negotiated another day with him. And she asked him to stock up on food, fuel and water after getting this tip from a colleague who reached the city earlier: "This is Armageddon. Don't bank on being able to get anything here."

Ruth had covered the war in Syria for a decade and lived in Libya in 2011. In her experience, reporting on major disasters or armed conflicts is easier and safer when you collaborate with fellow journalists. "You do occasionally meet correspondents who are cagey about what they're doing and keep to themselves. But they never do well out of it. Much better to team up with other people and share information."

So when Ruth went down to the reception desk to request a room closer to the fire escape, in case of aftershocks, and came across a French TV crew in the lobby, she saw an opportunity to make up for not having a translator. "I was so keen to team up that rather desperately I said, 'Oh, hi! I've got a driver, if you guys need a lift!'" It turned out that they were based in Turkey and didn't need her help. But they had a solid network of local journalists, so they put her offer on a WhatsApp group. "And, like manna from heaven," Ruth says, she received a message from Erin O'Brien, the freelancer working for the *Economist*. Erin was on a 16-hour bus trip from Istanbul to Adana and needed a ride to Antakya. And she spoke fluent Turkish.

When assigned a story, a reporter often wants to get out and start interviewing right away. If it's a sudden development, like the earthquake centered in Turkey, that will likely be the right move—get to the scene and talk to eyewitnesses. But for other stories, says NPR archivist Mary

Glendinning, "Stop and count to 10." Before interviewing people, find out what's already known and what's been said. "Investing the time is critical," Mary says.

In this early phase of research, known as reading in, the reporter has two main objectives:

- Find out what's already been reported so that you can pursue fresh angles and move the story forward.
- Have a base of knowledge so that you are sufficiently prepared for your interviews (more on that in chapter 6).

A lot of reporters start reading in by using search engines to uncover open-source information such as official statements, research documents and news reports. But that method will only go so far, since many news sites and academic publications place much of their content behind paywalls. There are databases, however, such as ProQuest, EBSCO and Gale, that provide access. If you don't belong to an organization that subscribes to these services, there are other possibilities: some resources are made available by journalist associations. Your local library may also have subscriptions.

One useful tool that search engines offer are alerts, says investigative correspondent Cheryl W. Thompson. "Investigative reporting is all about patterns. If I have an idea and I haven't quite developed it yet, I will set a Google alert with certain words." For example, she had heard that many homeowners associations across the country still had discriminatory property deeds, even though the Supreme Court outlawed them in 1948. So she set a Google alert with the words "racial covenants," which helped her team of reporters from NPR and several member stations track them and produce an investigative report.[4]

Building Sources

Reporters who regularly find original stories usually have one thing in common: they know lots of people who are privy to information that has not been made public yet. If you want to find stories no one else has found, you should start simply by talking with as many people as you can. And getting their contact info.

As an international correspondent in Africa, Eyder Peralta made sure to get the phone numbers of everyone he met, from "the guy who rents donkeys to the president of Ethiopia." A random person may seem inconsequential, but Eyder still got the number, "because you never know when you'll need it." Once while riding in a boat on Lake Victoria, he met the captain. Several years later, when there was a capsizing disaster on the lake, Eyder called that captain, who turned out to be a well-connected source.

FINDING OUT WHO OTHERS ARE QUOTING, AND NEW VOICES

Sources can help you find other sources. "I always ask my interview subjects who else they recommend I talk to," says climate reporter Rebecca Hersher. It can be a useful last question, rounding out an interview. And she is deliberate about building a network of sources that's diverse in terms of race, geography, age, gender and other ways. "If I know that my sources are skewed one direction, instead of saying, 'Is there anyone else I should talk to?' I'll say, 'I have plenty of voices that are like this. Can you recommend anyone who's different from that?' "

Sources can be anyone from a person you meet on the street and talk to only once, to an analyst or a business or government insider you come back to periodically. You'll want to cultivate a range of sources if you're developing a beat and trying to report original stories. When she was covering Silicon Valley, Shannon asked people to recommend essential reads, not just major books in the field but also work by journalists whose coverage stood out, to find out whom they were quoting. If a source is the go-to for other journalists, chances are that source is a good talker and is familiar with a reporter's needs and deadline pressures. But often the most widely quoted sources are not the most representative, and finding a reliable source who hasn't had much publicity yet is more likely to get you scoops.

If you have a beat, learn which government agencies, companies and organizations have a stake or interest in it. Ask yourself: "Who's making the decisions, where's the strategy coming from, who might be more interested in talking to the press than not?" says Mary. Scour academic and scientific journals but keep an open mind when you decide whom to pursue. "Maybe it's the lead author of a medical paper that's come out that's revolutionized something in health care," she says. "Or maybe it's not that primary author, but the fifth author."

Cheryl, the investigative correspondent, says that with law enforcement, she seeks out sources "who don't have much to lose" by talking to her. "Don't go to the police chief. They're not going to tell you anything negative about their agency. Not a street cop, either, because they often don't know what's going on." Instead, she says, find a regional commander or detective who has insights that you may need.

Building a network of sources, says Shannon, is a "slog at the beginning." But the effort is never really done. "When you're well established in a beat, it can be very easy to sit back and not continue to do that source building," she says. The most effective reporters keep seeking new sources while they work their existing ones, checking in from time to time to see what the sources are working on and the changes they see happening in their fields of expertise.

The best sources are ones you can regularly check in with and bounce story ideas off of. Try asking, "What have you heard? Who should I talk to? What's an angle I haven't thought of?" Cheryl says. Ultimately, Shannon says, you will "end up with people coming to you and saying, 'I have something for you' or 'I heard this,'" which is how she got the scoop on AI-generated LinkedIn profiles that she talked about in chapter 2. Then you will be the one with the sources that other reporters wish they had.

PEOPLE KNOW WHEN THEY'RE BEING USED

You get the most out of sources when there's mutual trust. "Get to know people first and let them get to know you," Cheryl says. "Developing sources really is about developing a relationship." When she arrived at the *Washington Post* in the mid-1990s, she went to all seven commanders in the District of Columbia and introduced herself. One of them became a top source for her. She met him every Friday in the same restaurant, where they would sit in the back and speak softly so that others wouldn't hear them. "It got to a point where the hostess who would always greet us—and you know, we came separately—she thought we were having an affair!"

They weren't. But meeting in a restaurant made it easier for the commander, because "people let their guard down when they're away from the office." Even if you meet them in their office, being in person is an advantage. "You get to see what's on their walls, or what's on their desk, or maybe that they have a dog or cat or they have kids," and you can show interest in those things. But, Cheryl says, it can't be an act, because people know when they're being used. You have to genuinely want to know about people. "I can tell you so much about my sources, that one of them likes to bake pound cakes and he likes to garden. And his granddaughter got straight A's in school and he bought her a new car. Over time, you develop a relationship with people and then it becomes really easy for them to give you information because it doesn't seem to them like you're prying them for information. It's more natural. It's more of a conversation."

Some sources are easier to cultivate than others. Being in public media brings credibility and respect, and many people are delighted to talk to an NPR reporter, says John Burnett, "because we matter." But while it opens some doors, it closes others. John's last assignment at NPR was covering the "divided America beat," where he also encountered suspicious views of public media. "And you tell people, 'I'm from NPR,' and they say, 'I'm not going to talk to you.'" He found that meeting them in person and looking for things in common helped win their trust.

That was also the case with sources in law enforcement. Covering immigration in Texas, John built a relationship with a Border Patrol commander in Washington, D.C., who hailed from a part of Texas Burnett knew well. The next time John was in Washington, he invited the guy out for pancakes. "I knew parks and cities and watering holes that I could mention to the senior agent so that he knew I was a real Texan." And from then on, he became a source. "You go out to eat, you shoot the shit, and then you have a cellphone number. It's not, 'OK, John, you may call me to confirm things.' It's an understanding. 'Don't use my name, but I'll be your reality check on shit that's happening with the Border Patrol.' It's old school, but it works."

Beware of sources you don't know well who are trying to get you to bite on their agenda in your coverage. "I really try to minimize people who are drones, who are media savvy and are just too glib," John says, adding that it's by no means easy. "Our worlds are populated with these people now who want to talk to us. And so you really have to kind of tune your antenna."

But when a source is good, invest in them. John says his best source was Letty Fernandez, the media relations liaison at the University of Texas Rio Grande Valley in Brownsville. A former TV reporter, "she just got what I do. Almost every story idea she threw at me was just fantastic." One was about a mariachi band whose gigs, because of the pandemic, were all canceled—except funerals.[5] Another piece, on the only parrot species native to the U.S., was full of squawking parrots—and one squawking biologist. The source "knew everybody in the valley. And we just hit it off. I bought her so many dinners, so many margaritas, I lost count. It was money in the bank."

You don't have to be a beat reporter to act like one. Many general-assignment reporters, editors and show producers find it beneficial to develop specialized areas of coverage, coming up with story ideas on topics they're familiar with. When he was starting out at *All Things Considered*, Alexi Horowitz-Ghazi's interest in documentaries initially compensated for his lack of depth in other areas of coverage. "As long as you can demonstrate one consistent piece of value, then you can get your foot in the door and buy yourself time and space to learn all the skills that you don't have and learn the terrain that you don't have," he says.

SOURCE DIVERSITY

Your sources should come from all walks of life and from all communities. It's important that your reporting reflects the true heterogeneity of the people touched by the subjects you cover. It's been a recognized problem that too many of our sources have been white, male, college educated, middle aged and living on either coast. NPR has been working hard to change that. Reporters and show staff are encouraged to make the extra effort to find sources we haven't quoted or had on our air. "I think a variety of perspectives in any story is worth the time to dig for it," says *All Things Considered* producer Jonaki Mehta. Anyone working on a story must track the diversity of their sources in a database that's used to make sure we're including all communities in our coverage. "It's a really important tool," says Tony Cavin, NPR's managing editor for standards and practices. "You need to force people to think twice about who they're talking to and what sort of opinions they're getting."

As a result of these efforts, we've made some progress. Whereas in 2013, 77% of the sources on our newsmagazines were white, by 2021 that figure had gone down to 61%, near the proportion of 59% in the general population. That still doesn't mean NPR's coverage was truly representative, because much of the increase in sources of color was due to their being quoted in stories about race, which increased after the murder of George Floyd.

While academic experts are indispensable to much of our coverage—their perspective often gives a story legs—an overreliance on them is problematic. From a diversity standpoint, you can get a distorted racial picture from quoting them, since white and Asian people are overrepresented on college faculties. And from a storytelling standpoint, having too many experts can make for pedantic journalism.

Burnett warned against this in 2023 just before he retired. "We're talking to too many consultants and too many academics and too many experts," he told me, though admitting that shoe-leather reporting takes more effort and guts. "I don't like it any more than anybody else does, having to walk up to somebody and saying, 'Hi, I work for NPR. We're doing this story on anti-trans bills in the Texas legislature. How do you feel about it?'" But you'll get better tape that way. The audience can relate better "to a real person like them reacting to an anti-trans bill," he says, than to a pro-trans organization.

Even when you do quote an expert, they don't need to feature prominently in your story.* Rebecca, the climate reporter, often begins her reporting by talking to scientists, but she uses those conversations to figure out who are the affected people she needs to find. "So, for example, if we're talking about the hazards of hurricanes, first I want to figure out, should I be looking for somebody whose house flooded because of storm surge, or should I be looking for somebody whose house flooded because of rain?" And once she finds that somebody, they usually become her opening scene.

Sometimes you don't need an expert source at all. Investigative correspondent Joseph Shapiro tries to let people with disabilities speak for themselves. In a long story about the high rates of sexual assault they experience, Joseph cited data from the U.S. Department of Justice, but all of the voices in the story belonged to people with lived experiences. "It was really powerful," he says, "because they told their own stories."

Rules for Dealing With Sources

When looking for a source, don't just think about what you want from them. Ask yourself, why are they talking to you? What's in it for them? If they are a random bystander or eyewitness, it's possible they just want to give you their perspective and stand to gain nothing from it. But in many cases, sources do benefit from the publicity you give them. There's nothing inherently wrong with that, so long as you remain in full control of your journalism and keep the relationship purely professional.

---

* Especially not in your audio story, where you really need to limit the number of personalities. In a digital story, quoting an additional expert can add to the authoritativeness of your reporting.

DON'T GET TOO FRIENDLY

Friendliness and charm are social lubricants and, as noted earlier, good reporters make the most of them. But there are lines you don't cross. "I never get too cozy with my sources," says Cheryl. "I don't let people buy me dinner, lunch, drinks, a piece of bread, nothing."

Brian Mann makes sure the parameters of the relationship are clear whenever he talks to someone. He'll say: "I know we've spoken three or four times, and you're calling me now, late at night, because you're heartbroken about something. But I am not talking to you as a friend now. I am a journalist. And we're on the record. And so I want to hear your story. I really think this is important. But I do want you to remember that you're not just talking to me. You're potentially talking to 3 million NPR listeners."

Speaking of which, be careful about using friends and family as journalistic sources. The *NPR Ethics Handbook* is very clear about avoiding conflicts of interest, and giving someone close to you publicity could amount to one.[6] On the other hand, our conversations with friends and acquaintances are often part of our reporting, and sometimes you come across a legitimate source. When FC Barcelona faced Manchester United in the 2009 UEFA Champions League European Cup soccer final, I interviewed a Spanish friend who was a Barça fan—because he spoke fluent English and described the team's style of play perfectly.[7] And when *All Things Considered* was looking for a Madrid resident to talk to host Robert Siegel a day after the 2004 terrorist attack, I connected them with a friend who lived a few blocks from one of the train stations that were bombed.[8] If you do go with someone close to you, there has to be a good reason. "The bar should be very high," says Rebecca. "We already have problems reaching a diverse group of Americans."

DON'T SHOW YOUR SCRIPT TO YOUR SOURCE

It's natural for people to care about how they are portrayed. Newlyweds, not the wedding photographer, select the best shots for their album. Parents, not the school photographer, do the same for their children. But it's the news photographer who makes the decision on a news photo, not the subject of that photo. Similarly, your sources don't get to choose how you quote them or what you write in your story.

At the same time, you might adjust your approach depending on who the source is. Shannon says she refused to let executives see her tech stories because usually they just wanted to argue about how they were being quoted. But if she reported on a controversial or complicated issue and wanted to make sure she quoted someone accurately, she might go back to them and say, "I just want to make sure I got this right. This is what you described to me. This is how I'm going to describe it."

PROTECT SOURCES THAT NEED IT

You should always try to get sources on tape, and fully identified. But sometimes there are good reasons to make exceptions: The source may fear physical or professional harm. The *NPR Ethics Handbook* spells out how we decide to let a source remain anonymous.[9]

If someone does need protection, and your editor has approved, it's up to you to make sure you and your source have the same understanding of the terms on which they are speaking, and how you will characterize the information they have shared with you. Cheryl tells sources that "on the record" means "I can use this information, I can use your name, I can use your title. Everything." But the categories of anonymity, she says, are less clear. "Is this off the record or on background? Don't ever let people define that for you. As the reporter, you define it. You determine, OK, 'background' means I can use this information, but I can't use this name. Or 'background' means that I can use this information only if I verify it with someone else."

Distinctions like that must be established upfront, Cheryl says. You don't want to be in the awkward position of having information that you thought was on background but the source thought was completely off the record.

Even when you're told something anonymously, it doesn't mean you can just go ahead and publish it. You have to determine whether a source may be dodging accountability for what they are saying. It's up to you to try to corroborate the information with other sources, whether it's people or documents, that can be identified. At NPR, anonymous sourcing needs to be approved by a newsroom manager to make sure it's in line with our standards.[10] We always want to be transparent with the audience and say why we have granted the request, whether it's for the source's safety or fear of retribution or whatever other reason was cleared by the newsroom manager.

And be as specific as possible in identifying what you can about the source, to establish their credibility. If your source is someone who attended a meeting where a pivotal decision was made, for example, you may want to describe them not just as "a source who didn't want to be identified" but rather as "a source who was in the meeting and who didn't want to be identified." But be careful: referring to that source's gender in a pronoun, if doing so would distinguish them from others, could give them away.

Using Your Microphone as Your Notebook

When Ruth was in Antakya gathering tape, she used her recorder as a notebook. As she and Erin headed into the city with their driver, and as they watched the effort to save a man who was buried under the rubble, she described everything she saw by talking into her microphone, as though she were describing it live on air in that moment. The technique is called the standup, and the beauty of it is that it takes sentences you would have recorded as tracks and turns them into scene tape that you can potentially use to liven up your piece. In Antakya, Ruth says, she "did way more standups" than usual.

At a training session a few weeks earlier, Ruth and I had talked about trying to have as much tape as possible in a story, almost as though you're putting together a non-narrated piece. While that's practically impossible to do in a news story—there's always something you have to explain—it's a good ideal to strive for.

In her story about Antakya for *Morning Edition*, Ruth recorded most of the tracks on location, including the ones below.[11] Show host Steve Inskeep introduces the piece:

STEVE: And let's go now to NPR's Ruth Sherlock, who's in the southern Turkish city of Antakya.

RUTH: Driving into Antakya now. We're just passing—building after building is flattened. One of the city's hospitals is literally on its side. The building's tilted, broken, rubble all around it, windows smashed.

(*Ambi of sirens.*)

RUTH: There are civilians digging through the rubble. There's just simply too many flattened buildings for emergency services to get to. So it's the civilians that are left to dig for their loved ones.

As listeners we are taken to the scene, not just by Ruth's vivid description but also by the sounds of a devastated city, with chopper blades whirling and sirens blaring.

At one of the collapsed buildings, Ruth taped rescue workers who were using an excavator to try to rescue a man under the rubble as his mother looked on, terrified that he wouldn't make it. "Very few people wanted to take the time to stand and do an in-depth interview on tape. People were digging through the rubble, trying to find loved ones," Ruth says. So she described what she was seeing, offering a kind of play-by-play account of the rescue effort. We hear her voice lower into a whisper: "They're telling everybody to be silent—try to hear for signs of life at this point." She then resumes in a normal tone: "They might have found him. But if that's the case, it's not good news. They've turned away, and they're walking back. Look[s] like they're going to talk to the mother now."

After a change of scene, Ruth does another standup from a location near the center of the city. We can hear from her breathing that she's walking:

> RUTH: Some people are walking out of the city with possessions in plastic bags, others just walking down the middle of the road in tears, lost. This earthquake happened in the dead of night here. Most people were in their beds sleeping. We can only begin to imagine how many people are under the rubble. (*Coughing*) So much dust and debris. One, two, three, four, five, six—six bodies lying on the street in front of a building—nobody is around—just left on the side here.

The standups are hardly polished writing, but it doesn't matter. They give a haunting sense of being there and experiencing the tragedy, especially combined with the ambi and conversations from the location.

It can be exasperating to do standups. When I did them, I would often become self-conscious recording myself in public. Other NPR reporters describe feeling the same way, including Ruth. But in Turkey, she says, "it was hard to feel self-conscious on almost no sleep for two and a half days at that point."

It's normal to have to do many takes. International correspondent Lauren Frayer estimates that the standups she files represent about 1% of what she's recorded. "I come home with 4 hours of tape and it's just me in the railway station, and it's aimless and it's dumb and it does not sound good."

But the 1% that does sound good brings the story to life. "I always tell reporters, try standups," says Northeast bureau chief Andrea de Leon. "Try several. If you hate them, you can just pretend you didn't do them and tell me you forgot." Even if you discard it, the standup will not have been in vain: What you say in your standups almost invariably includes details and descriptions you might forget once you get back to your computer. And since you're saying it out loud first, you are most likely formulating it in a script-friendly way.

In addition to describing what you're seeing, the standup is a great way to convey action, take a measurement or even do some counting. I'll consider those separately.

ACTION STANDUPS

In an action standup, you participate in or attempt to do the action that people you're reporting on are doing. For a story on cycling culture in Spain, I got on my own bicycle, went to a mountain range not far from Madrid, and hit Record:[12]

> (*Ambi of car zooming by.*)
> JEROME: So . . .
> (*Sound of me panting.*)
> JEROME: . . . here I am on my old mountain bike . . .
> (*Another car zooming past.*)
> JEROME: . . . and I'm trying to catch up . . . to one of the cyclists on this . . .
> (*Heavier panting.*)
> JEROME: . . . very steep road . . . as you can hear.
> (*Sounding almost out of breath, as I catch up with another cyclist.*)
> JEROME: ¡Hola, buenos días!

It was a little embarrassing to later hear myself breathing heavily on a nationally broadcast program, but it created an experience for the listener that brought them closer to that of a Spanish cyclist than anything I could have said.

MEASUREMENT STANDUPS

In 2021, Sam Brasch of Colorado Public Radio filed a piece on a major blizzard that knocked out power, closed roads and forced the cancellation of thousands of flights in the region.[13] Sam had a same-day turnaround for *All Things Considered*, and because Denver is two hours behind

Washington, D.C., he had to get the reporting done quickly. So, he stepped outside his front door and recorded this standup, which ended up being the opening track of his story:

> SAM: I'm out here in Denver, Colorado. The sun has finally come out. And we're going to do the ultimate test to see how much snow fell. And stick in that tape measure right into the snow, going all the way down. (*Pause.*) Wow! Looks like about 18 inches.

In a digital story, that whole paragraph might have been a simple piece of information: "It snowed 18 inches in Denver," but by inviting the listener outside with him and taking us through the act of plunging the tape measure deep into the snow until it hit the ground, Sam allowed the listener to experience it all with him. "What could have been a pro forma 2-minute weather update was brought to life by his standup with the tape measure in the snow," show host Mary Louise Kelly wrote in an email thanking the reporter after the show. "Drew me right into the story from the start and held me there."

"All we have are words," Mary Louise told me later. "But if you can paint me a picture where I can see it, where I feel like I'm standing there with you, that has as much value as all the great quotes, all the facts that you're going to throw out, because that's what's going to stick."

Investigative correspondent Joseph Shapiro used the technique to illustrate a central point in his story about "double-cell solitary confinement," the practice of putting two people in a solitary confinement cell.[14] Joseph's story uncovered an increase in homicides as a result of the practice. Here's his opening scene:

> (*Street noise runs under track.*)
> JOSEPH: Let's start this story with a quick trip to a mattress store.
> (*Bell rings as door opens.*)
> WOMAN'S VOICE: Hi!
> JOSEPH: Hey there. How are you?
> (*Ambi bed of shop continues under tracks.*)
> JOSEPH: I've come to measure the king-size bed.
> (*Ambi of Joseph extending tape measure.*)
> JOSEPH: So it's 6 feet 4 inches wide—
> (*Sound of Joseph measuring again.*)
> JOSEPH: —and 6 feet 8 inches long.

Because a king-size bed is just a little bit smaller than a solitary confinement cell at a prison I'm going to tell you about in Illinois, a cell where not one but two men live together for 23 to 24 hours a day.

And think about what goes into that prison cell: two bunks, metal, nothing comfortable like this mattress, a shelf, a sink, a toilet.

(*Ambi bed of shop fades out.*)

JOSEPH: Now imagine being locked in a cell about the size of this king-size mattress with a cellmate who's violent, a murderer . . .

Joseph could have just given the dimensions of the cell. Instead, he let us experience it by inviting us into the scene. A beautifully simple idea, though he remembers that it was not so easy to do it: "I think it was the third mattress store we went to before they would let us come in and record."

Sometimes all you have to do in a standup is record yourself counting. While doing a story about struggling automakers in Sweden, I wanted to give a sense of Swedish drivers' loyalty to domestic cars.[15] So I went out on the street and started naming the brands.

JEROME: I'm standing in the center of town in Gothenburg, and let's see what kind of cars are passing by. There's a Volvo. A Volvo. Three Volvos in a row. Here comes another Volvo. There's a VW and a Renault.

It may seem like a ham-handed experiment unworthy of a high school science class. But it got the point across without having to use a single percentage or statistic. And, I admit, the grammar is funky. But who cares? If I had gone back to my hotel and crafted a more elegant paragraph, it wouldn't have had the same effect as the image conjured up by my formulation.

Back in Beirut

Ruth spent a little more than a week covering the earthquake. In addition to filing dispatches from Turkey, she went into northern Syria and reported on the families displaced and the children wounded in the calamity. She filed videos for NPR's Instagram feed and did a recap on her reporting for the *Consider This* podcast.

As she prepared to return to Beirut, she was furiously working on her last story, which was about 11 people from a single family who died in the earthquake and another two people who were still under the rubble.

Many people think that when a reporter finishes a story, it no longer affects them. For Ruth, that couldn't be further from the truth. "The second I got to Lebanon, the moment the plane landed, I teared up," she says. When she arrived home, she was excited to see her husband and the kids and the dogs. "But I didn't feel myself for sure."

There were two children she couldn't get out of her mind. One was a 3-year-old girl whose leg was amputated at the site of her rescue and who didn't yet know that her mom had died. Another was a little boy whose parents and siblings had been killed in the quake. His nearest relative was a great aunt, and her house had been severely damaged.

"When you become a parent, it's very hard," she says. "These kinds of stories, they stay with you."*

* You'll read advice on processing traumatic events and caring for your mental health in chapter 7.

CHAPTER 6

Interviewing

Reporting is, to a large extent, about finding out what other people are thinking. The interview is the best way to do that. By engaging someone in conversation, you hear firsthand what they have to say, and you get an opportunity to gain extra insights, perhaps previously unreported, by asking follow-up questions.

One common misconception, though, is that the interview is about obtaining information. That's an important part of it. But because audio is all about the human voice, how someone comes across in an interview—not just how they talk, but also how they might hesitate or sigh or silently ponder the answer to question—can make all the difference. It can bring your story to life in a way that it might not in video or digital.

The best NPR reporters are conversationalists at heart: They really know how to connect with someone and find out what motivates them or why they think the things they do. "A lot of why I got into this business is that I love talking to people," says Sarah McCammon. "I really want to know what's in their head." No matter the person's views, Sarah always tries to bond with them as a human being, "being respectful, interested and curious. I think people respond really well to that."

"You have to be willing to go there with people" before they'll open up, says Adrian Florido. Especially when you're covering tragic news, people will only trust you if you're open with them. "I've cried with many sources and I've laughed with many sources. And I've gotten angry with many sources. I get emotional with them and that creates a bond."

Of course, that doesn't happen in every interview. What kind of bond you form will depend on who the source is. Are they a victim of a terrible tragedy, an elected official who is not responsive to their constituency's needs, a person who has been convicted of a serious crime?

Brian Mann once interviewed a woman who was involved in atrocities in Sudan. "I was probably sitting opposite a war criminal," he says. But that didn't change his underlying approach. "My goal was to say, 'This is a human being.' I want to understand not just the facts of what happened but, to the extent possible, how they got here, what their motivations are, what their feelings are about this thing."

During the 2016 campaign, political correspondent Asma Khalid endured shouted insults while on the campaign trail in New Hampshire and Ohio, and vitriolic attacks on Twitter as well. Asma is Muslim and covers her head, and the abuse got so bad she went into her editor's office and wept. Still, she insists on having conversations with all kinds of people no matter what their political views are. "The job of a journalist is to be curious, to be open minded," she says. "You have to view people as individuals and you have to come to them in interviews as individuals."

This chapter is about the reporter interview, where the questions are posed to elicit actualities and most of the time are not included in the final story. Host interviews, in which the audience does hear the questions, are discussed in chapter 14.

In this chapter, you'll learn how to talk to all kinds of people. For those in positions of power, I'll explain how to pose questions in a way that holds them accountable and gets them to go beyond the jargon and talking points. I'll also guide you on how to interview people who have endured terrible suffering so that they feel respected and are given agency in telling their stories. And I'll talk about how to approach people who are not used to talking to a reporter with a microphone. Remember that you're not just getting quotes, the way a print reporter might. If someone sounds stiff or unthinkingly rattles off facts, you will end up with bad tape, which sets both of you back. To help you avoid this, I'll suggest tips for putting an interviewee at ease so that they can express themselves in a way that grabs your audience. But before I get into all of that, I'm going to tackle an old debate in public radio that has been revived by advances in video connection software.

Remote or in Person?

During the COVID-19 pandemic, NPR closed its studios to guests and put the kibosh on in-person interviews. Like other broadcast media, NPR had been doing virtual interviews for some time using apps such as Zoom and FaceTime, which sound better than recorded phone calls. Suddenly, they were the only way to do journalism. Fortunately, the public was also getting better at connecting virtually, so producers could talk interviewees through the acoustics and the logistics: how to hold their phone, sync themselves on a second device and send the audio file from their device.

But despite an interviewee's best efforts, studio quality isn't attainable in someone's office or living room. You're always at the mercy of their technical aptitude and multitasking abilities if you don't have a field producer or reporter there holding the microphone and thinking acoustically. But the convenience of remote interviewing meant the shows could literally go on. And even after the pandemic subsided, many reporters continued doing journalism from their desks and living rooms. That practice upset many purists, who felt that it degraded the "takes you there" sensation of listening to NPR.

"NO SUBSTITUTE" FOR BEING THERE

"We're doing too many Zoom interviews," national correspondent John Burnett, who covered 20 hurricanes and a handful of wars, told me just before he retired in 2023. "We should get out of our cubicles and leave the house and go meet people." The reliance on remote interviewing "doesn't make for bad journalism, it makes for boring journalism," he said.

Adrian, who has covered major breaking news stories across the country and done in-depth reporting on race and identity, agrees. "There is no substitute for having the microphone in front of someone's face. You notice—when the conversation has been had in person—the longer breath, the pause, the reaction in someone's voice that might be triggered by the look that you as a reporter give them."

And there is evidence that inferior audio influences how much credence listeners give to a source.[1] "If lower-quality audio undercuts a person's credibility, we might be doing a disservice to our sources by settling for lower-quality tape. . . . It's another reason we should always make an extra effort to get the best-quality audio we can under the circumstances," Adrian Ma, co-host of *The Indicator from Planet Money*, said during an in-house discussion at NPR. He was pushing back against the notion expressed by some of the meeting participants that "often we are too precious about audio quality." But he added: "We don't have to be puritanical about audio quality. Interviewing a source through crackly phone tape is probably better than not getting the interview at all. And in some cases, it's germane to the story."

In *Serial*, a podcast by the creators of *This American Life*, much of the tape consisted of poor-quality calls to the protagonist, who was in prison, and the conversations were often cut short because of insufficient funds. Imagine, says NPR producer Brent Baughman, if the *Serial* production team had insisted on better-quality tape. "The whole thing was him on the phone, and it was amazing!" he says.

Another factor in how you meet a source is control. When you do an in-person interview, you walk away with the tape of the interview. But when you connect remotely, they may also be recording your conversation. National political correspondent Don Gonyea says he's careful about what he says, even if he's talking to a political scientist or some other source he's known for years. "Once we start the Zoom, I recognize that this is as if we're doing a live interview over the air. And in such cases I don't make casual remarks or engage in a lot of small talk, because anything can be taken out of context and misinterpreted."

To be sure, most reporters agree that when talking to eyewitnesses and people affected by violence or tragedy, it's usually better to be there. "But I would argue for some people, the second best, weirdly, is telephone," says climate reporter Rebecca Hersher, who reached elderly people on landlines in her reporting on the aftermath of storms. "I've gotten some really, really nice tape where they're in their favorite chair, they're looking out the window, their dog is in their lap and they're just talking to me about what's going on. It would be better tape if I was sitting right next to them. But it would not be better tape if I made them get up and figure out how to turn on their computer and download the Zoom app, and borrow their sister's phone so they could sync themselves."

So consider what medium people are most comfortable using. Rebecca was interviewing people from a generation that is used to talking on the phone, and she accepted that. "I know it's kind of heresy as a radio reporter," Rebecca adds with a chuckle, to say in effect, "Phone tape, that's where it's at!"

Whatever the merits of each medium, "sometimes the choice is made for you," says Ari Daniel. Although NPR had a team in Ukraine at the start of the war, and he was based in Boston at the time, editors asked him to do a story on Russia's bombardment of Ukrainian hospitals since global health was his beat. He did the entire story remotely, using an assortment of apps to do his interviews, but still looked for ways to make it feel as realistic as possible. While talking to a surgeon, Ari heard loud noises and asked him what was happening. It was artillery going overhead. And while Ari was speaking to a hospital director, her daughter, who was translating, started to cry over her mother's descriptions of the assaults on health facilities. Both became powerful moments in the story.[2] "When you're doing a remote interview, things can happen," Ari says. And being transparent and describing what's happening makes it a better story.

REMOTE INTERVIEWS AS AN EQUALIZER

When interviewing media-savvy sources like politicians or corporate executives, reporters might find it easier to challenge their statements in person. But not everyone does. A remote connection can serve as an equalizer when people in positions of authority try to intimidate reporters, especially young women and persons of color. "I do way more effective accountability interviews with government officials many times my age remotely," says Rebecca. When she was younger, she would find it hard in some cases to look a person in the eye and ask tough questions. Interviewing older white men in particular was challenging, since some didn't seem to take her questions seriously. "I really loved switching to remote interviews because I could turn off my camera and they couldn't see me," she says. "I was just 'a reporter from NPR.'"

Rebecca is not the only reporter who can get a little nervous. "I consider myself a shy person," says business reporter Andrea Hsu. "But I just think, when I'm in this job, I have to set that aside and go out and talk to people."

For a story on a threatened rail strike, she considered doing a remote interview, but decided instead to drive the two hours to Richmond, Virginia, to meet a rail worker named Reece Murtagh. "I'm really glad I went down and met him, because we just chitchatted a little bit longer than we would have on Zoom."

The intimacy of being there and being able to peruse a document together yielded some very visceral actualities in Andrea's story:[3]

ANDREA: . . . But deep in the report produced by Biden's emergency board, Murtagh found another reason the railroads are not offering more.

MURTAGH: It's on page 32, and I'm leafing through here to find it.

ANDREA: It's in the section that addresses the railroad's huge profits.

MURTAGH: Here it is. The carriers maintain that capital investment and risk are the reasons for their profits, not any contributions by labor.

ANDREA: Murtagh's first thought after reading that:

MURTAGH: Are you serious?!

Remote interviews do have one important advantage: They're quicker. When you're under deadline, which for some beats can be all the time, the rapid turnaround makes a lot of sense. But there are times when going out in person is faster. Andrea found that out while reporting on a spike in gas prices, when she first tried to find people to talk to online.[4] "I'd message people, 'Can I talk to you about gas prices?' And I got nothing. I was getting nowhere." So, she grabbed her microphone during the afternoon rush hour and went to a corner in Washington, D.C., that had three gas stations. Although many people declined her requests for an interview, she got what she needed relatively quickly. "It was actually much more efficient than messaging ten people and waiting for them to respond, and then not really [getting] the person you want," she says.

Be Prepared, but Don't Show Off Your Knowledge

Before an interview, familiarize yourself with the topics you'll be talking about. Research what your source has already said on the matter. Take note of their previous quotes on the topic, in case there's variance with what they say to you. But how much research should you do? How knowledgeable do you want to be? Is there an advantage in knowing less? And should you prepare a list of things you want to ask? I put these questions to some NPR reporters, and here's what they said.

CHERYL W. THOMPSON, INVESTIGATIVE CORRESPONDENT

Before I go into an interview, I read everything I can get my hands on about the issue. Because you don't want to repeat a story that was done a week ago or a month ago. You want something new and fresh and different. And then I come up with questions of, "OK, where do I want this story to go?" I always write the questions down because it helps you stay prepared and on track. Interviews can get out of hand really quickly. They can get unwieldy.

Another reason I write my questions down is because you don't want to be thinking of another question while somebody's answering. You really want to focus on what they're saying. You really want to listen to their response, because their response oftentimes leads to a question that you don't have written down.

BRIAN MANN, CORRESPONDENT, NATIONAL DESK

I don't want to say, "Don't go into an interview with lots of research," especially if it's going to be a difficult interview. You should be as prepared as you possibly can. But prepare to be naive, to say, "I do not understand" or "I thought it worked this way" or "My understanding was this. Can you explain how I've got it wrong? Can you walk me through this? Because I don't remember it working that way." Those are perfectly valid things to say.

REBECCA HERSHER, CORRESPONDENT, CLIMATE DESK

I go in with questions written for every interview. I don't want to read the questions, but I want to have sat down and really put some thought into the different ways in which we might talk to each other, me and this person, whoever they are.

The questions don't have to be deep. If I'm prepping for an interview with somebody, and all I know about them is that their friend told me that their trailer flooded and that they might be a good person to talk to, my first question is, "I'm sorry, your friend actually just gave me your first name. What is your last name?" But I'm going to write it down. It's an exercise for me to imagine what that conversation might be like, and how I might put that person at ease, and to tease out any goals that I have and also pitfalls that we might fall into.

I did a number of interviews with a woman who had a terrible tragedy happen to her after a storm. She had a child die. It was awful. And I prepped for those conversations intensely, in part because I wanted to be really sure that this interaction she was going to have with a reporter was not going to be traumatic for her. And that I wasn't going to say something off-the-cuff that was either hurtful or dismissive. Or forget about something we'd already talked about before.

Maybe everyone has their own style, but the goals are the same. Some people might say, you're trying to build a human connection, so just go in with a blank slate and be a good listener. But not everyone is naturally a good listener. Am I always the best listener when I am just listening to my family? No. Might I benefit from having sat down and really thought about what it is I can do to be a good listener in this moment? So professionally,

I'm going to prep by writing down questions. I often won't have them in front of me. But the exercise of doing it helps me be more prepared.

SARAH MCCAMMON, NATIONAL POLITICAL CORRESPONDENT, WASHINGTON DESK, AND CO-HOST, *NPR POLITICS PODCAST*

I'm often, not unprepared, but a little raw, because of the kind of interviews I often do—for example on abortion and reproductive rights, where you're talking to people about very sensitive things. I really just try to be myself and be very natural and just talk to them. I go into it, obviously, having prepared in the sense that I've done some research. I found this person for a reason.

But I'm not afraid to get a little bit messy and rambly. I'll ask a question by saying, " 'Hey, you know, I've been thinking about this,' blah, blah, blah. 'What do you think? Am I getting that right?' " Sometimes when you're talking about really sensitive stuff, just being a little exposed and a little not organized, helps the other person feel more relaxed. I don't want to approach somebody who I'm talking to about their abortion the way I would if I were doing a big formal interview with Dr. Fauci (the former presidential adviser). It's just a completely different tone and purpose.

GREGORY WARNER, *ROUGH TRANSLATION* HOST AND FORMER NPR CORRESPONDENT

I am definitely of the camp to have questions in advance. I also believe that there's an art to it and you don't want to have five pages of questions, even if you may have five pages of questions in your head. Because you don't want to just robotically go through each question; you want to be responsive to what they're saying.

What works for me is a one-page document that really gets at why we're curious about this. What are we trying to have answered? And even specific scenes that we want them to go slowly on.

One of the big differences with podcasting at this length versus the news format is that you often need the person to tell the complete story, which means asking, "What did it look like when you arrived? And then you walked in, and what did you see next? What did it smell like?" And even some sensory detail.

It's really nice to have thought about it in advance and on paper. So you know why you're asking them to tell you what something smells like while there's 17 other important questions to ask.

Strategic Ignorance

Sometimes you might need to show that you've done your homework. Earlier in her career, Sarah reported for member stations in Iowa, Georgia and Nebraska. When she covered the agriculture beat, she interviewed a lot of farmers and had to work to gain their respect. "Sometimes I felt like I had to first make sure people know that I was smart," she says.

But demonstrating knowledge can backfire, especially with sources who like to hide behind jargon and technical language. Nathan Rott, who is on the environmental beat, says a lot of scientists are like that.

"Everything has to be so precise," he says. For example, they might say, " 'Well, 78% of the forest is likely to be gone in the next 10 years.' And I'll say, 'Well, that sounds really bad,' " and they'll repeat, "Well, it's 78% of the forest."

When a source gets too deep in the weeds, Nathan tries to get them to shift mental gears. "I'll say, 'Hey, that's really, really complicated. Could you explain that to me the way you explain it to your kid?' " Or he just tries to be himself. "I kind of am a schlep and I just own it," Nathan says. That self-deprecation can work to his advantage during interviews, he says, if people feel they're smarter than he is. "Who is this idiot from Montana? He doesn't know nothing! I gotta explain it to him in the easiest way I can."

Health correspondent Rhitu Chatterjee also finds highly educated professionals to be a tough bunch. "Sometimes you need to get these very highly skilled, very well-known scientists to come down out of their ivory tower and be the humans that they're trained to not let show." Rhitu does that by having them imagine they're in a less formal setting. "We're at a party, and I'm like your neighbor, or your aunt or uncle. And I haven't read your scientific papers and attended your seminars."

Chatterjee has a master's degree in biotechnology and was a correspondent for *Science* magazine before coming to NPR. But she rarely tells people about her science background. "I have to understand the science, and that's also part of the interview. But if I understand it well, I can often tell it better than the scientists. What I want them to do is tell me a story."

A scholarly person may bristle at being asked to talk in terms a layperson would understand. "'Well, it is a complicated thing' and 'How dare you try to dumb down the thing that I'm explaining?'" Nathan says. "But your job at the end of the day is not only to make sure you're accurately representing what people are telling you, it's to make sure that people understand" what those scientists are saying, he adds.

No matter whom you're interviewing, Brian says the job of the journalist is to be "naive in their curiosity and vulnerable in their ignorance and open about it. You're not trying to impress a politician or an actor or an activist. Your job is to inform your audience."

Still, it's tempting for a journalist, especially an insecure one, to show off how smart they are. "When you're young or have some anxiety about whether you actually belong in the space, it's hard to put aside the feeling that you need to prove that you know what you're doing," says Rebecca. But don't get fooled into thinking it will change anyone's mind. "Asking a question in a way that makes them know you're smart is not going to work if they've already decided that you're not."

In fact, when she doesn't let on that she understands something, Rebecca feels as though she's in control. "When you ask the question the way you want to ask it, the way that serves the audience best," she says, "that is an act of power."

The Approach

You've probably noticed that a lot of NPR stories begin with a reporter meeting someone. In this story, Ari Shapiro greets Phil Spagnuolo, a local political candidate in New Hampshire:[5]

> (*Ambi of footsteps, knocking on door, and door squeaking open.*)
> SPAGNUOLO: Hello!
> ARI: Hey! Are you Phil?
> SPAGNUOLO: Yes, I am!
> ARI: Hi! I'm Ari.

Showing up to an interview with tape rolling is a relatively easy way to make sure you have ambi to establish a scene that sets up the interview, in case other sound opportunities don't materialize. If you plan to do that, let people know ahead of time. When he calls to arrange an interview, Nathan tells people: "I'll be the dude standing out there, probably wearing a hat, with a big microphone in my hand." And most have no problem with it. Nobody will say, "I'd prefer if you don't record until we establish a rapport," Nathan says. Advance notice is advised, because a microphone can frighten someone. Especially shotgun mics, which, as the name suggests, are long and look like a weapon. Even with a less intimidating microphone, it would be a pity to arrange an interview over the phone, only to show up at the guest's office or lab or factory and have the person say in astonishment, "I didn't think you were going to record this!"*

When reporting on the street it's a little trickier, since you are usually talking to people for the first time. You may be rolling tape in order to gather natural sound when you see someone and strike up a conversation. Just don't go in full throttle, even if other news organizations are doing it. "One of the things I've noticed, especially on a lot of breaking news stories," says Adrian, who covered the aftermath of the mass shooting in Uvalde, Texas, "is that there's a lot of really insensitive reporters who are willing to just stick a microphone or a camera in someone's face without asking any permission and just exploit people's pain."

Interviewing people after traumatic events is best done with gentleness and compassion. In one of his early assignments as an NPR producer, Alexi Horowitz-Ghazi covered the high school shooting in Parkland, Florida, with Brian, and he remembers Brian's sensitivity as he approached the situation. "He just had this very gentle, very empathetic way of walking up to people, starting a quiet conversation and checking in with them as people," before eventually continuing with, "'Here's what I'm doing. I'd like to talk to you about this part of your experience' or whatever, and doing it in such a thoughtful and nuanced way," Alexi says.

Before taking out his microphone, Adrian tries to spend time with people, getting to know them and feeling their pain. It builds trust. "People can feel when you actually care. They can definitely feel when you don't care. I don't ever want a person to feel like they have been exploited."

---

* Another reason to be rolling from the start is for documentation. "It's a very good idea to record interactions with sources from approach to departure, being careful to note what's on- and what's off-record, in order to protect oneself and document professional encounters," says Brian.

Vulnerable Interviewees

Rhitu Chatterjee had been reporting on burnout among health care workers when a source put her in touch with the roommate of a nurse who committed suicide. The roommate, also a nurse, was clearly struggling with the sudden loss of his friend.[6] "He was late for my call and obviously anxious about it," Rhitu says.

For all the strides society has made, there's still a taboo on talking about suicide, depression, anxiety and other mental health conditions. Finding people willing to talk about their struggles is hard. And even when they do, talking to a journalist risks reliving the trauma. Rhitu treads carefully. She follows a protocol that's designed most of all to protect her sources. It's based on the following principles:

MAKE SURE THEY'RE READY. The person has to be in a state of mind where they can reflect on their experiences. "Nobody who's in the midst of a crisis is able to give an interview. When you're in the midst of something, you don't have the perspective." She also tells them, "I want you to know that you don't have to answer anything you're uncomfortable with."

EXPLAIN WHAT YOU'RE DOING. In the first interviews, Rhitu does most of the talking and outlines her goals. "In telling the story of someone who is in a vulnerable place, I want to help others," she says. But it can take a while for someone to open up. "These are intensely personal narratives. I almost need two or three interviews to even confirm a narrative."

SHOW COMPASSION. She approaches the person with empathy, and shows she cares about their well-being. "Conveying to them that I know they're hurting right now. And just asking, 'How are you doing? How are you coping?' They don't get that from a lot of reporters. Just a little bit of humanity helps."

DRIVE HOME THE PERMANENCE OF A PUBLISHED STORY. Rhitu makes sure the person has no qualms about having their story out there. "Once it's out there, it's out there. It's not going to be taken back," she says.

Rhitu says she is always ready to drop a story, or try to find another interviewee, if she thinks harm will come to someone as a result of talking to her. "The person," she says, "is more important than my story."

Old-school journalism dictated that if someone says something in an interview, no matter what their mental state, it's fair game to publish it.

While the public needs to be informed about violence and tragedy, and it can be therapeutic for those who've experienced it to talk about it publicly, journalists can cause survivors a lot of harm if they're not careful.*

"I'm very straight-up about it," says international correspondent Eyder Peralta, who has done extensive reporting in war zones. "I tell people, 'You don't have to talk to me.'" But if they do want to talk, he treats them with honesty and compassion. "You don't have much time to create a long-term intimacy" when covering armed conflicts, he says. "But you can at least create a short-term intimacy by just being respectful and giving people dignity."

Several NPR reporters I spoke to felt it is important to give people agency, especially those in vulnerable situations. Florido tells people that what they say in the interview doesn't have to be final. "I never do this with politicians or public officials, but when I'm talking to everyday people who generally don't speak to the press or are a little hesitant about whether they want to talk to me, and who don't really have any obligation to talk to me—I'll say, 'If at some point you decide that you're not comfortable with me using this, just call me and we'll talk about it.'"

It's no guarantee that he won't quote them, but he'll consider that if warranted. In the aftermath of the Uvalde school massacre, he got a scoop from a local resident who told him that Eva Mireles, the fatally wounded teacher, called her husband, a policeman, from the classroom after she was shot. Adrian was about to include that detail, and tape of the interview, in a live two-way. "And we were 15 minutes away from airtime. And she [the local resident] called me and said, 'I don't want to be the person telling you that Eva Mireles called her husband.'"

He told his editor that Uvalde is a small town and that the source worried that people would turn against her. "And so we took it out," Adrian says. Although the information eventually became public, he still believes that protecting his source was the right thing to do. He says that giving people agency leads to better interviews because people feel more comfortable talking to you. "It's a way to very quickly establish a level of trust with someone you just approached on the street, and that you're meeting in a moment of great pain." And even though they didn't get that initial scoop on air, the person "kept talking to me and I got other information," Adrian says.

---

* The Dart Center for Journalism and Trauma (dartcenter.org) offers extensive resources on covering conflict and tragedy.

Adrian Florido on Approaching People on the Street

My approach is to be as subtle as possible and blend in. It's important for people to know that I'm a reporter; I don't want people to be fooled into thinking that I'm a random person. So I always identify myself, but I first try to connect as a human being.

I was in Puerto Rico, reporting in the aftermath of the earthquake in early 2020. And I was going around the emergency shelters that the Federal Emergency Management Agency had set up for people whose homes had been damaged or who were too afraid to sleep inside because of the aftershocks. These were people who were poor and living in tents for weeks. And I would approach them as they sat on their cots.

In situations where people might not be super-receptive to talking to a reporter, I try never to have my recorder out or even my notebook out. I don't like it to be visible because I don't want people to feel pressured. So it's like I'm just walking around, and usually what I do is, I ask people if I can introduce myself, as opposed to going up to someone and saying, "Hey, can I interview you?" or "Hi! I'm a reporter!"

It's important, because this is Puerto Rico, and in Latin American culture, respect is very important. Formalities too. And if people feel like you're approaching them with respect from the very beginning, it's like you've earned a point.

So I'll say, "My name is Adrian Florido. I'm a reporter with NPR, National Public Radio, in the U.S. Do you mind if I sit down?" And before I say, "I'd like to interview you," I just strike up a conversation. And then at some point I'll say, "Well, I'm working on this story, and I'd like to interview you, because what you're saying sounds really interesting and important and I'd like to include your voice. Would that be OK?" And if they say yes, then that is usually when I'll take out my notebook and recorder.

It's a way to approach someone that is human. At the end of the day, we're there to take something from them, but that doesn't mean there can't be a real human connection.

Transparency

Not everybody knows how journalists do their jobs. There are plenty of misconceptions out there. So no matter whom you're talking to, you may want to explain what you're going to do with their comments. Rebecca has a spiel that goes something like this: "Hey, I'm a reporter. I do stories that go on the radio. The radio goes all over the country. It also lives on the internet forever. We have a website. We also publish photos. I know that's weird, ha-ha, a radio network with photos! We also have podcasts. My job as a reporter is to file things for all of those places. I don't always have control over where the things go, which show they air on, or when they air. I won't always know or be able to give you any warning."

Even sources she talks to frequently can benefit from a shortened version of the spiel: "Hey, I'm interviewing you. I'm pretty sure this is going on *All Things Considered* tomorrow, but I'm not 100% sure. It might go somewhere else. I'm going to start recording now. Is that OK?"

Rebecca also doesn't promise an interviewee that they will be in her story. In fact, she tries to preempt disappointment by saying that she may use the interview solely to inform her writing. She used to promise to let the person know when a story was going to air, but found it hard to follow up. "I'd inevitably forget somebody."

It's all about setting expectations. Cheryl Corley sometimes does it with a bit of humor: "I tell them, when you hear it you're going to say, 'Was that all she used?!'"

Essentially, what you're doing is telling people, in just a few words, how journalism works.

Respect for Their Time, and Your Own

An interview can take just a few minutes if you're calling someone up with a very focused question. On some podcasts, it can go on for hours and involve a producer, a reporter and maybe even an editor and a host. "I've been in interviews where the interviewee is messaging me, 'When is this going to end?'" says correspondent Lauren Frayer, who has collaborated on podcast episodes with *Rough Translation* and *Planet Money*.

Doug Mitchell of Next Generation Radio, which trains college students and early-career professionals in public radio, often has them do a non-narrated audio project over five days. And right at the beginning, the participants are told, "Don't interview anyone for more than 45 minutes. Because if it's longer than that, you really aren't listening."

How long should your interview be? It depends on how much material you need, how central this person is to your story and the sensitivity of the subject matter, especially if you're building up to a controversial question.

But the Next Generation standard is a good one: Don't waste anyone's time. You don't need a recorder to get all the facts and figures; you can find those on the web, or get them in a pre-interview. When you're in a taped interview, your objective is to get good actualities. TV reporting used to be good training for this. Back when cameras were heavy, the photographer *really appreciated* having an interview end before their shoulder started falling off.

Not wasting people's time also means making sure you've asked all the essential questions and don't have to return for an additional interview. It's not like print reporting, where you can ask another question by phone or email, since the sound quality will be different. "Radio is a lot harder to do," Nathan says, "because all of a sudden you realize, 'Crap, I'm not sitting in that room with that air conditioner buzzing. I could call them but the tape is going to sound different, and I don't want to have to try to explain in the story why we're talking to them again in a different setting.'"

When Rebecca prepares her questions, she organizes them in order of importance. "I don't want to be wasting my time on the side questions if we actually only have 20 minutes together, and they have a hard stop time because they're picking up their kid."

Even if she has agreed on an interview length, she always asks the person at the outset how much time they have. They may have more than they thought, and you can ask follow-up questions. Or they may have underestimated their commitments. One interviewee told Rebecca she had no time constraints, but Rebecca says she should have known better because the interviewee was at a conference. "After 40 minutes, I asked her a question about the morality of climate economics in general. And she said, 'I am so late right now!'"

Ways to Make Them Comfortable

No matter how long an interview is, you'll want your subject to feel relaxed and ready to talk as quickly as possible. Author and journalist Isabel Wilkerson talks about "accelerated intimacy" that journalists need to create.[7] It's an extraordinary challenge in audio journalism, both because of the scary microphone and because listeners will hear if someone who should be comfortable in the interview is not.

Here are some tips that will help put your interviewee at ease.

PRACTICE USING YOUR GEAR AT HOME. The more you fumble with your recorder and microphone, the more distracting it is for the interviewee. And the less you will be able to think about what you need for the story. You won't always be going out with a producer, so get to know the equipment. Try starting, pausing and stopping the recording. Adjust levels, mic-to-mouth distance and angle and so on. Interview your children, partner, parents, friends and pets. "If you're comfortable with your equipment, it melts away for you," says Eyder, "and it melts away for everybody."

DON'T LOOK AT YOUR QUESTIONS. It will interrupt the flow of the interview. Study them beforehand so that the main ones are in your head and you can maintain eye contact and manifest your keen interest in the interviewee. "There's nothing more awkward," says Nathan, "than reaching in your back pocket and trying to pull out a notepad with questions that you've written before." If you need to, you can say at the end of the interview, "Let me just check if I missed something," and take a glance at your notebook.

FOOL AROUND WITH THE MICROPHONE IF SOMEONE SEEMS INTIMIDATED BY IT. "If they've got a kid," says Nathan, "give it to them for a little while, let them play with it. Put it in front of the dog. Scratch your back with it. Do those things to make it not seem scary."

DURING THE INTERVIEW, KEEP THE MICROPHONE OUT OF THE LINE OF SIGHT. Angle it to the side. It's better for audio quality, and after a while, the interviewee will likely no longer notice it. "If a person forgets it's there," says Nathan, "then you're gold, because they're going to feel like they're just talking to you."

MAKE SMALL TALK. Chitchat about sports or the weather or whatever while you're checking the recording levels, and before you launch into the questions on your list. When she's traveling, Cheryl often asks for restaurant recommendations. "It's not only just chatting them up, but giving you good information for later!"

START WITH UNCONTROVERSIAL SUBJECTS. You can put people at ease by asking them about things they are knowledgeable and excited about—how they got into their field, what they're working on at the moment, and so on. Beginning with simple subjects will help interviewees feel they can talk without fear of making a mistake or saying something they'd hoped not to divulge.

TELL THEM NOT TO THINK OF IT AS AN INTERVIEW. The idea of an interview makes people stiffen up and sound stodgy and formal. Instead, say it's a conversation. "I find that being supercasual and setting the tone as supercasual puts people at ease," says Nate. "It makes them more likely to just talk."

SHARE YOUR OWN STORIES. If appropriate, talk about how you or someone close to you has been in a similar situation and how you reacted. "I will totally reveal parts about myself," says Nathan. "It makes you a human. You're not just like this autonomous journalistic machine there trying to get information from them."

TAKE CONTROL. Tell the person where to sit, to turn off a noisy appliance if it's obtrusive or to find a better room if necessary. It shows you know what you're doing. "Don't be afraid to be bossy," says Cheryl. "And you can even say it. That's how I do it. I say, 'Well, I'm going to be a little bossy here because we need to get the best audio. Can we go someplace else?' "

And if there's a loud sound during an interview that doesn't fit in the story, stop the interview, because it will sound jarring in the middle of an actuality. "Let's just wait for this thing to pass overhead," is what Nathan says when an airplane flies nearby. Or if an air conditioner is clicking on and off, he might say, "Can you do me a favor, and just turn that off for the next 20 minutes?"

"It's really important that you own the space," he says.*

AND DON'T FORGET THE BASICS. Start the interview by asking the person to identify themselves. This is a good way to ensure that they know their name will be used, so that they don't say at the end of the interview, "I'd rather not be named." Have them say and spell their name on tape so that you have the correct pronunciation for your audio story and the correct spelling for digital.

And double check that you're recording! Especially if you're new to audio, or a little absent-minded (like me!), you can quickly find yourself engrossed in a conversation that yields no tape. Also, some recorders—infuriatingly, if you ask me—require you to press the record button twice: the first time just gives you levels, and only the second time starts the tape rolling.

---

* There are ethical considerations in creating a suitable environment for recording in the field. See Lauren Migaki's comments in "Noisy Environments" in chapter 4.

Brian Mann on Putting Interviewees at Ease

What I have learned over the years is that with my body language and with my tone of voice and my whole approach, I can establish a very human social contract pretty quickly.

When possible, I have my recorder out from the beginning because I want them to be comfortable that this is part of what we're doing. I don't make small talk, and then suddenly, right before we do the interview, pull out my device. Instead, I have it out from the moment I arrive. It's part of the room. It's there. It's, "This is not that big a deal. Yeah, it's on. We're just talking."

And I am also explaining my process. I'm saying, "Here's what's going to happen. Here's what we're going to talk about. And it's going to be pretty relaxed. We're just going to be having a conversation." And I'm really meaning it. I am not there to spark some quick, juicy, sexy quote.

And I'm sitting in a way that's as relaxed as I can be. And I'm engaging with them in as human a way, I'm sitting in a place where we're both comfortable, and I try to hold the microphone so it's outside their sightline. It's close, but it's not right in front of their face.

Closeness

In 2021, Eyder, who was based in Nairobi and covering Africa, got an interview with Ugandan president Yoweri Museveni.[8] He drove to the president's ranch, located five hours from Kampala. It was the height of the COVID-19 pandemic, and Museveni was terrified of catching it, so his handlers tried to keep Eyder as far from their boss as possible. "They wanted to use their microphones and have him on a speaker from the other side of a field." Eyder responded by saying, "No, we're not doing the interview unless I use my microphone, and unless we sit close enough that I can hear him."

Eyder got his way. A few months later, he worked on a very different kind of story, but where it was also crucial to be close to his interviewee. It was in the war-ravaged Tigray region of Ethiopia. Eyder visited a school that had been turned into a safe house for women who had been raped during the conflict. He interviewed one woman who talked about what she had endured:[9]

EYDER: Eritrean soldiers, she said, hit her with their guns. They insulted her. They tied her up and dragged her to a military base.

(*Woman speaking in Tigrinya.*)

INTERPRETER: For about a month, she was chained. She was chained—both her feet and her leg was chained. And she was gang raped both vaginally and anally by soldiers.

EYDER: There were other women there, and the soldiers took turns. They would rape them for days at a time.

(*Woman speaking in Tigrinya.*)

EYDER: They assaulted her so bad, she says, she wouldn't stop bleeding. Some of the women begged to be killed. And eventually, the soldiers drove her out to a field and left her for dead.

While listening to her horrific account, Eyder positioned himself as close as he could. He, the woman and an interpreter sat in a triangle, their knees almost touching, as he miked each question, answer and translation.

"When somebody you don't know is suffering, the human thing is to pull away," Eyder says. But as an audio journalist, you can't do that. "What you're doing with the microphone is actually getting really close in." And you're an active participant in the conversation. "As a print reporter, you can get away with sitting in a corner and not doing anything," says Eyder, who worked at the *Miami Herald* and *Houston Chronicle* before joining NPR as a producer in 2008. "As an audio reporter, if you're 20 feet away, it sounds like you're 20 feet away."

Like a Date

Imagine you're on a first date, and it's going well. The person you just met is pretty neat; there's potential for a relationship. How are you acting? What's your body language? Are your eyes wandering toward your phone, perhaps? Are you checking out other people in the café? Is your mind meandering, thinking about a show you haven't streamed yet? Something your mother said?

Or are you hanging on this person's every word?

An interview is not unlike a first date. Just as when you're meeting a romantic interest, you should be in a heightened state of awareness, fascinated by this person and engrossed by their stories, trying to find out all you can about them. "You have to be utterly present," says Neva Grant, who helped found *Weekend Edition* in 1985 and has worked as a producer on *Morning Edition*, the *TED Radio Hour* and *How I Built This*. Your interest in the interviewee, she says, can't be an act. You shouldn't *pretend* that they are the most important person in the world at that moment. "You have to believe it," she says, because if you don't, you won't be able to hide it. "They'll know if you're drifting."

With clogged calendars, bleating phones and bulging inboxes, we all struggle to give anything our undivided attention. "For journalists, it's really toxic," says Brian. "I'll be in a media scrum and I will see other journalists holding out their recorders, while they are literally tweeting or reading texts from their editors."

That's why you shouldn't even be looking at your notebook as you are listening to someone talk. When Brian is in an interview, he treats it as "sacred space." "It is not texting space. It's not social media space. It's not thinking-about-my-editor space. It's not even thinking-about-my-audience. It is, 'I am now in a human relationship with another person where they are my focus.'" Even when he's on deadline, Brian tries to block out all the sources of his stress and concentrate on the conversation. He describes it as a physiological transformation. "I'm bringing my heart rate down. I'm bringing my pace of thought down. I'm setting aside a lot of the buzzing thoughts in my head."

Now comes the high-wire act. Because while you're in this heightened state of awareness, with the interviewee thinking they've got your undivided attention, you've actually got a few synapses attuned to your surroundings. After Hurricane Katrina, Neva was on a host trip with Steve Inskeep in New Orleans, visiting the home of a family they'd interviewed before. They had no water and were getting power from a generator. During the interview, Steve asked if they could go upstairs. That led to a rich scene—climbing a ladder and looking out over a dark and empty neighborhood, as the owner waxed lyrical about how peaceful it was to be able to see the stars.[10]

While immersed in conversation, Steve was also thinking about how the interview might figure in the story. "While you are all-in and authentically engaging," Neva says, "there has to be some tiny little corner of your brain that is doing what Steve did in that moment," and asking yourself, "How can we make this even more vivid?"

Another reason to focus part of your brain on your environs is to pick up on things that could endanger your safety, or even just your work. In 2013, Nathan was deep into an interview with the fire chief of the Granite Mountain Hotshots in Arizona. The chief had just lost 19 firefighters in a wildfire and was getting emotional. "The guy's just bawling," Nathan recalls. "And I was so engrossed in it, I was so focused on him, I did not hear that there was a golf cart backing up behind me." When Nathan got back to his hotel, he realized that the interview was useless. The backup alarm would be too distracting for the listener. "What they're going to hear is the most grabbing sound, which is a golf cart going *beep, beep, beep, beep!*"

OK, so maybe on that score, an interview isn't entirely like a date. And definitely not, in the way you have to be a bit annoying and ask follow-up questions.

The Follow-Up Question

Some reporters, when starting out, may feel intimidated by famous or powerful people. They are so grateful to have landed the interview that when they get a vague or misleading answer to their question, they don't challenge it. They think the problem lies in themselves. If this person, who has extensive experience communicating with the public, chose to formulate it the way they did, surely the audience will get it. Right?

Wrong. You are a journalist. You chose to do this story, or your editors assigned it to you, on behalf of your audience. And if you have trouble understanding the answer, chances are so will your audience. If you need a better answer, or more information, or you think there's a falsehood or inconsistency, or just something that needs elaboration, the time to bring it up is in the interview, not when you're writing the story or going over it with your editor.

That's why the follow-up question is so crucial. It can be as broad and simple as asking why. Or it can be a specific statement that pushes back on a claim, like this one that Sarah McCammon asked a legislator who was claiming that there were voting irregularities in Georgia in 2020: "Your Republican secretary of state, Brad Raffensperger, reviewed the entire vote count three times, including Fulton County's, and said it was legitimate."

This is another reason to not look at your prepared questions, so that you can focus on listening closely. "The follow-up question is the surgical tool of our trade," says Brian, explaining how he comes up with one. "When people say things, I stop and I think. And I play it back in my own mind. If it's something that I think is exciting or confusing or controversial, I'll play it back for them. I'll just say, 'Hey, you just said that you legitimately believe *x*. I don't understand that. Can you explain how you came to feel that way or, you know, or what does that mean for you on a daily basis?'"

An insightful follow-up question, one that comes from listening closely and with interest, often lands him the "money quote." "If you come back from that human interaction with a powerful moment, a meaningful, thoughtful moment, which is almost always triggered by a follow-up question, the rest of your story will write itself," Brian says.

A follow-up question can expose a lie, but for Brian it's more often about getting beyond the spin and superficiality. "This is not gotcha journalism. I just want them to be human and talk about this policy idea that they've put forward. I want to understand human-to-human why they care about it, what's driving their commitment to it."

Brian, who started his public radio career in Alaska, says covering its fisheries was a great training ground for asking follow-up questions. "It's one of the most exciting industries in the world in which the people who do it love to talk about it in very boring ways. If they can talk about bycatch and quotas, they're excited. And I didn't want to hear about bycatch and quotas. I wanted to hear about the actual experience of the thing. And so I really learned to let some of that stuff go by, and then get to the questions that felt like we were really having a human conversation, about what they're experiencing and what they're going through."

If a person refuses to engage meaningfully with a follow-up question, it can be just as illuminating. There was a period when Brian was doing a lot of coverage on abortion. And when speaking to people who were against reproductive rights, he said they were very good at talking about "the first beat of their argument." Then he would try to get them to think through the ramifications of their positions. "'You say a woman can't get a legal, safe abortion,'" he would say. "'Let's talk through what happens next, and let's really go into it.' And often the resistance to that was stubborn and awkward and objectively ill informed." So that became the story. "The crux was not, 'I oppose abortion.' The crux was, 'I really have no great interest or policy ideas for what happens next.' And I found that to be a much more interesting story than yet another piece about a March for Life, which I'd already done and many other journalists had already done."

Don Gonyea on Interviewing Voters

APPROACHING SOMEONE ON THE STREET

I'll drive by a place, and I'll see somebody sitting on their front porch, and then I'll be about three blocks past and in my head I'll go, "That person is just sitting on their front porch, I got to go talk to them." And then I'll have this little argument with myself, why I don't need to go talk to them, because I'm kind of an introvert—I'm not good at walking up to people just cold. But next thing you know, I'm turning the car back, I'm parking it and walking up their street. And I figure out a little icebreaker. "Is there a diner around here? Is there a place I can get a cup of coffee?" And then if they answer and they give me some information, I go, "Yeah, I'm a reporter, I'm interviewing people. I could talk to you?" And next thing you know, you're talking to them. Sometimes they give you the thing that's magic; more often than not there's nothing particularly great, because the odds are pretty low that any given interview is going to be all that great. You might have to talk to a lot of people. You have to approach 10 people to get one to talk to you. And of the ones that talk to you, maybe only half—if that—say anything that's really on point.

So basically it's all about the numbers—it's simple math—the more people you talk to, the more likely you'll get something that fits well into the story. And it's not that I'm looking for them to say something specific, something I've decided in advance that I want. I'm more than open to being surprised. In fact, that's the best thing. But mostly I need it to be interesting and on point, even as it can be either predictable or unexpected.

APPROACHING PEOPLE AT TRUMP RALLIES

I've got a microphone, and I've got my gear on, and I identify myself. But if I want to talk to somebody waiting in line to get in, I don't walk up to them and say, "Hello, I'm from NPR. Can I do an interview with you?" That can be off-putting. Instead, I might walk up to them and I go, "Whoa, you guys are first in line! What time did you get here?" They want to brag about it! And next thing you know, we're 30 seconds into this fun little encounter where I am truly interested in how they got to be first in line, and just having that kind of a casual start to an encounter increases the odds probably tenfold that they're going to have a real conversation with you.

Before the Rally

I've started talking to people on the sidewalk outside these rallies as early as 5 a.m. Maybe they've been there since the night before, staking out their spot in line, so they can get as close to the front as possible when the doors open at 5 p.m. or 3 p.m. or whatever it is at any given place. That works in my favor. Because they're just sitting there and they've got another seven hours to wait before they open the doors, and they can't exactly say, "I'm too busy to talk to you." In fact, they're generally happy to talk to me because it's going to help them kill 15 minutes or a half an hour. I have had long conversations with people in line, long enough to get past the talking points. We talk about their families and how many of these rallies they've been to and what is it they like most about them. This is all stuff I want to hear, even if most of it is never going to make your story. Maybe you get the sound bites you're going to use in the first minute, or—and this is why patience is important—maybe you get the sound bite you're going to use 29 minutes in. And if you hadn't had the first 29 minutes of whatever you talked about, you weren't going to get what they said that was the most interesting and the best tape.

After the Rally

It's no news to anyone that during his speeches Trump regularly vilifies the journalists gathered on the press risers. In such moments he calls reporters "enemies of the people," or "so dishonest" or "fake news," all the while prompting loud booing, jeers and other gestures from the audience.

We get used to all of that. But I've also had the somewhat jarring experience where people I had long, friendly conversations with before the rally have looked at me with contempt and refused to talk to me at all when I bump into them again after the rally. I recall one time in Cleveland in 2016. In the hours beforehand I interviewed lots of people waiting in line to get in. One couple was particularly nice, answering my questions, making small talk. Then I bumped into them again heading out of the event after the rally. I asked what they thought of the speech. The only reply was a curt, "Tell the truth." They wanted nothing to do with me or my microphone.

I recall thinking that it was an instant example of the effect this candidate's words can have on an audience that is clearly enthralled. I was eager to get this couple's post-event thoughts. But their sudden contempt for me made it clear that we weren't going to have a friendly follow-up to our pre-rally interview. Had they been willing to talk I'd have been happy to talk to them. You never know what you'll get, and I certainly saw it as a follow-up conversation worth having. Alas, it didn't happen.

I should add that in this instance I felt no reason to be concerned about my safety. The post-rally crowd was fired up by the speech, and that included their disgust for the reporters present. But that doesn't mean this was a dangerous place to be. As a reporter you're always mindful of the circumstances and of what's going on around you. At times, I have been in some dicey and volatile situations, including at some political rallies where there were clashes between rally attendees and protesters. Sometimes there are people who show up just to make trouble. But none of that was present at this particular rally. Still, it's important to always be alert and aware of your surroundings, especially during the rally and while walking to your parked car afterward. The situation can change quickly.

FACT-CHECKING IN REAL TIME

It's not easy to be a fact-checker in real time, especially when you want to actually hear where they're coming from and what motivates them to think what they're thinking. Because that's the really important thing when you're talking to regular folks and regular voters. But it's also good to think ahead to when you're actually writing your story. Fast-forward to when you're sitting at your laptop in your hotel room. You've got your taped interviews and you're on deadline and you've selected something to put on the air. And that's when it would be nice to have a quick on-tape exchange with you and them, where they will often tell you Trump won the election because there were ballots stuffed into boxes across Ohio or Michigan or in whatever state he lost. And you say to them, "There's no evidence of that." And then they assert, "Well, it happened." It's just a quick moment in the story, but you didn't just let the falsehood stand. You said something to the person, and it's on tape. It's not that you'll change their mind. You won't. And nothing is really served by getting into an argument with them in that moment. But you didn't just let the falsehood fly past unchallenged.

WHY YOU MIGHT NOT WANT TO FACT-CHECK IN REAL TIME

Sometimes it's a situation when you may not want to push back and fact-check in real time, to keep the interview going and not have it end instantly, or not have it be your last interview in the room. Because you know that as soon as you're done talking with her, you're going to go 10 feet over there and approach somebody else and talk to them. And if you get into arguments with the first two, there's not going to be a third, because the others in the room will hear that this guy is here with an agenda and he just wants to make you look bad. So you're making conscious decisions on the fly to allow you to keep gathering tape. But you are obviously beholden to the truth and accuracy. So if you do put them on the air saying something that's utterly, provably false, you say that it is false in your script.

What If You Get Stuck or You're Not Getting What You Need?

If someone isn't answering questions in a way you think will work for your story, there are strategies you can follow.

PERSIST

"I'm going to ask the same question, six different ways," says Alexi Horowitz-Ghazi. That's often necessary with scientists. Rhitu Chatterjee might pose a question one way and then again a few minutes later: "I know you said that, could you try and say it a little differently?" Sarah McCammon also keeps at it. "If I feel like somebody is really nervous and giving really wooden answers or is just overly complicated, or if I don't think the audience will understand it, I'll just be very transparent and say, 'OK, I already kind of asked you this, but I'm going to ask you a different way because I'm just trying to understand a little more precisely.'"

An audio journalist will often know better than the interviewee how they will sound in this medium. So the aim is not to coach them, just try to get them to sound more human.

"I never go in with a script and say, 'OK, here are your lines,'" says Ari Daniel. But he will try to prompt them in a way that gets them to tell their story. "They're going to the trouble of talking to me. I'd like to go to the trouble of, you know, empowering them with questions and interactions that will allow them to be present in the piece."

TAKE A BREAK

Get a cup of coffee, or have lunch together, and then try again. And use that time to build trust.

If mic jitters are keeping someone from sounding human, Nathan says he sometimes addresses it directly. "I will sometimes straight-up stop an interview. I won't stop recording, but I'll point the microphone away and I'll say, 'Hey, it seems like you're kind of nervous?' And then we talk about it, and usually in the talking they become less nervous."

Sometimes people freeze up during an interview. In those situations, former special correspondent Melissa Block would say, "'Let's just start over,' or 'Let's just take stretch, or have a cup of coffee.'" She says it helps show that you are really interested in a conversation rather than an interview. "That's when people really open up the most, when they feel like they're not talking to a reporter who's jotting down every word they say, but they're talking to somebody who's interested in hearing something."

A break can help, even if you're not stuck. Lauren Frayer had started interviewing a family about the 1947 Partition of India when they invited her to have pizza with them at their house. They talked about soccer and other things unrelated to the interview, and it was time well spent; when she started recording again, the interview went much better because of the new rapport they had established.

"There have been some gems that I probably didn't capture because my mouth was full of pizza and I wasn't recording. But it's a gamble. You have to gauge whether your time is better spent building trust with the tape recorder off or whether you need to capture every moment."

GIVE UP

But not entirely. The reporter put in the effort to interview the person, so they should finish the interview but give the person a more limited role in the story. And, if necessary, cut the interview down drastically. "You might do a 15-minute interview, or a 20-minute interview with the person," says Nathan, "where, in your actual piece, you use 15 seconds of tape."

Next, they might find someone else to interview. "I'm going to have to figure out how to book my way out of this," Alexi Horowitz-Ghazi tells himself. For the synthetic drugs story I mentioned in chapter 2, Alexi interviewed a professor whose scientific research inadvertently played a role in the rise of the multibillion-dollar synthetic drug market. Alexi had hoped the professor would be the central figure in the story, but when he talked to the 86-year-old professor, it turned out he was "not quite reflective enough to fully flesh out some kind of emotional arc to carry the whole story." Alexi used part of the interview for a little scene in the middle, but found other people to tell the story following a different arc.

Andrea Hsu had a similar experience. She went to Martinsburg, West Virginia, for a feature story on President Joe Biden's plan to increase funding for child care.[11] She had set up an interview with the owner of a day care center, but the interview didn't live up to its promise. "I was really discouraged because I thought, 'Oh my gosh, I just drove all the way here and I don't really think I have a story.'" When she got out of the interview, she called up other day care centers in the city and reached the owner of another one. "She said, 'Yeah, come on over!' I was able to go to the second place, get more sound, talk to her, talk to a parent. It was just fortuitous that she was willing to talk, and she was a lot more forthcoming."

Other Ways To Get Great Interview Tape

Your interview is less likely to run aground if you make sure your interviewee is relaxed and immersed in the conversation and you're well prepared and listening intently to their answers. But if you want to get really great tape that will make your story come to life, here are some tips.

"MAKE SOMETHING HAPPEN"

Just because we don't manufacture scenes, doesn't mean you can't get someone to react to something, especially if that reaction offers a window into their views or their personality. In 1990, special correspondent Susan Stamberg went to Nyack, New York, to interview the legendary actress Helen Hayes. In her just-published memoir, Hayes wrote that she loved M&M candies, so Susan bought a bag and handed it to her while rolling tape. It became the opening of her story:[12]

> SUSAN: Helen Hayes does not disappoint. Almost 90 now, her eyes are as twinkling in person as they are on screen. Her skin is a soft nest of wrinkles. She's thrilled with a small gift, prompted by a chocolate habit she reveals in her book.
>
> HAYES: Oh, M&Ms, I adore them!
>
> (*Ambi of unwrapping candy.*)
>
> HAYES: Oh, bless your heart, aren't you sweet!
>
> (*Continues unwrapping.*)
>
> HAYES: M&Ms, my blessed thing. Well, you know when I like to eat these? At night, sitting up in bed, reading.
>
> SUSAN: Ms. Hayes lives in a big white Victorian house in Nyack, New York . . .

Susan recalls: "It was such a revelation of character in her, and something that wouldn't come out in an interview but added a lovely moment to the story." And it didn't violate the rule against manufacturing sound, because this was something really happening. "You can't fake a thing. But I wasn't faking anything. I was handing her this thing." Any interaction you have,

Susan says, is "totally recordable." If your interviewee is a musician, you can tell them you'd like to hear them play. If they are an author, get them to read from their book. If they are a baker, tell them to show you their baking techniques. And if they are a chess master, try playing a game against them.

What you ethically can't do is ask them to pretend they are doing an activity or to just go through the motions. You also can't give a gift of significant value, or something that gives an interview the appearance of a quid pro quo.

Susan read Hayes' book before the interview and saw the possibility of an interaction by offering her something small, a bag of M&Ms. Totally fine. So long as you do it within the bounds of ethical journalism, take Susan's advice and "make something happen."

GET THEM ON THEIR FEET

When people who work at desks agree to an interview, they will usually want you to interview them in their office. That's a recipe for a stale interview. Try to get them out of there.

Lauren says that in the beginning of her radio career, she brought a mic and tripod stand to her interviews. "I would go into their office and sit down and put the tripod in front of them and sit next to them. And it was horrible because they would talk for an hour and drone on. And then when we were leaving and I had packed up the mic, they would say the most amazing thing."

Now Lauren always tries to get her interviewee out of the office and on their feet. Many resist. Especially businesspeople and politicians, who feel comfortable and authoritative behind their desk in their office. When that happens, rather than offend them, Lauren does an interview and tries to keep it short, around 20 minutes.

"But then I will say, 'Can you walk me around your office and show me these photos of whatever we're talking about?' Or, 'Let me walk you to your car. Could you just reiterate what you said back there while you're walking?' "

Her experience is that people loosen up and say the best things when they're on their feet, walking or even just standing up. "So I threw out the tripod and I've never used it again!"

Jane Greenhalgh recalls trying to set up an interview with a farmer at his farm. "And he couldn't quite understand why that was necessary. 'Why can't you just interview me over Zoom, and then you can go to the farm and get some mooing of the cattle?' " She didn't go for it. "If you're actually there observing them as they're doing what they do, it's more immediate. It's more compelling."

TAKE THEM BACK TO THE MOMENT SOMETHING HAPPENED

Rhitu says she tries to get people to recall how they felt or what they were thinking at a pivotal moment. In a story about the origins of agriculture, she talked to a German botanist who had the chance to examine grains found on a farm dating back 12,000 years.[13] "She was telling me technical details," Rhitu says, but she pressed her, asking, "How did you feel? What went through your mind?" That prompted the botanist to talk about how excited she was when she received the samples, because it's so rare to get plant samples from such ancient sites. "Taking somebody to a moment, to a place, to a time and helping me, and through me the listener, to be there with them, that's going to be key."

GET THEM TO FIND ANALOGIES

"You need very visual descriptions of things," Rhitu says. And an effective way to get them is to have people use analogies. Rhitu did a story about a study into how a certain species of bat uses its tongue to slurp nectar from flowers. The scientist started explaining the mechanics of it, but Rhitu stopped her and asked if there was something comparable in everyday life. So the scientist compared it to a mop. "Not a sponge mop, but a stringy mop," the scientist said, conjuring up an image of a bat sucking up liquid using the hairs on its tongue.

Nathan says it's OK to come up with an analogy yourself and test it out on your source to see if it works. For example, a situation may be "like if you put three cookies in the cookie jar, but then somebody is stealing a cookie every night, or whatever.' And if they say 'Exactly!' That's very clear, boom." You can put the analogy in your story, in your words or theirs.

GET THEM TO DESCRIBE OR REACT TO WHAT THEY SEE

In chapter 4, I recommended thinking visually, the way Lauren Migaki did in telling the drought story in Brazil by describing an empty marina. You can employ the same idea in interviews, by asking a person to look around and describe what they see. "Everything you're seeing with your own eyes, you want your subjects to say that on tape," says reporter Elissa Nadworny. Working with Lauren as producer, she did that when reporting on a teachers strike in Los Angeles, where she toured an empty school with the principal.[14]

ELISSA (*recorded in studio*): Only about a half of the student body came to school on Tuesday. Principal Joe Nardulli leads us down what's usually the eighth-grade corridor. Today classrooms are empty, hallways quiet.

ELISSA (*asked during interview*): If there wasn't a strike right now, what would this hallway look like?

NARDULLI: O-o-o-o-h, this would be a full place! You know, about 410 students roaming the hallways, passing. And this is the location that students would have instruction.

As you can see, the answer isn't always a direct description. Often people are really *reacting* to what they see. So you may still have to describe the scene in your tracks. But their reaction can be very powerful. When Adrian Florido was reporting in Uvalde, Texas, he went to the cemetery where the victims had just been buried. In his report, he described the scene like this:[15]

ADRIAN: You look out across this section of the cemetery, and there are just fresh mounds of dirt everywhere. There's Jacklyn Cazares' grave. And then just a short walk away, Jose Flores and Alithia Ramirez. I can see Makenna Elrod's grave from here. Annabell Rodriguez and Xavier Lopez, who were sweethearts, were buried right next to each other. There's a huge, massive mound of flowers covering their graves, so you can't tell that there are two, but there are two here.

Adrian then talks in the story about how he met the two sisters of the murdered teacher at the cemetery. He asked one of them, Sandra Sanders, what it feels like to see those fresh graves, and here's her response:

SANDERS: It's almost like the aftermath of a war zone. You look around. We're in this little beautiful town . . . and it's, like, did we have a war? Did we send our children to war and this is what has happened?

Adrian could have described the scene by saying it looks like a war zone, but it's much more powerful coming from Sanders.

AVOID YES-OR-NO QUESTIONS

A simple yes or no may be enough for a print reporter, but it's hard (though not impossible) to use a one-word actuality. So imagine the kind of answer you hope to get, and tailor your question accordingly. Instead of asking a Social Security Administration official, "Will Social Security will be viable when current workers retire?" you might ask, "How will you make sure that Social Security recipients will receive full benefits when they retire?" or "How will Social Security have to change if you find that there's not enough money to pay out promised benefits when current workers retire?" Sometimes, as this example suggests, all you need is a "How" at the start of your question.

SOMETIMES IT'S OK IF THE ANSWER IS MUDDLED

Nathan Rott was doing a story on the Biden administration's pledge to protect 30% of the nation's land and water by the year 2030.[16] It was the most ambitious effort to stop biodiversity loss that the country had ever undertaken, but Nathan did his homework and knew that the Interior Department did not have a clear method for measuring progress toward the goal.

And so, during an interview with Kate Kelly, an Interior Department official, he recorded himself asking, "OK, so how do you measure that? Like, what counts as a percent of conserved land?" Kelly's reply was bureaucratic and evasive: "Yeah. It's not easy. There are a lot of complex questions that go into considering what counts." The way the official answered the question spoke volumes. "If it's a person who is in an official capacity," Nathan says, "who should know the answer to the question, I don't have to say in my reporting, 'They don't know what the hell they're doing.' " Listeners can hear that for themselves without the reporter having to say so explicitly.

KNOW WHEN TO SHUT UP

In conversation, people tend to interject words or phrases that don't add any meaning but lubricate the conversation by showing interest, such as "Uh-huh," "Yeah, I know!" or "That's interesting!" Or, maybe just a laugh or an "ugh!" But if you do that in an interview, it can spoil an actuality. "I can't tell you how much tape I've ruined, where a person is saying something cool and you hear me in the background, saying, 'Oh, cool!'" Nathan says.

Instead, use silent gestures to give feedback and encourage the person to continue telling their story. "I've mastered the art of quiet laughing and nodding my head and being very expressive with my face," producer Christina Cala says. "That goes a really, really long way in getting them to open up."

"If they say something that surprises me," says Adrian, "I just look at them with a surprised face or raise my eyebrow or puff myself up a little bit to indicate to them that 'What you said there is really interesting, say more about that!'"

The silent gestures may be the only part of your interview that might not feel like a natural conversation, but if they're done right, the interview will sound better.*

Getting Vox

Sometimes reporters and producers are sent out to get "vox"—shorthand for vox populi (the voice of the people). Rather than asking one person lots of different questions, the reporter poses the same question (or maybe the same few questions) to a lot of people. For example: "Should the U.S. provide Ukraine with fighter jets? Why or why not?"[a]

You may be asked to get vox for a story you're working on, or for someone else's reporting. Try to find a busy place so that you don't have to go searching for people—preferably a place that is relevant to the story. Is the vox question about politics? Stand outside the state capitol or the city council building. Is the question about food? Head to the grocery. About war? Take your pick. Just find somewhere that you can collect tape from a diverse group—diverse in age, race and income level.

Brace yourself for rejection. When Andrea Hsu went to a gas station for her story on gas prices, lots of people gave her funny

* You shouldn't be silent all the time during an interview. A well-placed vocal reaction can make for a good exchange, if it helps you tell the story.

looks or just said no. "This is just part of my job," she had to remind herself.

As Don Gonyea observed earlier about interviewing, you may need to approach dozens of people to get enough answers. If they agree to talk, start by asking them to say their full names and spell them. And, if relevant, find out their age, where they live and what they do for a living. Then help people respond to you in full sentences, as in, "I think private Social Security accounts are a good idea because they let me control my investments," as opposed to, "They're a bad idea." Avoid unclear pronouns! You might have to ask interviewees to back up and start over in a full sentence. Consider using the "rule of three": Ask three short questions at once, and your interviewees will likely respond in full sentences (for example, "What is your name, what did you see and how do you think the police handled that situation?"). And feel free to ask follow-up questions to get a complete story.

Record at least one minute of background sound at each location. Begin by saying where you are so that the producer can identify the proper ambi for each cut and gather location details that might be written into copy. (For example: "This is the street corner outside city hall where I talked with Lou, Susie and John. Lots of traffic passing by. It's lunchtime, so there are tons of people around.")

a. This section is drawn from the excellent training guide written by my predecessor Alison MacAdam, available at https://training.npr.org/2015/10/07/a-guide-for-gathering-vox-and-doing-it-quickly/.

CHAPTER 7

Safety

Eyder Peralta was frustrated. It was 2017, and he wanted to go to South Sudan, where a civil war was raging, six years after the country won independence with U.S. support. Reports at the time suggested hundreds of thousands of people had been killed. Eyder wasn't able to apply for a journalist visa. The government was unhappy with his reports of atrocities based on the testimony of refugees fleeing into Uganda. So he and his fixer, Tong Akot, decided to get a business visa. "We weren't going to do any reporting," he recalls. "We were just going to show up and say, 'Please, we're here now. You might as well accredit us so we can do some reporting.'"

Before leaving for South Sudan, Eyder sent his editors a note with a long list of contacts: everyone from the military generals and government officials he hoped to meet to make his case for accreditation, to an employee at the hotel where he was staying. Once in the capital Juba, after doing the rounds, he got a call from the media authority. "Hey, I've got good news for you," the official told him. "We're going to give you accreditation."

He was instructed to go to a hotel, where he met two officials. Eyder was so upbeat he ordered a round of beers for all of them. "And five minutes after the beers arrive, dudes with guns come in and say, 'You're coming with us.' And I said, 'There's no fucking way I'm coming with you! I don't even know who you are.' And then another 20 dudes with guns come in. And I said, 'OK, I am coming with you.'"

The Grim New Normal

Reporting in the field has always had its hazards, with journalists at risk of being kidnapped, abused, arrested and even killed while working in war zones and countries with authoritarian regimes. They've also faced dangers while covering natural disasters and severe weather events. But it's a sad testament to how bad things have gotten that the previous edition of this book, published in 2008, didn't mention journalist safety. Not only had no NPR journalist ever been kidnapped, but that was also before the deaths in 2016 of David Gilkey, an NPR photographer, and Zabihullah Tamanna, a freelance journalist in Afghanistan who had spent several weeks as an interpreter and local guide with an NPR team. Both men were killed while traveling with the Afghan National Army when a convoy they were riding in on a remote road in Helmand province was ambushed by Taliban fighters.

Concerns about safety used to be directed mostly overseas on reporting on armed conflicts. That's no longer the case. America has become a markedly less friendly place for the press. Journalists have been both verbally and physically assaulted while covering events during Donald Trump's presidential election campaigns. And public radio journalists trying to cover nationwide protests after the murder of George Floyd were assaulted and arrested by police.

The proportion of Americans who tell pollsters they don't think journalists act in the public interest has risen; so has the proportion of the public that sees political violence as justified.[1] With this in mind, journalists and the newsrooms they work in cannot ignore the threats to their safety.

What follows is not a comprehensive safety guide. For that, I'd refer you to the organizations listed in appendix 1, which offer plenty of excellent resources. What I will address are the specific vulnerabilities journalists face in the field. You'll read about our close calls and other experiences and get specific advice on staying safe.

What Saved Eyder

"No story is worth a life," says Caroline Drees, NPR's senior director for field safety and security. "No story is worth you getting injured." Not only should you never knowingly put yourself in danger to get a story, she adds, but news organizations should do everything they can to keep you safe. Find out the security protocols at your organization, or if you're a freelancer, at the outlet you're filing for. Some of the organizations listed in the appendix provide specialized support to freelancers, such as medical insurance while traveling.

NPR provides its journalists with training on working in hostile environments and dealing with online threats. What may have saved Eyder's life was the contact list he had sent and the regular check-in schedule he had established with his editors. When he was detained, he said, "I missed a check-in, then I missed two check-ins. And then the editors started calling the list of people I'd left them." The hotel employee gave them a tip. He told NPR that South Sudanese national security had been to the hotel and searched Eyder's room. "What that told them was that I was being held by the authorities," Eyder said. That was decisive.

NPR got in touch with the U.S. consulate, and Eyder's treatment improved. He was released after four days of detention. But it took two more weeks for the South Sudanese authorities to release Tong. And after that, he wasn't able to work as a fixer again, and ended up moving to the Netherlands, where he got a job as a nurse. "It's one of so many examples," says Eyder, "of the heightened risk our local staff face when they are working on the same stories we are."

Risks for Audio Journalists

Audio journalists face particular risks when reporting in the field.

YOUR MICROPHONE DRAWS ATTENTION

At a protest, part of your job is to capture the sound, but you also need to minimize your exposure to risk. Using a shotgun mic will allow you to record from the periphery. You'll want to do your interviews there anyway, where the shouting doesn't drown out your conversation. Or get some sound close up and then move to the margins of the protest. If you think protesters will be hostile to a reporter, be discreet. A shotgun mic, as the name suggests, can resemble a weapon. In certain situations, you may want to use a smaller interview mic, the internal mic of a pocket-sized recorder or, as a last resort, your phone. "A microphone is an attention grabber," says producer Lauren Migaki. "You have to make choices and analyze each situation: Is it right to blend in here? Or is it better to have really good audio quality?"

On the other hand, a larger microphone can protect you from police or authorities when you're covering a protest in a place ruled by a repressive regime. "It can make sure that everyone knows that you are recording," Lauren says.

YOUR HEADPHONES MAY BLOCK OUT NOISE YOU NEED TO HEAR

Remember the beeping golf cart in chapter 6, the one that ruined Nathan's interview? Using a directional microphone while wearing headphones or earbuds limits your situational awareness. Things that make noise around you may not only affect your recording, but also your physical safety. "I never, ever put two headphones on when I'm in the field," says Lauren. "I always have one ear free." Another option is to not wear headphones at all and keep an eye on the recorder's level meter. But that could limit how much you see around you. And you have no guarantee of audio quality.

Whatever you do, make an extra effort to monitor your surroundings, and be prepared in case things get out of hand. Also be careful not to raise the volume of your headphones so high that you risk damaging your eardrums.

Safety Tips

Because of the nature of the stories we cover, NPR producers and reporters have had occasion to put our safety training into practice. Here are some of our experiences and insights.

BE A JOURNALIST ABOUT SAFETY

As a journalist, you need to get the lay of the land quickly for your reporting. Staying safe is not all that different. "What you're doing for safety is you're gathering as much information as you can. You're trying to understand the situation," Eyder says. Recalling the hotel manager's role in winning Eyder's freedom, he adds: "And it's about sourcing, in that you've spoken to the right people before you go."

"The kind of person who should have been a journalist 100 years ago, and the kind of person who should be a journalist today are kind of similar," says Caroline. "You've got to be inquisitive and curious and not afraid of the legwork. All of that helps you on the security front too."

IF YOU THINK IT CAN'T HAPPEN, THINK AGAIN

In early 2022, all-out war in Ukraine seemed unimaginable, even to many Ukrainians. It's no surprise that the NPR team there didn't expect it either.

But Caroline insisted on having a plan in case Russia attacked the capital city. "Everyone was laughing at her," Lauren recalls. "They said, 'Kyiv is not going to get bombed!'" Nevertheless, they made a plan and agreed on a rendezvous point far from the capital. Sure enough, when the war started, every correspondent was in a different city, and they all met up at the appointed place. "It's really lucky," Lauren reflects, "that someone was thinking this through!"

TREAT ANXIETY AS YOUR FRIEND

On a trip to Brazil, Lauren and reporter Lulu Garcia-Navarro were walking through the Amazon rainforest with "rubber tappers," workers who extracted latex from rubber trees. When a motorcyclist started coming toward them, the tappers took out their weapons and opened fire. It was part of a running conflict between the tappers and illegal loggers in the area. "It was not a scene that I expected to get violent. And we were miles away from our driver." No one was hurt in that incident, but the experience taught Lauren that she had to game out all the possible risks. "Anxiety is kind of your friend, because you have to think about both what could go wrong with every recording" and what could go wrong in terms of safety.

And no matter how much training you've done, brush up on your awareness of the particular risks. "When there's a flood after a hurricane, you're not supposed to go in the water because it could be electrified," Lauren says. "But you're not going to remember that until you're facing that situation. So, as much as you can, keep that information fresh."

FEEL FREE TO SAY NO

People on a reporting team, regardless of seniority, should be able to voice their security concerns and be comfortable with the agreed arrangements, made in consultation with the security team, before proceeding. "Even the most junior team member should have a veto right, and no one should ever be pushed to go on an assignment or into a situation they don't feel comfortable with," Caroline says.

LEARN FROM OTHERS

When going abroad, or to any unfamiliar place, seek out veteran journalists from your news organization and others who have experience in that part of the world. "We all learn through stories. That's why we love public radio, right?" says Lauren. Before she went out into the field, Lauren sat down with as many producers as she could. "Everyone has a story of how they've messed up. And to me that was so helpful." That's why Lauren wants to share this story with you, about not taking unnecessary risks.

In 2014, she was on a reporting trip in Crimea with *Morning Edition* host David Greene and a Ukrainian interpreter. It was before the Russian annexation, and they were on what was then the Ukrainian side of the border, when a guy in camouflage carrying a Kalashnikov rifle walked over from his military vehicle. She shoved her memory cards down the car seat and took out her microphone as he approached.

"Are you recording?" he said.

"Yes. I assumed you understood that," Lauren answered.

That displeased him and he made her turn off the recorder. Then another guy in military fatigues came over and told them they broke the law being so close to the Russian border. He took their passports and left them standing outside their vehicle for about an hour. They eventually were allowed to leave with their passports, and thanks to the hidden memory cards, they didn't lose the audio. But Lauren says she should have been more careful. Holding on to the memory cards was not worth the risk, especially given the revelations of brutality that emerged in later fighting. "Knowing now what I know about Russia today, I would have given it all up."

SOME RISKS HAVE NOTHING TO DO WITH YOUR STORY

If you're covering the aftermath of a hurricane, you may risk illness from infectious disease or from eating spoiled food. If you're covering a wildfire, you may walk by a tree that looks healthy but could fall on you. If you're in a rainforest, you need to know which snakes are poisonous. "Traffic accidents are my biggest threat," correspondent Lauren Frayer told me back when she was based in Mumbai. She would check in repeatedly with her manager and with security when she was on the road, telling them, "We've passed that part of the road that's famous for bandits or that's famous for landslides, or whatever."

OR THE RISK MAY NOT BE IN THE OBVIOUS PLACE

When I was based in Spain, anti-American sentiment swelled over the U.S. invasion of Iraq. There were massive anti-war protests, and people were furious that their prime minister, José María Aznar, was one of President

George W. Bush's staunchest allies. And when I would identify myself, people assumed National Public Radio was a state broadcaster like Radio Nacional de España. Still, I was never threatened at the rallies. But on a completely unrelated story far from the capital, I started talking to a man outside the town hall. When I told him I was with "la Radio Pública Nacional de Estados Unidos," he screamed "Yanqui!" and lunged at me. Fortunately, others I had just interviewed helped restrain him and sent him away.

In the U.S., political correspondent Asma Khalid, who wears a Muslim headscarf, didn't worry too much about her safety while covering the Trump rallies during the 2016 presidential campaign, despite Trump's anti-media incitement and his proposal to ban immigration from Muslim-majority countries. "You have some protection when there's a lot of other journalists around you and there's a lot of other people around you," she says.

But it was riskier away from the stump. Once, while Asma was following a canvasser in Ohio, a woman started yelling at her to get off her property, until the woman's daughter sent her mother inside. Asma often responds to distrust by emphasizing another part of her identity. "When I do interviews in the Midwest, I always bring up the fact that my family's from Indiana. You start talking to them about Indiana, and the demeanor softens."

"YOU'RE WHERE YOU BELONG"

Veteran journalist Cheryl Corley, who was NPR's Midwest correspondent for many years, didn't cover the Trump 2016 campaign. But she did go out on a number of stories where safety was a concern, including the 2014 riots in Ferguson, Missouri, after the shooting of Michael Brown, and the aftermath of Hurricane Katrina in 2005. "You were working in a lot of situations where you didn't know exactly what the atmosphere was like or going to be," she says.

Cheryl, who is Black, describes her approach to safety as "acting like you belong, but also taking a sense of where you are, getting the pulse of where you are." She has this advice for women of color: "You're a reporter. You're there to do a job. And you're where you belong."

WORK YOUR ADVANTAGES

Women can be targeted sexually. In one of the most notorious cases, CBS journalist Lara Logan was assaulted while covering the pro-democracy protests in Cairo in Tahrir square in 2011.

"But as a woman," says Caroline, the NPR field security director, who worked for Reuters in the Middle East, "I feel like I've had a lot of advantages as well. I've had people come to my assistance who might not have had I not been a woman."

"A lot of people ask me if I've felt disadvantaged as a woman in journalism," says Lauren Frayer, "and I have only seen advantages." In some conservative places she can cover herself completely and "blend in on the street in a way my male colleagues couldn't." Being a woman also gave her additional access. "In India, I can meet with the men and then go into the kitchen and interview the women," she says.

PRACTICE ONLINE HYGIENE

Online, it's another story. Much of the worst abuse is directed at women and people of color.* "And that can be everything from being soul destroying, to being physically dangerous, to being emotionally completely destructive," says Caroline. During the 2016 presidential campaign, Sarah McCammon endured heaps of abuse from people, but she reminded herself that she just symbolized what made them angry. Her strategy? Don't poke the bear. "I don't want to minimize a real threat, and everybody should use their best judgment. But it's not valuable to live in fear about every angry person who's mad at you and then probably goes on with their life."

In addition to not engaging with trolls—and blocking them, if need be—Caroline says that you should be very conscious of your privacy and security settings on social media and take anti-doxxing measures to protect yourself.[2] Using strong passwords and encryption on your computer and communications may seem banal, but it is crucial to online safety. If you really want to make sure no one is intercepting your communications, consider old-school methods like meeting a source in a park. And, Caroline says, be careful what you share online. "Oftentimes we put things on there like a picture of our kid at a playground, and there is location metadata [in the photo] that tells somebody who's savvy where that playground is."

* A survey among women and gender-nonconforming journalists by the Committee to Protect Journalists showed that 90% of respondents in the United States considered online harassment the biggest threat they faced. You can read more about the survey here: Lucy Westcott, "'The Threats Follow Us Home': Survey Details Risks for Female Journalists in U.S., Canada," Committee to Protect Journalists, Sept. 4, 2019, https://cpj.org/2019/09/canada-usa-female-journalist-safety-online-harassment-survey/.

READ THE SITUATION

In 2011, while reporting on the pro-democracy protests in Cairo, I went to a loyalist rally with a TV crew from Voice of America. At the time, President Hosni Mubarak's regime was disseminating conspiracy theories about foreign press helping his enemies.

When we got there, our car was surrounded by a mob, all men, some with rocks in their hands. They shouted and started rocking the vehicle back-and-forth. Quickly but cautiously, I opened the door and told the cameraman to follow me out the door with his gear. I pleaded with the Mubarak supporters, using the Arabic words that I knew. "We're here to interview you. We want to know what *you* think!" The next moment felt like an eternity, but one of the guys told us to wait. He disappeared into the crowd and came back with one of their leaders. We interviewed him, and when we were done, the men thanked us for coming and let us go on our way.

As Caroline says, what I did was "read the situation." Sometimes you have to show your own vulnerability and that you're not a threat. "And sometimes," she says, "you might have to drive through."

Covering Protests and Other Crowded Situations

CAROLINE DREES ON PLANNING AHEAD FOR SAFETY

If you have the luxury of being there a day before, do reconnaissance of the site. What's going to be your way out if it gets ugly? What's a dead-end street, and which streets will actually let you get out? If you're going by car, where can you park and what's going to be a good exit route? How can you make sure that you can actually get away, whether you're using a car, a bicycle or your feet? Always have a plan B.

If you have the luxury of time, get in touch with the organizers in advance so that you know how to reach them, and vice versa. Maybe there are phone numbers you can gather of people you can reach out to if there's a problem among the organizers. Or find out who the local police chief is, or the PR officer of the local police, if you're new to that town, and say, "I'm going to be covering this protest tomorrow. Do you have anything going on?" You'll want to know if there is any information the police would like to share with you that could be relevant for your work: scheduled media briefings, press releases, safety zones for journalists, curfews or dicey incidents have already occurred.

Wear the right footwear. I have a colleague who went to Iraq in flip-flops—she regretted it! Don't take more with you than you need because you could lose it or it could get stolen. Make sure you have your journalist ID on you, in a place you can easily reach to show the police, if needed. And carry it in a way that your lanyard can't be used to strangle you.

Depending on your job, stay on the sidelines if you can, and only dip in and out of a protest to get what you need. It's very easy to suddenly be sucked into a crowd and hard to extricate yourself from that, which is both a security issue and a news issue, if you have to report live. Try to stay on the sidelines and have your peripheral vision going. Make sure you're not so focused on the person you're interviewing in front of you that you don't notice the two guys in balaclavas behind you, or the ones over there who started a fistfight. It may have nothing to do with you, but you might walk into it!

And if you can, go with a colleague or someone else. Having a second pair of eyes can help each of you spot dangers—say, if one of you is focused on doing an interview, gathering sound or working with equipment. Situational awareness—basically, being aware of your surroundings—is critical to staying safe.

SARAH MCCAMMON ON GETTING IN AND OUT

The thing that makes me feel personally most comfortable in those situations is just being very strategic about where I go and for how long. They tell us in hostile-environment training, "Don't get in the middle of it if you don't need to." And you probably don't need to. There's nothing wrong with going to the periphery of an event and keeping your eyes open, ideally having somebody with you who can also help be eyes and ears. And having very clear communication with editors and colleagues about where you're going and for how long, so people know where you are and if you might need help. If I'm in a hurricane zone, I always put a tracker on my phone and send my location to my editor and usually to at least one person from my personal life.

And once you're in the situation, approach things cautiously, keeping eyes open. This is one of the areas where having a lot of experience as a reporter or correspondent comes in handy. When I was starting out, I would interview everyone and spend so much time—report, report, report. And that's a good thing when you're learning the ropes. But when you're in a potentially dangerous situation, and you've been doing this for a little while, you kind of know what you need to get a three-minute story on the air. You go and get a few quotes, get out, take some notes and that's all you need. There's no medal for having the best sound bite ever at a rally where people are toting long guns. That's my biggest advice: Get in, get out.

TAKE CARE OF YOUR MENTAL HEALTH

While covering the Syrian civil war, Ruth Sherlock spent lots of time with other war correspondents and aid workers, and many times they became close. She remembers hanging out with Peter Kassig, a former U.S. Army Ranger, at parties where they'd "have these meaningful conversations about what we wanted to achieve in our lives." Kassig was beheaded by the terrorist group ISIS, which showed the killing in a video. So were James Foley and Steven Sotloff, both of whom Ruth had gotten to know while covering the Libyan civil war. Ruth's husband, Paul Wood, a former BBC correspondent, was kidnapped in 2013. A week or two after he escaped, Ruth's close friend, Syrian-born journalist Susan Dabbous, was abducted while shooting a documentary for Italian TV. Ruth also lost her mentor, Marie Colvin, who had given Ruth her first byline in a British national newspaper, and another friend, Kayla Mueller, whom Ruth had once sent flowers to on behalf of Mueller's Syrian boyfriend. Both women were killed in Syria.

Ruth used to push such things away, telling herself: "It's not your place to be affected by this really, because those around you are the people that are directly affected. You hide behind the mic or the pen and the paper, and you report and then you go home." But her approach changed as she got older and started a family. "I had to do a lot of soul searching. And now I have a different approach, which is that I recognize that these things are awful and they do have an impact on your mental health. And that the first step to handling it is accepting that, not trying to push it away," she says.

National desk correspondent Brian Mann has a tough beat: addiction. He's covered the opioid crisis and the havoc it wrought on people's lives. He's also reported on heavy stories beyond his beat, including mass shootings, natural disasters and the war in Ukraine. His self-care includes pitching joyful stories that "tell the full story of life: people who are having babies and starting cool art programs and helping educate people about really cool, nuanced things."

One aspect of mental health that too many journalists don't think about is how little they sleep. Even a minimal amount of rest can be elusive when you're on the road, reporting and filing all day long. It can also be hard to get when you're back home and dealing with late deadlines or early-morning broadcasts. *Morning Edition* host Bob Edwards used to wake up at 1 a.m. every day to prepare for the show, which begins broadcasting at 5 a.m. After Steve Inskeep took over in 2004, he worried about the health effects of overnight shifts, which research suggests can be harmful.[3] "I would like to have a semi-ordinary human life and also not die," he says. "So I've adjusted the hours and feel really strongly about taking care of myself." He gradually moved his wake-up time to 3 a.m., which allows him to get seven hours of sleep and spend more time with his family.

Your mental health can suffer even if you get enough sleep and are not covering violence and human suffering. Ordinary interviews can be stressful and psychologically exhausting, and so can working on breaking news where you have to come up with a concise, coherent paragraph in no time. Going live on the air and talking without flubbing can be nerve-racking. "Journalism is hard work. It's emotionally draining," says Caroline. "You will not be able to do the work of being a journalist if you don't feel strong and don't feel good about what you're doing."

There are many options for dealing with the stress and the emotional toll it takes, and often the right way is very personal. Mindfulness might help, or listening to music. Some people find prayer restorative. "Figure out what works for you to make you genuinely decompress," says

Caroline, adding that you should also give yourself time to regroup after an assignment.

How Far We've Come

In 1994, I flew to Yemen for an international news agency to report on the fraying of the country's post–Cold War reunification. It was supposed to be a geopolitical story, about a crisis once again rending a nation formerly split between the pro-West north and the Soviet-allied south. My flight arrived in the capital, Sana'a, after midnight, and during the ride to my hotel, I thought it was strange that it was dark everywhere and that my taxi driver and I had to negotiate multiple armed checkpoints to get into the city. But when I arrived at the hotel, I brushed it off and went to bed. About two hours later, I was awakened by the sound of gunfire. Outside the window, the sky was illuminated with anti-aircraft fire and tracer rounds. I hid under my bed and cried. Finally, I got a hold of myself and filed a report on the outbreak of hostilities, and then I told my editor I'd come back as soon as possible on one of the planes evacuating foreigners. "No, you won't," he responded. "You're going to the front." And so I went. I hitched rides in Jeeps with northern Yemeni forces and flew in rickety Soviet-era helicopters to the front lines, where artillery rockets slammed into the ground, sometimes just a few yards from me and the other reporters who, unlike me, were not accidental war correspondents.

Such a story would be unthinkable today. Most news organizations, including NPR, have come a long way in ensuring safety, with training and resources and respect for their staff. In retrospect, if I had known how dangerous that reporting trip would turn out to be, I wouldn't have gone. On the other hand, if I'd been trained on how to minimize the risks, I might have—which is the way it ought to be.

CHAPTER 8

Writing for the Ear

During the "golden age" of radio, listeners gave their undivided attention to the stentorian voices coming through the mahogany-encased, vacuum-tube-powered sets in their living rooms. We may now be living through another such period—a golden age of audio—mainly because technology has made it portable and adaptable to modern life. People can take their favorite show or podcast with them as they jog through the park or drive the kids to their piano lesson.

But it also means listeners are easily distracted. When people multitask, they are only listening with a part of their brain. When a driver cuts them off at an intersection, or when the kids in the back seat get into a fight, the listener will be distracted. So, if your writing requires too much mental effort to follow, if the prose is dense or has too many facts and figures, you will lose people along the way, no matter how well you can turn a phrase.

Newspaper writers can cram lots of facts into a single sentence, and many do. Have a look at this *New York Times* lede from the 2023 banking crisis:

> In an extraordinary effort to stave off financial contagion and reassure the world that the American financial system was stable, 11 of the largest U.S. banks came together on Thursday to inject $30 billion into First Republic Bank, a smaller peer on the brink of collapse after the implosion of Silicon Valley Bank last week.

Nobody would say a sentence like that in a normal conversation. Hearing it on the radio or in a podcast, listeners would be put off by the jumble of facts.

Now, consider the way *Morning Edition* hosts Steve Inskeep and Sacha Pfeiffer started the same story:[1]

STEVE: Some of the country's biggest banks are sending money to a bank in trouble.

SACHA: First Republic Bank is headquartered in San Francisco.

That's not far from the headquarters of Silicon Valley Bank, which failed a week ago.

And although First Republic appears healthier, it's been facing some of the same pressures, and a lot of anxiety.

So other banks passed the hat.

And it's a fairly large hat—30 billion dollars.

STEVE: NPR's David Gura joins us now . . .

The *Morning Edition* story has about the same amount of information, but it's spread out in smaller, bite-size chunks. It takes six sentences to convey what the *New York Times* story did in one. It got listeners interested in a very complicated financial story with writing that was conversational and a pace that was not overwhelming. Before we talk about how to write this way, let's drill down on what makes writing an audio script different from other forms of news writing.

What Makes Writing for the Ear Different?

When writing for listeners, you have to keep in mind what your story doesn't have:

HEADLINES, PHOTOS OR VIDEO. Without a headline that proclaims what the story is about, you need to get right to the point and do that with your opening lines. And you have to make up for the absence of images by creating pictures with your words.

INDENTS OR PUNCTUATION. Without visual cues that demarcate thoughts and ideas, you need to write simple, clear sentences that deliver the information at a pace the listener can handle.

EASY WAY TO REVIEW. If listeners miss something you said on the radio, it's lost. They can't go back and reread the paragraph, the way they can in print. With a podcast, listeners can scroll back, but it requires effort, and you don't want to spoil their jog—or make them take their eyes off the road. So tell a story they can follow and don't overwhelm them with facts and figures.

EASY WAY TO DO RESEARCH OR FOLLOW HYPERLINKS. In a text story, people who don't know the background can search the internet; all it takes is a few keystrokes. But in an audio story, you need to include enough context so as to make it understandable, but not so much of an explanation that the people who already know it are bored or, worse, offended that you're spelling out something they already know.

Writing for the ear may require a fair amount of deprogramming: casting off what you learned in creative writing classes—or while working at newspapers or TV stations—and listening carefully to the way you, and other people, talk. Because you're writing in what is essentially a style that mimics conversation, only more intelligent and more articulate. And it's precise, despite the more casual tone.

How To Find Your Voice

"My simple rule for writing for radio," says national political correspondent Don Gonyea, "has always been: if it wouldn't come out of your mouth in conversation, it shouldn't be on your page." He's absolutely right. It's why I do the following exercise in my training sessions on writing for the ear. Try it with a friend or colleague:

Take something you've written—a news article or an expository work such as a term paper. Read it again, at least the first few paragraphs. Now, set it aside and, without looking at your text, tell the other person what you wrote in those opening paragraphs. Once you've done that, read the paragraphs out loud to them. Ask them to do the same for you.

How were the two ways of communicating different? Were there words you used in writing but not in speaking? How did your sentence structure differ? Was it simpler in the spoken version? How much did you go into detail in each case?

My training exercise is aimed at helping people recognize that the way they talk and the way they write could almost be different languages. Some people may express themselves more eloquently than others, but rarely is their writing and their speech constructed identically.

At the start of your public media career, you may be writing scripts that others read, so pay close attention to how people around you talk. Record your conversations, and then analyze both your speech and that of your interlocutors. How do people begin sentences? What vocabulary do they use?

The point is not that you should learn to write exactly as people talk in a conversation. We're often a little messy and can be a little incoherent when we talk off-the-cuff. But you want to learn the syntax, the rhythms and the vocabulary of speech, and then maybe elevate it so that it sounds how you—or the person you're writing for—might, on your smartest day, perhaps after a cup of coffee.

No Rules, Only Guidelines

Most days after hosting *Morning Edition*, Steve leads an internal discussion in a Slack channel titled "writing-notes," in which he reflects on that day's show and invites comments. But he's not categorical about the craft.

"There are no rules," Steve says. "There are guidelines, and anything that I take as a guideline is something that I will also violate at some point."

Still, you can't ignore the guidelines without understanding why they're there in the first place. Good writing begins with a mastery of the most important ones.

WRITE SHORT SENTENCES—BETTER YET, IDEAS

Susan Stamberg, one of NPR's founding mothers, was an English major in college. When she joined NPR in 1971, she says, she tried to imitate the writing in the great works of literature she had studied.

"I wrote sentences that were much too long," she says, recalling that she even tried to include "some obscure word that nobody would know, because I wanted them to pick up their dictionary and look up the meaning."

But she quickly realized her mistake, and transformed her writing style. "For radio, you write haiku. You write really short, punchy sentences that will get what you want to say across clearly and quickly," she says. In fact, she often doesn't write complete sentences, "just a handful of words to get you to the place you need."

Steve tries to keep his sentences within one line in NPR's editing program, which fits about 10-15 words before spilling to the next line. When that happens, he doesn't always actually cut, but makes sure anything longer than a line does "earn the time."

In fact, some reporters don't even write in complete sentences. They write one thought or idea at a time, as Elissa Nadworny did in this early draft of a story:

ELISSA: In the city of Kharkiv—in northeast Ukraine—there is a kindergarten classroom. With pastel-green walls—with little beds and little chairs.

(*Tape of video from before war—video of kids marching*)

ELISSA: Once—it was a magical place where children played chess and grew flowers and learned and laughed.

(*Tape: pop video before war—kids marching down the hall*)

ELISSA: Where each day—for 27 little 6 year olds—started with a hug.

Elissa says she rarely uses punctuation other than em dashes, which separate her ideas and tell her to pause in her delivery. Listeners can't see commas and periods, so sentence fragments are just as good as sentences so long as they express an idea—or, as Susan says, get you where you want to go.

AVOID COMPLEX SENTENCES

Imagine that someone starts telling you an idea, then suddenly diverts to another idea, before returning to complete the original idea. Sound confusing? Well, that's essentially what complex sentences are. Many journalists, especially those who write for the web, love them because they seem sophisticated.* But on the air, complex sentences can come across as stilted and even pretentious, because it's not the way we normally speak.

---

* Next time you read a digital news story, note how many dependent clauses appear. Often, the more prestigious the publication, the more intense the use of clauses. Consider this sentence from a Vox story: "The measures, which include sanctions targeting Russia's financial system, the wealth of powerful individuals, and Russian fossil fuels, are designed to punish Vladimir Putin and the oligarchs who support and depend on him, and hobble the Russian economy." Newspaper obituaries often employ the longest clauses. Here's one from the *New York Times*: "Bushwick Bill, who helped inject vivid

Let's do a quick grammar refresher here:

- A complex sentence is one that contains a subordinate clause.
- A subordinate clause, which is also called a dependent clause, is sometimes marked off by one or two commas.

For example, the words between the commas in the previous sentence, "which is also called a dependent clause," form a dependent clause. Whenever you see a comma, consider turning it into a period and making two sentences instead. Susanna Capelouto, NPR's Southern bureau chief, gives this example when training interns and new reporters:

> The storm, which will hit South Carolina this week with lots of wind and flooding, may take the roof off of Susanna's house.

This sentence starts listeners thinking about the storm, but the clause then gets in the way, forcing them to hold two thoughts in their head at once and increasing the chances they'll get lost. It's better to take the dependent clause and turn it into a separate sentence:

> The storm will hit South Carolina this week with lots of wind and flooding. It may take the roof off of Susanna's house.

Now try reading both versions of Susanna's example aloud. Which is easier to say? Which is easier to understand?

START WITH YOUR SUBJECT

In much of the prose that readers come across in journalism, the writer uses an introductory clause to set up a sentence, like the one I wrote at the beginning of this sentence. Or consider the whopper in the *New York Times* story quoted at the beginning of this chapter: "In an extraordinary effort to stave off financial contagion and reassure the world that the American financial system was stable . . ." People don't begin sentences that way in conversation; they get right to the point. Pay attention to the

---

psychological horror and lightly morbid comedy into Southern hip-hop storytelling in becoming one of the genre's most recognizable characters, died on Sunday at a Colorado hospital."

syntax you use when you talk casually with people, and you'll find that most of the time it is subject-verb-object.

So instead of writing a sentence like this:

> Over the years, Freedonia's president has drawn global criticism for his crackdown on press freedom.

Write it like this:

> Freedonia's president has drawn global criticism over the years for his crackdown on press freedom.

ATTRIBUTE FIRST

"Nobody in conversation ends with attributions," says newscast editor Elizabeth Wynne Johnson. "'You'd better be home by 11, said my mother.' That's not how we talk!" It works for a written story, because the brain processes words on paper, or on a screen, visually.

> About 60% of the U.S. is under winter weather advisories, the National Weather Service says.

You just read that sentence and it seems fine, right? But read it out loud and listen to how ill-suited it is for audio. This is better:

> The National Weather Service says about 60% of the U.S. is under winter weather advisories.

Another reason the attribution at the end doesn't work is that the listener's mind hangs on to the last thing it hears. "Whatever really interesting thing you've just said," Elizabeth says, "you then step on it with an attribution and it takes all the energy out of it."

FAVOR ACTIVE OVER PASSIVE

In 1987, President Ronald Reagan tried to obscure his administration's role in the Iran-Contra affair by saying "mistakes were made." The passive voice may be good politics, but it's bad journalism. If you say, "Eleven people were killed," when you know who killed them, or worse, when you could find out, you are not doing your job as a journalist. Write "The assailant killed eleven people" or "The train crash killed eleven people." Use the active tense, or at least try to, so that it forces you to look for the facts.

AVOID CLICHÉS AND JOURNALESE

Do you ever hear sentences like these in regular conversation? "I am poised to go on vacation, after an increasingly difficult week at work." "My boss slammed the proposal I put forward." "She met many revelers at the party last weekend." I certainly don't. But "poised," "increasingly," "slam," "revelers" and similar words creep into news copy all the time (search for "journalese" online for more examples). So does the expression "stave off," as in the *New York Times* story above. There are also constructions such as "oil-rich," "tech-heavy" and "cash-strapped" and clichés such as "shocked the sports world" or "it remains to be seen" that are rarely, if ever, uttered by real people and should be avoided in all news copy.

News stories also use unorthodox formulations, such as "President Biden today told members of Congress" or "a world increasingly grappling with the effects of climate change." You probably wouldn't say "I today went shopping," rather "I went shopping today." You also wouldn't say "I ride a bicycle increasingly showing signs of wear and tear," but rather you might say, "I ride a bicycle that has more and more signs of wear and tear." Again, if you wouldn't say it, don't put it in your script.

Another tip from Susanna, the editor, is to avoid words ending in "ing" because they don't sound conversational. For example: "Susanna says playing the guitar is a lot of fun" vs. "Susanna says when she plays guitar she has lots of fun."

SPELL OUT TITLES, NUMBERS AND STATE NAMES

When writing for someone else, such as a host or a reporter, don't make them work out abbreviations. You may think something is universal that is not. Does "MI" stand for Mississippi, Missouri or Minnesota? Actually, none of the above! It stands for Michigan. Here are more rules to remember:

- Spell out state names, such as Pennsylvania, Massachusetts and California.
- Write out professional, military, clerical and royal titles, such as doctor, colonel, lieutenant, saint, reverend and his royal highness.

- For numbers, write out one, two, three, through nine. Use numerals for 11 through 99. Spell out hundred, thousand, million and so on. And don't use symbols such as $ or %.

So, $4,415,922 would be four-million-four-hundred-15-thousand-nine-hundred-22 dollars. But unless you have a good reason for using such a precise number, it's better to round off: "about four and a half million dollars."

MAXIMIZE SHELF LIFE

Use the present tense as much as you can. And avoid what senior newscast producer Carol Anne Clark Kelly calls the "little landmines" in wording that make a story unusable. Suppose it's the night before an election and you're writing a newscast spot to be used the next morning. You might be inclined to write, "Voters will decide . . ." But the next day, when the spot needs to run, Carol Anne says, "voters *are* deciding." So, she says, "keep the 'wills' out."

It may seem counterintuitive in writing about news happening now, but focus on the context and the wider issues. And put anything that might change or get dated in the intro, which show producers can update. Notice how the following spot was written so that it could run before the NASA panel convened, as the meeting was starting, or even for a good time afterward. Newscast anchor Lakshmi Singh introduces the story.[2]

LAKSHMI: A NASA panel charged with studying Unidentified Anomalous Phenomena is meeting today in Washington, D.C. As NPR's Geoff Brumfiel reports, they hope to bring some science to the debate over strange things in the sky.

GEOFF: From flying saucers to little green men, the government has dealt with claims of what used to be called UFOs for decades. But a lot of the old reports are pretty low quality. Astrophysicist David Spergel heads the NASA panel . . .

SPERGEL: Fuzzy pictures that people took in the 1950s are not terribly valuable.

GEOFF: The 21st century has brought some better data, but its collection and study is still ad hoc. Spergel says NASA's panel will try to lay out a more scientific approach looking at U-A-Ps, as they're now known. At today's meeting, they'll also respond to public comments and concerns. A final report is expected later this year.

Geoff Brumfiel, NPR News, Washington

By writing the spot this way, Geoff saved himself from having to write another one. And the newscast team could count on it being able to run for several hours, or even later in the day. Although the example here was a newscast spot, the same principle applies to pieces for the newsmagazines, which need to stay fresh while being rebroadcast to stations across the country over the course of as much as six hours.

Another way to maximize shelf life is to write the story as a "now what" spot. Suppose a city elects its first woman or person of color as mayor. "The person coming out and giving their victory speech," says Carol Anne, "can sound very 'last night.' But that first mayor of color or that first female mayor, that's a story that's just starting with the victory speech. The best spot that's going to run the next morning is really, 'What are the challenges this person is facing now?' And you can actually even use some of that same tape you got the night before."

EDIT YOURSELF

The best writers are never satisfied with their writing. They will go back and edit themselves, always looking for a more precise or evocative word, and getting rid of what doesn't need to be there. There comes a point where you have to stop, and your deadline should determine that. But until then, keep improving your prose.

Even as his show is about to go to air at 5 a.m. every day, Steve Inskeep is doing just that. "I believe in going back again and again and again to look at things I am constantly struggling with, in the context of *Morning Edition* and *Up First*, to find even five more minutes to reflect on the scripts, to reflect on the things that we've written, and to make them the best that they can be," he says.

But don't panic if your script looks simplistic. That's what written conversation—a succession of short, fragmentary sentences, as in Elissa's script on kindergartners in Ukraine—looks like. Getting used to it is part of the necessary deprogramming.

READ YOUR SCRIPTS OUT LOUD

This is the one rule that should never be violated. Read aloud anything you or someone else will say on air, and time yourself. If you are on deadline, work in the time to do it.

And don't just move your lips or quietly read off your computer screen. Say it with your radio voice. That forces you to listen, which will alert you to problems that might come up later in tracking, or live on air. You will get a more accurate time. And you'll catch unfortunate rhymes, words that repeat too often and sentences you can't breathe through.

If possible, read your script to another person. "If someone hears your sentence and says, 'I don't get it,' you need to take that seriously, even if they don't know your story very well," Steve says. "Because the audience probably will not know your story very well."

Even better is to try telling your story before you write it. If you're reporting in the field, you can do that by composing as many parts as possible as standups, and then transcribing them later. Even if you're working at your desk, you can start by telling your story to a friend or colleague, or call up your mother and tell her about your story (I'm sure she'd be delighted to hear from you!). Shannon Bond practices talking out loud and then writing the way she speaks. "After a decade as a print reporter, my instinct is writing for print," she says.

Elissa Nadworny on Writing With Voice Memos

This voice memo technique has been very helpful for me, especially when I first started at NPR. You basically explain the story out loud. It's as if you just got all your tape, you reported your story, and you call your mom and she's like, "What are you working on?" And you're like, "Oh, well, I went to this school and I met this person and they told me this. And it's really interesting because it actually relates to this big idea. That was the reason I went to the school in the first place."

You can do it with an actual person, a friend, your mom, or you can just do it alone in your room. You're basically telling a story orally, the way that we do on the radio. Not a whole cohesive story, but the way I might start the story, my favorite moments and certainly the nut graf or the main idea: why it matters, what's at stake.

You record it on voice memo, and then you plug in your headphones and you go sit at your computer, and you type up verbatim what you recorded yourself saying.

> And the more you do this, the more you begin to write like you speak. You will pick up on your cadence and the word choices you use and how you emphasize certain words. I did it enough times that I can hear myself as I write my radio scripts.

WRITE FOR ONE LISTENER

Don't refer to the audience in the plural, by saying "our listeners out there," or "everyone in our audience." It's true that we have millions of people tuning in and downloading, but you should talk as though a single person is listening. Don't ruin the experience of intimacy.

DELETE ADJECTIVES

Channel Ernest Hemingway and liberate your sentences from adjectives. Many adjectives that appear in news stories are there to "make the speaker sound concerned," says Steve. "I will call it a deadly shooting, a tragic event so that I sound like I care. But my opinion of this event is not important. What I want is for the listener to feel what the story is like, and me telling them how I feel gets in the way."

An adjective can be a sign that something needs explaining. Instead of talking about "a strategic town," explain why it's strategic: "Capturing the town would give Russian forces control over the whole region."

Don't delete all adjectives. Well-chosen modifiers can add color to a piece, like this sentence from a story by former Texas-based national correspondent and NPR veteran John Burnett: "Her gray hair is cut sensibly short. A blue apron is fastened around her waist." But get rid of adjectives that are superfluous. Would you be reporting on the devastation if it wasn't widespread? Can a murder be anything but brutal?

AVOID MEANINGLESS ATTRIBUTIONS

Terms such as "officials," "analysts," "critics" and "experts" are overused in journalism. And they're problematic: They often suggest a consensus that doesn't exist. Sometimes a reporter will use one of these terms after talking to just two people in the category.

And people don't use these terms in everyday conversation. Has anyone with small children ever said, "Parents say you should stop messing around and go to bed now"?

In the name of transparency and storytelling, cite the expert or critic you interviewed by name. If you are quoting officials for a company, government agency or other entity that has a unified position, do what's known in grammar as metonymy: let the thing speak for the people it represents. "The White House says" rather than "White House officials say." Or "Apple says" instead of "Apple executives say."

COMPRESS TITLES

In a story that's meant to be read, it's OK to give someone's title first: "USDA under secretary Chavonda Jacobs-Young" or "Los Angeles schools superintendent Alberto Carvalho." But that's not the way we talk. I wouldn't say, "I just talked to NPR education correspondent Elissa Nadworny yesterday about the story she's working on, and she said that NPR *All Things Considered* executive producer Sami Yenigun told her . . ."

So beyond "president" or "senator," when a person has a long title, use one sentence to identify them and another to say what they said. "Chavonda Jacobs-Young is a USDA official. She says . . ." And remember that "Dr." is used only for medical doctors.

WATCH OUT FOR ERRORS

Some writers like to get their thoughts down on paper first, and then go back and make sure everything they've written is accurate. That's a risky strategy, especially in journalism today, when online collaboration is widespread and an error can take on a life of its own. So make sure you've checked every fact, and make sure what you're reporting is true, before—not after—you write it down. If you need to follow up later, make a note of what you're unsure about and then be rigorous about going back and double-checking all facts and figures and verifying sources. Beware the mindless slip-ups: If you wrote "billion," make sure you didn't mean "million," or that a Tuesday rally didn't actually take place on Thursday.

Spelling and grammar should also be correct in audio scripts. Don't mix up "their" and "there." Parts of scripts can end up live on the web.

And your subjects and verbs should agree. "The appeal from the economists *calls* on the Treasury Department . . ."

SOMETIMES, IGNORE THE GUIDELINES

Any of these guidelines can be overridden in the interest of better writing. You can write a sentence brimming with dependent clauses, Steve says, "if it has a purpose and expresses a meaning, and if it adds to the rhythm of the piece and works with the piece."

Consider this line in Frank Morris' story on the victory parade for the team that won Superbowl LVII: "After walking much of the parade route high-fiving fans, Kansas City's quarterback Patrick Mahomes got on stage, about as humble as a liquored-up Superbowl MVP in an enormous, bejeweled professional-wrestling-style belt can be."

It's an awkward sentence with overwrought adjectives and an introductory dependent clause. "And of course, it's great," says Steve. "Frank cheerfully violates every 'rule' of broadcast writing," but it works because it creates an image, gives a telling detail and carries a single thought to the end.

Write Pictures

"Radio is your most visual medium." This counterintuitive assertion by Ira Glass of *This American Life* couldn't be more accurate. It works like this:

1. Audio lacks pictures.
2. You create pictures in the listener's head with your writing.
3. The listener's imagination becomes a co-author in your storytelling.[3]

In courses on writing for broadcast, radio is often lumped in with TV. But it's not the same thing. TV reporters tell a story by weaving pictures with graphics and words. And writing that is too descriptive risks getting in the way of the pictures. "In radio," says John, who has won numerous awards for his reports, "we have to write vividly, because that's what is going to make the story come alive."

TV news scripts are often very sparse in detail and color, so as not to distract from the pictures. But in audio scripts, words create pictures in the listener's imagination. That's why your prose should be vivid. "Every time you use a verb, it's an opportunity to surprise and delight your listener with an interesting word. Every time you use an adjective, it's an opportunity to create a scene," John says.

Here's how he did that in a report from Costa Rica's Monteverde Cloud Forest on the threat to its biodiversity:

> JOHN: At 5,000 feet, Monteverde straddles the continental divide. Rains and cloud mists from the Pacific and the Caribbean bathe the mountain in near-constant moisture, producing a cool, lush, shrouded setting that verifies our dreams of what a cloud forest ought to be. Plants grow on plants that grow on other plants in an orgy of fertility and adaptation.

Push yourself to elevate your language, says John. Don't settle for the sentence you just wrote. Rewrite it. Use a thesaurus; find a better word. "Throw it out there, try to make it interesting, try to make it funny, try to make it memorable."

And make use of metaphors. In reporting on NASA's mission to knock an asteroid off course, Geoff Brumfiel didn't convey the rock's size with measurements or bland adjectives such as "big" or "huge."[4] Instead, he said the asteroid was "the size of an Egyptian pyramid," and that it was struck with a spacecraft that "splatted like a bug on the windshield" of its target.

You can also do it in two-ways. In his conversation with *Morning Edition* host A Martínez on Federal Reserve interest rate hikes, Scott Horsley compared its monetary policy to the bathroom shower in his "creaky old house":[5]

> SCOTT: You know, you turn the knobs, but the water temperature doesn't change right away. And if you're not careful, you can overcorrect. And then suddenly, whew, you get hit with this icy blast. You know, the Fed has been turning the cold-water tap on the economy with these rate hikes now for the better part of a year. And we're only now starting to feel the effects. Consumer spending is slowing. Job growth has cooled a little bit. Ideally, inflation will settle gently back down to the Fed's 2% target. Turn the knobs too far, though, and we wind up shivering through a painful recession.

Vivid writing doesn't have to be complicated. As she begins her story on the Ukrainian school, Elissa is already painting a picture with her words:

> ELISSA: In the city of Kharkiv—in northeast Ukraine—there is a kindergarten classroom. With pastel-green walls—with little beds and little chairs.

"A short sentence can pack a lot of power if it's well written," says Steve, who doesn't let a deadline keep him from trying to write well. "I'm constantly aspiring to be elegant and poetic in the language that I choose, even though I'm just doing a news story that I only have two minutes to write."

This chapter has been focused on writing scripts. But I'm not quite done. There's a whole other aspect to writing that depends on how you use tape in your story. You'll learn about that in the next chapter, as I zoom out and look at the structure of audio storytelling.

Reporter Two-Ways

Not every audio story needs to be a meticulously produced piece. A conversation between a host and a reporter, known as a reporter two-way, is a great option if you have just interview tape and no ambi.[a] Years ago, NPR did reporter two-ways only for fast-breaking stories. But as the news cycle accelerated, they became a go-to, frequently appearing as lead show segments at the top of the hour and half hour.

The reporter two-way can be pulled together quickly, requiring minimal tape and allowing the reporter to be the expert. And for many topics, it is easier to digest than a piece. It's conversational and the listener doesn't have to follow a storyline.

A typical reporter two-way runs between three and four minutes. There's an intro that sets up the story and a script with three to four questions. It's almost always on the reporter to script both the intro and the questions.

The reporter two-way works when it sounds like a conversation. Think of it as an update after you've done your reporting, like the conversation you have with your editor before sitting down to write the story. Except that what you're telling the show host is confirmed fact and your analysis is thought through.

Here are some tips on how to prepare a reporter two-way and make it sound good:

IT'S NOT AN INFORMATION DUMP. It's tempting to want to show off all your reporting, but don't do it. Be judicious about how much you say. "You're lucky if you can leave your listeners with more than two or three things that people are going to really take away," says disinformation correspondent Shannon Bond.

"You don't want to kind of clutter it up with too many facts, too many numbers, too much information."

FOLLOW A CONVERSATIONAL ARC. Begin with a question that offers an on-ramp to the story. How would you start telling it to your mother, friend, neighbor? This is not a print story, where all the news has to appear at the top. The questions in the middle can be about reaction or context, or drill down on some interesting detail. The last question can be a summary, analysis or look forward. Here's how Shannon comes up with questions and thinks about the conversational arc:

> I write the host intro first, because that's the who, what, when and where. The host intro should be snappy and get somebody to want to keep listening. And then I try to think out the questions. The first could be, "I just told you what happened; tell us more about this." Or the first question could be, "Why does this matter? Why is this such a big deal?" Or, "How should we interpret that this has happened?" And then maybe the next question is, "Let's step back. How did we get here?" And then the third question can be, "How are others responding to this?" And then it's, "Where do we go from here? What happens next?"

DON'T READ YOUR ANSWERS. Write them as bullet points and fill in the rest as you talk. If you've done your reporting, you'll know your stuff and it will come out conversationally. If you've never done a reporter two-way before and you're afraid you'll freeze up or stumble, then write full answers the first couple times. But at least delete the conjunctions and the articles: the "thes," "ands" and "buts," and fill those in in real time. And each time you do a two-way, write less.

Don Gonyea sometimes writes out the complicated bits, but he'll add little signposts on the fly. "If I'm on with Inskeep, what I might say is, 'And Steve, here's the thing you've got to hear.'"

PACE YOUR QUESTIONS. In real-life conversations, most people vary how much they say. "I put a lot of thought," says health correspondent Rhitu Chatterjee, "into not having every answer and question be the same length, and having a little bit of an exchange, where possible, that's quick and short before getting into something longer."

LEARN HOW TO BUY TIME. Our brains sometimes need a split second after the host asks a question to gather our thoughts, and for the listener to adjust to the change in voice. It's OK to think for a second, or use filler words. But don't start every answer with the same word, like "Absolutely!" Change it up. Or avoid filler words altogether. One trick I found useful to buy my synapses a little time was to start by repeating part of the host's question. I did that in a two-way with Andrea Seabrook on a U.N. climate panel's meeting:[b]

ANDREA: Did the panel make any recommendations?
JEROME: No, it didn't make recommendations . . .

And again in a two-way with Renee Montagne during my reporting on forest fires in Portugal:[c]

RENEE: And elsewhere in Portugal, how much damage is there?
JEROME: Most of the damage is in . . .

PAUSING IS GOOD, RIGHT? Filler words are not signposts. Whether it's the trendy "right?" or the old standbys "like" and "you know," they don't help the listener. Sometimes, they're uttered out of a fear of silence at the end of a statement. Other times, it's insecurity over sounding authoritative. But it's an illusion that these words help. On the contrary, the implication is that you're pushing people to agree with what you've just said. And that, to be honest, is a little sneaky. Instead, let your words speak for themselves and use pauses. You will sound smarter and more trustworthy.

GIVE A VISUAL. As Russian troops were massing on the Ukrainian border in early 2022, Steve Inskeep asked Daniel Estrin, who was doing the two-way from Kyiv, what people there thought of a speech that Russian president Vladimir Putin had just given, accusing the U.S. of trying to lure him into a war. Daniel began his answer with an image: "Well, I saw a dartboard here with Putin's face on it yesterday, so that gives you a sense of what Ukrainians think of him."[d] A detail like that will convey much more than quoting what an analyst or politician said.

THROW TO TAPE. A two-way is a conversation between two people. When you play a clip of tape, you're bringing a third person into the conversation, so you have to introduce them. You can say something like, "I spoke to the head of the labor union, Charlene Watson. Let's listen to what she said." And then play the tape.

BUT DON'T HAVE TOO MUCH TAPE. If you have a lot of voices, do a piece.

KEEP THE QUESTIONS SIMPLE. Don't load them down with quotes and preambles. Just write it as simply and straightforwardly as you can, and let the host embellish it by putting it in their own voice. But . . .

AVOID "STUPID HOST" QUESTIONS. The reporter two-way is a conversation between two smart people. "The host," says health correspondent Rhitu Chatterjee, "can't just be saying, 'Oh, and then what? And what else? And why is this important?'" Put some of the story in the questions, so that the host comes across as "somebody who's well informed, just not up to date on the details," she says.

BE EMPHATIC. Imagine you're on a date, and you're trying to impress the person sitting across from you. You want them to be interested in something you find fascinating. That's how it should be in a two-way. "You have to be enthusiastic," says John Burnett. "If you sound bored with the material, God knows our listeners are going to be."

a. At NPR, the term "two-way" can refer to a host conversation with a reporter or a conversation with an outside guest. Some stations differentiate the two by calling the former reporter debriefs and the latter two-ways.

b. Andrea Seabrook, "U.N. Science Panel Sees Faster Warming of Earth," interview of Jerome Socolovsky, *All Things Considered*, NPR, Nov. 17, 2007, https://www.npr.org/2007/11/17/16393441/u-n-science-panel-sees-faster-warming-of-earth.

c. Jerome Socolovsky, "In Portugal Another Season of Forest Fires," *Morning Edition*, NPR, July 27, 2019, https://www.npr.org/templates/story/story.php?storyId=4813410.

d. Steve Inskeep and Daniel Estrin, "Russia's Putin Accuses the U.S. of Trying To Drag Him Into War," *Morning Edition*, NPR, Feb. 2, 2022, https://www.npr.org/2022/02/02/1077522522/russias-putin-accuses-the-u-s-of-trying-to-drag-him-into-war.

CHAPTER 9

Building the Story

You've done your interviews, you know how to write for audio, so you sit down to write your story. Where do you start?

Put down your pencil. (Yes, some journalists still use pencils! But I should probably say, take your hands off the keyboard). If you start writing now, you would be giving your voice primacy. Instead, go back to your interviews and start by looking for the tape that captures what you want to convey. "Our job is to let people tell the story," says *All Things Considered* producer Jonaki Mehta. "That's why I love radio so much."

I pointed out earlier that in print quotes, the reader is getting the information only from words that someone said. But in audio, it's a multidimensional experience: The words, the emotion, the timbre, the pacing and the tone of voice all tell you something about the person and what they are saying.

Science reporter Ari Daniel compares a script to "a flattened balloon," like that character in a cartoon who's been run over by a steamroller. "And then he gets reinflated, and his eyes pop out and he's real," Ari says. "The voices in a piece, they breathe life into it."

In this chapter, you'll learn about how to breathe life into your stories by highlighting those voices. I'll discuss the kind of tape you should be looking for and how to write around it. I'll also talk about scenes and story structure and creative ways of telling news stories. But first I'd like to focus on the host intro. Because while it's not technically part of your story, in that you as a reporter or producer won't voice it, you are responsible for drafting it.

The Intro

The intro must grab the audience's attention. It's the host's pitch explaining why the listener should care about the story coming up, whether it's a reported piece or an interview with an on-air guest. The intro can vary in length.* For newscast spots, it typically comprises two sentences of copy, where the first summarizes the news and the second names the reporter and foreshadows the content.

> JEANINE HERBST (ANCHOR): After more than three pandemic years the federal government's declaration of a COVID public health emergency is ending.
>
> NPR's Selena Simmons-Duffin reports on what's changing and what's not.

For newsmagazines, intros mostly run 20-30 seconds, although some that include tape can go up to about a minute. Either way, the aim is to write the intro as short as possible to give the story or conversation as much time as possible.

In podcasting, there's more variation. You might have a vignette with several pieces of tape that teases the theme of the episode before the first break. You may want to end it with a cliffhanger, or try something a little avant-garde that intrigues the listener. Whatever you do, you want to give your listener an on-ramp to the story.

TWO QUESTIONS TO ASK YOURSELF

The intro is the audio equivalent of a newspaper headline and lede rolled into one. It's during the intro when many listeners decide whether they want to keep listening. The intro has to whet the listener's appetite for the piece that's coming up, explain why the news event or issue is worth paying attention to and why it's relevant to the listener. It should set the scene for the upcoming report, by providing the time reference, the location, and perhaps even the main characters of the story. As much as possible, it should be visual; it should give the listener something to see in their mind's eye. It should make people care about the upcoming story.

The graph on the next page shows analytical data from the NPR app, which tracks when listeners click on the skip button. This shape, with a sharp peak at the extreme left, is typical of many of our stories on the app. It shows that

* There is no intro in newsmagazine stories that follow the host-narrated format known at NPR as tape and copy. But the first few lines need to grab and hold the listener's attention in more or less the same way as an intro.

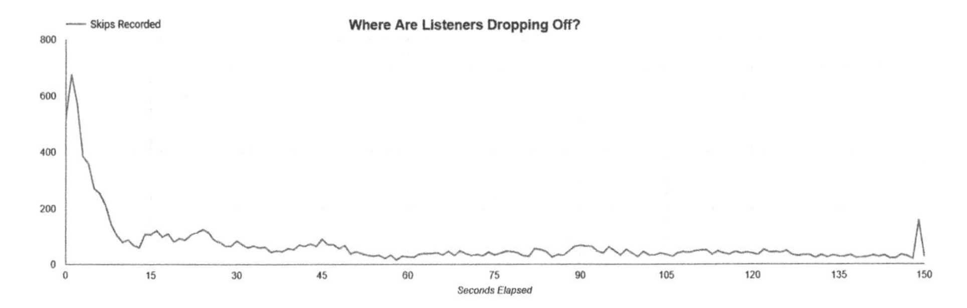

Graph showing when listeners hit the skip button on the NPR app.

most users decide within the first 10-15 seconds whether to listen to a story, whereas the typical intro length is 20-30 seconds.

That's why intros should start by giving people a natural way into a story, while quickly piquing their curiosity. Start crafting the intro by asking yourself two questions:

- What does the audience already know?
- What will my story tell them?

For some topics—presidential elections, the housing market or certain celebrities—it's reasonable to expect much of the audience will have a familiarity with some of the broader themes. For others—including many foreign stories, certain financial developments or underground culture—listeners are coming in cold. Either way, your first sentence has to hook them. Consider this intro to a guest interview read by A Martínez on *Morning Edition*:[1]

> A: A lot of us dream of the stars. Very few of us, though, have been able to travel a little closer to them. But a small group of civilians got that chance in a history-making journey last week.
>
> (*Ambi of rocket blasting off and flight control team announcing liftoff.*)
>
> A: The crew of the Inspiration4 spent three days in a SpaceX capsule. It was the first spaceflight in history to orbit earth with a crew made up entirely of private citizens. One of them, Sian Proctor, became the first Black woman to pilot a spacecraft, and she is here with us now.

Imagine a web-based news article on the same story. It might start like this: "For the first time in history, a crew made up entirely of private citizens orbited the earth in a spacecraft." For a reader, it works to get them interested, but a listener needs something a little more conversational and inviting.

A's first sentence, "A lot of us dream of the stars," is written to be heard. It doesn't deliver any news. Rather, it's a sweeping statement that's hard to disagree with. And that's the point. It's the way we talk. If someone tells you, "It's sweltering today," you'll respond, "Tell me about it!" Or, "Actually, I don't really mind it." Or something else. Whatever the case, the conversation is already underway.

That doesn't mean all intros should start with sweeping statements or shouldn't contain news. But what you do want avoid, if possible, is starting in the past: *It used to be that way, but now this is happening*. So instead of an intro that starts like this:

> For the past 16 years, German Chancellor Angela Merkel has led her country's rise to assume greater leadership of Europe. On Sunday, German voters will elect her successor . . .

It's better to start with the news:

> Germany elects a new leader Sunday. The winner will replace Angela Merkel, who has spent 16 years in power. And during that time, her country rose to assume greater leadership of Europe . . .

Earlier I compared an audio intro to the headline and lede of a web story. In fact, it's more. Frequently it's also a nut graf, because it tells you why you should listen to the story. The significance should be clear in the intro. But it shouldn't tell the whole story. That's why cliffhangers work, or anything that catalyzes curiosity. Notice how Geoff creates intrigue in this intro:[2]

> GEOFF: Saudi Arabia is building its first nuclear reactor. It's small, and nuclear-powered electricity is an important part of Saudi Arabia's plans for its future. But as NPR's Geoff Brumfiel reports, there may be an ulterior reason for the interest in nukes.

WRITING THE INTRO

Some NPR reporters write their intro first to help them establish where they're going with the story. I think there's something to that. I would at least take a stab at an intro, often by copying and pasting my pitch, which usually contains nut-graf-like information, and then rewriting it as an intro. And when I'm done with the story, I'd go back one more time and try to improve it, or throw it out and write an entirely new intro. You will probably develop your own method, which is fine, but keep the two questions above in mind, and take the following tips into consideration.

DON'T LARD THE INTRO WITH FACTS AND FIGURES OR WITH NAMES THAT ARE HARD TO PRONOUNCE. Show hosts and newscast anchors already have a lot on their minds, so don't make them struggle with a foreign name or give them numbers they may stumble over. Save the name and the numbers for the body. When I was in Spain, the government was led by José Luis Rodríguez Zapatero, a name that was a handful for non-Spanish speakers. I always made sure to put it in my tracks as a second reference, and let "the Spanish prime minister" suffice for the host or anchor.

DON'T ECHO. Second references to proper nouns in the intro should be nonrepetitive. If you have a sentence like this in the intro:

> President Biden is traveling to Atlanta, where he will deliver a speech at Ebenezer Baptist Church.

Then in the body, especially in your first track, say "the president is planning," or "Biden is planning," or "at the church where Martin Luther King Jr. ministered." Don't repeat first references identically.* Also, if the host or anchor's intro says what happened, don't "layer-cake" by telling it again in different words. Instead, continue the story with another thought.

MAKE SURE YOUR STORY DELIVERS. An intro is a promise to listeners, telling them what they will learn in the body that follows. If an intro says, "federal documents show the coal mine has a record of serious safety violations," the story needs to describe that record. An intro should also manage expectations by letting listeners know if the report is an investigative piece, a profile, a chronology of events and so on.

SAVE THE BEST FOR FIRST. Everyone wants to sound good. But if you save the best writing or tape for your story when it might work in the intro, listeners won't stick around for your story. Let the host have their moment on the stage. Give them the opportunity to connect with the audience and sell your story.

---

* NPR's style omits a sitting president's first name.

AND MAKE IT FUN, WHEN APPROPRIATE. OK, so you've written a top-notch intro that follows everything I just talked about. If you have a few minutes, set the intro aside and unleash your creativity. Try to write a nonstandard intro. You might come up with something like this, read by *All Things Considered* hosts Ari Shapiro and Audie Cornish:[3]

> ARI: With all the financial turmoil this week, here's a simple sentence that it might be useful to repeat. Audie?
> AUDIE: The stock market is *not* the economy.
> ARI: Let's try that together.
> ARI AND AUDIE: The stock market is NOT the economy.
> ARI: NPR's Sonari Glinton explains why.

As with so much in this book, there are exceptions to almost every rule. But the aim is always the same: You want to entice the audience to listen to your story. And just as there is no one right way to structure an intro, the possibilities are infinite when it comes to your story. I will discuss some basic ideas that should be helpful.

Scenes: The Basic Unit of Structure

Audio stories are built around scenes. They take the listener to the moment and place where something happened, or is happening, either through the eyes of the reporter or a witness or affected person recalling an event. In its most basic form, a scene consists of a piece of ambi, followed by a bit of descriptive writing, and then an actuality. A story can include one or more scenes, and the most common place you'll find a scene is at the beginning, right after the intro. Look at how international correspondent Joanna Kakissis sets up the opening scene of a story about wildfires on a Greek island:[4]

> *(Ambi of birdsong and running water, posted then faded under tracks.)*
> JOANNA: Athina Zioga often plays this video she recorded earlier this summer.
> It's from her favorite hike in a forest above her village.
> You can hear songbirds and a babbling brook, and a breeze she remembers as scented with wild oregano.
> ZIOGA: I spend a lot of time here. Every day, I was coming here with my dog, and it was my oxygen. It was my power to live.

AMBI, TRACK, ACTUALITY AND OTHER VARIATIONS

The rest of the story is a succession of scenes, each starting with a different piece of ambi: church bells ringing, a goat bleating, footsteps, a car door closing, people sweeping up debris. Each ambi is followed by tracks leading into actualities. Note that the ambis are the kinds of sounds we talked about gathering in chapter 4—the kind that immerse us in the reporter's experience.

Not all scenes follow the ambi-track-actuality model. You can have one piece of ambi leading to a series of actualities, or a series of ambis leading to one actuality. I turned it around once and put the ambi after an actuality from a 91-year-old war veteran named Mo Fishman, whose recollection set the scene better than any sound or writing could have:[5]

FISHMAN: When you're 21, there is no bullet that was made for you, and here was a chance to do something about fascism. Some men were running away from bad marital or love situations, but what united all of us was that we hated fascism.

JEROME: During the Spanish Civil War, tens of thousands of young men and women came from around the world to fight against the military uprising led by General Francisco Franco.

(*Ambi of Spanish Civil War veterans singing fight song.*)

In fact, you don't even need ambi to set the scene, if a person's spoken recollections do the trick. In this episode of *Rough Translation*, host Gregory Warner began with LaTasha Barnes' memories of growing up in Richmond, Virginia:[6]

BARNES: I grew up in a multigenerational household, as we say. It was my mom, my grandmother and my great-grandmother and myself, all in the same house.

GREGORY: Growing up, LaTasha Barnes spent most of her time with her great-grandmother.

BARNES: My great-grandmother was a cook. One of the few cooks, few Black cooks especially, that ran her own kitchen in the city of Richmond.

GREGORY: She ran the family home too.

Barnes creates the scene, and listeners are instantly immersed in her household without needing to hear any sounds. Which is fortunate, because, as with so many recollections, there's no time machine that would enable a producer to go and record a scene that already happened.

MAKING THEM HUMAN

There are three things you should aim to do in a scene:

- Create a visual
- Introduce the person, object or idea
- Give the context

Each of the scenes above paints a picture—of a bird-filled forest, of enthusiastic war volunteers, of a multigenerational household—that introduces the listener to people and provides the context necessary to understand the story. There's a lot of information in just a few lines of script. But it's important to do scenes well, because they embody that notion we keep coming back to, that radio is a visual medium. "It's like really zooming in on an opening shot of a movie, and you can see everything in the shot and you know what's happening there," says Meghan Keane, founder, editor, and managing producer of NPR's *Life Kit* podcast.

When you use a scene to introduce an individual, present them first as a human, before getting to them being a parent, judo champion, biologist, trombonist or whatever. Look at how *Planet Money* host Kenny Malone, when he was at member station WLRN, introduced a local government official at a traffic nerve center:

> KENNY: Brian Rick squints at a massive wall of live traffic feeds from all over Miami. He's a spokesman for the Florida Department of Transportation but looks like a nature show host—khakis, hiking boots, prominent mustache.
>
> RICK: I've alternated between mustache and goatee. When I was 23, I had a full beard, long hair down to my shoulders, and I wrote poetry on the beach.

We get to know Rick as an ordinary dude, because it's much easier for listeners to relate to him, or at least picture him in their mind, that way than as a transportation official. And, mind you, all this humanizing takes place before hiker-type Rick says anything about the actual subject of the story: dynamic congestion toll lanes.

Adora Namigadde of member station WBEZ humanizes her main character before connecting him to the tension at the heart of her story about a Black church rejecting homosexuality:[7]

> ADORA: Reverend Don Abram dresses his slight frame neatly, with navy chinos and a crisp button-up shirt.
>
> The 28-year-old grew up worshiping at a church he describes as—
>
> ABRAM: —a handclapping, toe-tapping Black church on the South Side of Chicago.
>
> ADORA: He even started preaching at that church when he was 14.

It's not until Abram's third actuality that the listener hears the money quote: "The church that loved me, and that raised me and reared me, nearly turned its back on me." Compare that to how stories tend to begin in other media formats. In a web-based story, you want the tension soon, so you're likely to want the money quote right after the lede. In video, you try to start with your most emotional moment. So again, that money quote needs to be high up. When I worked in TV, a colleague started a story about a refugee by showing him weeping.

That's not our approach to audio storytelling at NPR. In your interview, you've most likely had to work up to the point where the interviewee is in tears or tells you their most intimate thoughts, so you should work up to that moment in your story as well. You may be talking to the person on the worst day of their life; it shouldn't define them. They are a human being first, and the beauty of audio is that presenting them as such is the best way to get listeners interested in what they have to say.

Overall Structure

In an article for the NPR Training website, former *Planet Money* host Robert Smith discusses a study he did of four minute-long NPR stories.[8] He finds that the best stories have "three distinct sections," like acts in a play, each with a change of scene, character and focus. A story about a political rally or press conference, for example, might be told in three chronological acts:

Act 1 (before the event): Describes what's at stake
Act 2 (during the event): Tells what happened
Act 3 (after the event): Discusses changes, if any, as a result of the event

A story about a social or economic problem might structure the acts differently:

Act 1: Introduces a person, object or idea in connection with a problem
Act 2: Quotes an expert talking about the problem or someone attempting to solve it
Act 3: Returns to the main person, object or idea for a resolution, prediction or closing thought

These are just two examples of the three-act format. Stories may follow a hybrid of these structures or an entirely different one, but the arc will be there.

A STORY IN ACTS

Consider a real story that was told in three acts. Early in its war with Russia, Ukraine depended on U.S.-supplied Javelin missiles, and journalist Frank Morris reported about it in a four-minute story for *Morning Edition*.[9] Act 1 starts in the host intro, with Steve Inskeep describing the problem:

> STEVE: One of the best weapons Ukraine has in its war with Russia is Javelin missiles. But supplies are low, and restocking them will not be easy. Frank Morris of our member station KCUR explains why.

The problem is now evident but not the story's object, which is a weapon. Morris introduces it and provides the context in his opening tracks:

> FRANK: Every conflict has its iconic weapons—tanks in World War II, helicopters in the Vietnam War. And Mark Cancian with the Center for Strategic and International Studies says the war in Ukraine has distinguished an American-made, shoulder-fired, precision-guided antitank missile—the Javelin.
> CANCIAN: We've seen pictures of Saint Javelin. We've heard Javelin songs.
> FRANK: That's right. T-shirts, murals, even songs, venerating a missile.
> (*Ukrainian singer singing "Javelin! Javelin!"*)

Remember my advice in the preceding section, about making a person human? Frank is doing the equivalent here with an object. By playing the song and telling us about the mural and the T-shirts, he's able to make even a weapon of war relatable. Then in act 2, Frank has experts talking about the problem's causes (lack of qualified workers, outdated factory, component shortages) and the proposed solutions (training, investment, reactivation of old machinery).

Finally, in act 3, he comes back to the object and makes a prediction.

> FRANK: So U.S. factories will soon be producing a lot more weapons, including Javelin missiles, the iconic weapon of the war in Ukraine.
> For NPR News, I'm Frank Morris.

There you have it: a complicated story, boiled down and presented in three parts. For longer feature stories, the structure can take on extra acts. For example, you could do a four-part news feature, framing it this way: problem → solution → complication → future. Robert offers a hypothetical example in his training article:

> There are too many potholes on Route 9! (*Sound of bumps.*) And so citizens have started to fill them in on their own. (*Scene with shovels.*) But this created a different problem: people with shovels slowing down traffic in the middle of Route 9! (*Mayor complains.*) And so now the city is trying to work out a deal with the neighbors.

Mapping out a structure helps you plan your coverage and organize yourself when it's time to build your story. Because, as Robert says, life is complicated, and the real world usually doesn't fit into an obvious structure.

> There is rarely a clear protagonist or villain. There isn't always a beginning, middle and end. And if the story you are covering doesn't have a natural structure, you will need to create one; as a sort of life preserver to keep you from drowning in information.

So take that life preserver that is being thrown to you.

BEATS

Many podcast producers at NPR structure their stories around "beats," the essential components of the narrative. "Here's what happens, and then this is what happens," says Meghan. "And then this is going to lead to this idea. This is where I want the climax of the intro to be."

When she was a producer at *Invisibilia*, Meghan used beats to build an episode about a man who owned 10,000 Disney artifacts.[10] He began collecting them after his brother died. Here is what she wanted to convey as she introduced Richard Kraft to the listener:

1. Kraft and his brother were close.
2. They bonded over the world of imagination and movie soundtracks.
3. The soundtracks helped them escape something about their reality.
4. Soundtracks were also a metaphor for how Kraft's family dealt with emotions.

Here's how she begins sketching out beats 1 and 2:

MEGHAN: When Richard Kraft and his older brother, David, were kids, they used to sit at home and listen to movie soundtracks. Their bedroom walls were lined with sagging bookcases stuffed with their vinyl collection. And they really loved this one soundtrack.

Kraft then talks about how he and his brother would listen to the music from *The Ten Commandments* and feel like they were there on the bottom of the Red Sea, crossing with the ancient Israelites. "Nothing in our real world felt like that or looked like that, but the music made us feel that way," Kraft says. With that actuality, Meghan establishes beat 3, that this ritual was a form of escapism. Then in beat 4, Meghan shifts to how the family dealt with emotions:

MEGHAN: Soundtracks have a kind of special power. They're a great shortcut to making you feel an emotion. It's kind of their whole purpose. You know what else soundtracks are good for? Not feeling emotions.

KRAFT: I can stop it. The record's going to end. I'm not going to be drowning in my tears for eternity; I'll probably be drowning in my tears for the next three minutes and 58 seconds, as the counter counts down, and I know when this thing's going to end.

It's important to note here that Meghan chooses one visual for the beat. "I didn't say, well, they also like to ride their bikes together. They also went to the ice cream store together with their family," Meghan says. "I chose one specific thing that I thought best illustrated their relationship." And she did that with the help of structure, because in an audio story, you can't tell "every single thing that happened," she says.

Meghan is the editor and managing producer of NPR's *Life Kit*, a how-to podcast that gives listeners advice on a range of topics, and she uses beats in that podcast too. Rather than tell a continuous story, *Life Kit* offers takeaways on each episode's topic. For example, the takeaways in an episode on giving speeches sounded a lot like some of the advice in this book: "one idea per sentence," "say it aloud," "use more contractions." "Within those takeaways, we typically have time for about three ideas. As an editor, I know then we highlight the context of the takeaway, an example, and a caveat or a nuance to the takeaway," Meghan says.

Once you have an inkling of your story's structure, it's time to find the tape to build it around.

Climbing the Wall

In offering tips on how to use quotes, Roy Peter Clark of the Poynter Institute says a journalist should work from the assumption that they are the primary storyteller: "Begin with the idea that you are the writer, that you can write it better than the source can say it. When that is not the case, use the quote."[11]

I would argue that that advice doesn't apply to audio. Actually, it's the other way around. Your writing is secondary to the tape, and your job, as Jonaki Mehta noted earlier, is to let people tell their own stories. "You're not writing the story and filling in the gaps with the voice," as you might in a story for a website, she says. "You're letting people tell their story, and you're really filling in the facts around it."

Try this: Take a news story from a website and remove all the quotes. Chances are you will still understand what the story is about. Now do the same for a radio or podcast story. "If you can take your script, remove the tape cuts, read the script and still get the full story, then you've failed," says *Morning Edition* senior producer Barry Gordemer. Writing for audio means "not just keeping the tape around to effectively say, yep, that's right." Every actuality should "build on the thought you've made or take you to the next idea in the story."

That's why veteran producer Neva Grant, who teaches audio storytelling at George Washington University, compares audio scriptwriting to scaling a wall in a climbing gym: "The tape is like those little studs sticking out, and you're building the story on them."

TWO TYPES OF ACTUALITIES

So as you go through your interview tape, looking for the climbing studs on which to hang your story, listen for two kinds of actualities:

HOT TAPE. Brief utterances such as "I couldn't believe it!" or "It's not going to happen," or basically any short, pithy statement that may contain just a fragment of a thought. Hot tape works like ambi in that it is mostly about establishing a mood, and also like ambi it can be the start of a scene.

EXPLANATORY TAPE. These are longer actualities that, despite the name, can express a point of view. The key is that the speaker is expressing a complete thought, and the meaning of the words is the main thing you're trying to get across, although how the speaker says it also matters.

In this story by Iowa Public Radio's Katie Peikes, about the rising price of farmland, see if you can identify the hot tape and the explanatory tape:[12]

(*Auctioneer proclaims, "Here we go! We have an online bid of seven thousand . . ." and his voice fades under the following track.*)

KATIE: At this farmland auction in Jesup, Iowa, there's more than two dozen people attending in person, and lots of others are online and on the phone. Just a couple years ago, the final sale would have been near seven or eight thousand dollars an acre. But now it's a lot different.

(*Auctioneer's voice fades up for this post: "Eighteen thousand nine hundred nine—now 19 thousand!"*)

KATIE: The winning bids topped more than 17 and 19 thousand dollars an acre.

RIENSCHE: I was astounded.

KATIE: Jesup farmer Ben Riensche came to check out the prices. He didn't bid this time around, but recently at another auction, he bought 40 acres for 15 thousand an acre.

RIENSCHE: It was way more than I ever hoped to pay. But we'll find out in a decade whether I was too high or too soon.

KATIE: Farmland prices have rocketed, especially in the Midwest . . .

You are correct if you identified Riensche's first actuality ("I was astounded") as the hot tape, and his second actuality ("It was way more . . .") as the explanatory tape. As this example demonstrates, using both types of actualities, and some good ambi, can really bring a scene to life.

A WORD OF WARNING ABOUT TRANSCRIPTION TOOLS

Transcription software has made the work of logging interviews so much easier for journalists across platforms. For audio, the text it produces can help flag potential actualities, but it rarely, if ever, gets every word right. So you always have to check the tape, because that's what the listener will hear. And words, by themselves, can be deceiving. So a quote that looks amazing on your computer screen may be a dud on tape.

That's why I suggested earlier noting down your best actualities as soon as possible after each interview, while they are fresh in your mind. And here's another hack: follow the waveform in your audio editing program. When people get animated during interviews, they tend to speak a little louder, so skip to the highest mountains in your interviews and see what the person said there.

For more tips on managing large quantities of tape, see "Organizing and Labeling Tape" in chapter 4.

Writing Into Tape

Many NPR reporters follow this procedure: They select their best tape—ambi and actualities—put them in the order they think will best tell the story, and only then do they start writing, filling the gaps with their words. Writing into actualities, in particular, should be done with care, because you want to ensure a seamless transition that keeps the storytelling momentum going as the voice alternates between the narrator and the interviewees. Let's look first at techniques for writing into tape.

THE STANDARD SETUP

The standard way to write into tape is with a sentence or two that identifies the speaker and gives just the right amount of context the listener needs to understand the actuality. Here's how science reporter Nell Greenfieldboyce wrote into an actuality in a story on a study of mosquitoes:[13]

NELL: Lindy McBride is a mosquito researcher at Princeton University. She says the results are exciting.

MCBRIDE: Finally we have evidence that there's some sort of pathway in the sense of smell that is required for mosquitoes to like us.

Nell keeps it simple. She could have said more about the study or about how mosquitoes smell. Or she could have paraphrased:

NELL: Lindy McBride is a mosquito researcher at Princeton University. She says scientists have a better understanding of how mosquitoes are drawn to people.

MCBRIDE: Finally we have evidence that there's some sort of pathway in the sense of smell that is required for mosquitoes to like us.

But she didn't. Which makes it a much better experience for the listener. And that's one of the cardinal sins for writing into tape: paraphrasing or repeating.

Avoiding repetition means you don't always have to give the context. Look at how former *Morning Edition* host Rachel Martin set up the actuality in this tape and copy on how New York City libraries were abolishing fines:[14]

RACHEL: Dennis Walcott is the president of Queens Public Library, one of New York City's three public library systems.

WALCOTT: We want you in our libraries, and we want you using our resources that are available to you for free.

It works because the context was given earlier in the story. And it makes for an elegant variation.

Speaking of variations, I mentioned in chapter 8 that most of the time, you should start with your attribution. But if it gets in the way of the narrative, try putting it at the end, the way I did in in this story on social distancing:[15]

JEROME: It's not mandatory to wear a mask in Washington, D.C.; neither is it in most U.S. states. And that frustrates many Americans who worry about how easily the coronavirus spreads, says Northeastern University health law professor Aziza Ahmed.

AHMED: In these moments of uncertainty, we all suddenly feel the need to police our own behavior—but also police other people's behavior, as well.

You could, in theory, write a whole story with the first kind of setup, giving the name, then context, then the tape. But that would give little joy to the listener since it would be repetitive and predictable. Such stories are disparaged as "just acts and tracks." That's why you want to vary the way you set up tape with the other methods discussed in this section, especially the more artsy ones that I'll get into now.

FINISHING A THOUGHT

When you have an actuality that is a sentence fragment, you can lead into it with a track that completes the sentence. That's what *All Things Considered* host Mary Louise Kelly does in this tape and copy about a jet suit for paramedics, having already identified the speaker, Richard Browning:[16]

> MARY LOUISE: He floats gracefully over rolling hills, grass billowing beneath him. He says the feeling when you're up in the air—
> BROWNING: —is pretty hard to describe. It is the most free, liberating, kind of dreamlike state you get.

This is an effective way to maintain the momentum of the narrative as you change voices. In some circumstances, though, you may want to avoid this technique. One is if you have a piece of tape that expresses an opinion, because introducing it this way suggests that you or whoever is tracking it is endorsing it. The other is if the narrator's and speaker's voices sound similar enough that a listener might not pick up on the change.

BACK ID

Up to now, the stories you've read about usually identify people before quoting them. Now I'll discuss the "back ID" or "tease," which starts with a short actuality, and names the person afterward. You saw this technique earlier, in the farmer's auction piece from Katie Peikes:

> RIENSCHE: I was astounded.
> KATIE: Jesup farmer Ben Riensche came to check out the prices . . .

This method works best with a piece of hot tape. You don't want the listener to start wondering who's speaking because the actuality is going on for too long. And since the first bite functions as mood tape, you'll most likely also need an explanatory actuality after a brief script, as in the continuation of Katie's piece:

KATIE: . . . He didn't bid this time around, but recently at another auction, he bought 40 acres for 15 thousand an acre.
RIENSCHE: It was way more than I ever hoped to pay. But we'll find out in a decade whether I was too high or too soon.
KATIE: Farmland prices have rocketed, especially in the Midwest . . .

BUTT CUT

If you play one person's actuality after another's, you get what's called a butt cut, in that you're butting two pieces of tape up against each other. Here's an example of the technique in a story by Stephanie O'Neill, on church gatherings during COVID:[17]

STEPHANIE: For now, Harvest Rock will continue in-person services, says attorney Mat Staver. He represents the Pasadena church and its nearly 170 affiliate congregations, and he's confident of a win in California.
STAVER: At this stage, it's a no-brainer.
STEPHANIE: Because the Constitution guarantees the freedom of assembly and worship, Staver argues, in-person church services should be exempt from any ban on gatherings.
STAVER: People need fellowship, and they need support. You can go to warehouses and big-box centers and spend hours there with no limitation. But you can't spend an hour in a church socially distanced. Frankly, if you go to a grocery store, you're encountering more risk.
CHIN-HONG: I wish that were true, that it was safer in a church environment. But unfortunately, it's not.
STEPHANIE: Dr. Peter Chin-Hong is an infectious disease specialist and a professor of medicine at UC San Francisco.
CHIN-HONG: You go into a store, you're generally in for a very short period. And you keep your mask on at all times.

Staver was introduced the standard way. And then his tape flowed right into Chin-Hong's first cut of tape, which comes before he is identified. Essentially, this second piece of tape is a back ID, but that's not an imperative: Both voices can be introduced before the butt cut. Either way, the butt cut is a great way to vary the flow of a piece, but there are several considerations to think about before employing it:

- Butt cuts can work as transitions, to reinforce a point or demonstrate a contrast.

- The two voices, as well as the reporter's, should not sound similar, or the effect is null.
- If you are using the butt cut as a contrast, make sure the person speaking in the second cut was responding to the precise point the person in the first cut of tape made, otherwise you'll be quoting them out of context.

THE REPORTER'S QUESTION

If you're struggling with how to write into tape, listen back to your interview, to see what the person said just before the actuality. That might give you an idea for how to lead into it. And if what came just before the actuality was your question, you may want to try setting it up by playing your question, the way Nathan Rott did in the conservation story in chapter 6.

> NATHAN: OK, so how do you measure that? Like, what counts as a percent of conserved land?
>
> KELLY: Yeah. It's not easy. There are a lot of complex questions that go into considering what counts.

This technique is great for pushback, because listeners get to hear exactly how the speaker responded to your question. If you did the interview outside or in an environment with background noise, it might also be a way to create a scene, since your asking the question would essentially be a short standup.

That sums up the main ways of writing into tape. Now I'll look at how you can move the story forward as you write out of your tape.

Writing Out of Tape

In a story for the web, it doesn't matter all that much how you come out of a quote. For example, in this *Wall Street Journal* article, the writer goes from quoting one person to another:[18]

> "These financial metrics that have traditionally signaled adulthood don't apply anymore," says Julie Lythcott-Haims, former dean of freshmen at Stanford University and author of the book "Your Turn: How to Be an Adult." "And that is not their fault, but society is telling them, 'There is something wrong with you.'"

> Kelly Williams Brown introduced the world to the term "adulting" with her original blog and 2013 book, "Adulting: How to Become a Grown-Up in 468 Easy(ish) Steps."

Such a change of voice is too abrupt for audio. Imagine someone getting distracted for a second and not hearing that it was Lythcott-Haims speaking in the first paragraph. The second paragraph would sound like a back ID to them, and they would think Williams Brown was saying, "And that is not their fault." Even without distractions, it's a little overwhelming to jump from societal expectations to a neologism, and from one book to another.

In the interest of full disclosure: The article has a subhead between these two paragraphs. But that actually strengthens my point. You need a little help when switching voices, and in audio, there is no subhead.

What you need is a transition. It can be a signpost, like this: "Now we're going to hear from another author who has written about becoming an adult. Kelly Williams Brown introduced the world to the term 'adulting.' "

Or you can build on the actuality you're coming out of, by summarizing the same person's thoughts on something else. Asma Khalid does this in a story on President Biden's economic policy. See how she writes out of an act by factory owner Drew Greenblatt:[19]

> GREENBLATT: But right now it's wildly unfair to build in America compared to China because we have so many things stacked against us.
>
> ASMA: *Greenblatt wants Biden to go further.* He wants less regulation and a tax break for R&D. In the end, he says, supporting American manufacturing is smart policy and smart politics.

You don't even have to stay with the same voice. You can follow the thread of the thought they just shared, as Asma does later, after an act by Biden adviser Brian Deese:

> DEESE: We are, for the first time since really the 1960s and in many cases earlier than that, using targeted public investment over multiple years to try to crowd in private capital.
>
> ASMA: *They're doing this in a few ways.* They're giving out subsidies for semiconductor plants and electric vehicles. They're also keeping Chinese products out by maintaining the Trump-era tariffs and imposing sweeping export controls to limit China's access to technology.

And there are a few transitioning tricks that can spice up your script. Here's how Asma continues the story:

ASMA: And they're doing this all out in the open, which is even more unusual, according to Dani Rodrik. He's an economist at Harvard.
RODRIK: Among economists and mainstream policymakers, they think "industrial policy," for a number of decades now, has been kind of a dirty word. And I think that sort of has completely changed now.
ASMA: *It's changed because* politics on the right and the left have changed, and politicians have decided China is a common foe.

By echoing the word "changed," Asma not only makes the jump from act to track, but also underscores that part of the actuality for her listeners. Nathan does something similar in this *Short Wave* episode, where he's in a boat on a reservoir with environmental advocate Eric Balken:[20]

BALKEN: People are like, oh, this place is so beautiful. And, like, if you were to build a dam in the Grand Canyon like the bureau wanted to, you know, that would be a beautiful reservoir too, and it would also be a crime against nature.
NATHAN: *A crime against nature.* Balken thinks that's what happened here at Glen Canyon, and he's not alone in that sentiment. But to understand why . . .

A "crime against nature" is a harsh judgment. But the speaker's intonation didn't match the import of the words, so Nathan comes back to it with his own emphasis, hanging on each word, and then using it as a bridge to his next thought.

This is called the repeat, or echo, technique. And another way to use it is in contradiction. Culture reporter Elizabeth Blair uses it after musician Divinity Roxx describes how she felt about the idea of auditioning for Beyoncé's band:[21]

ROXX: And I didn't think anything about it because Beyoncé is Beyoncé. She can call anybody, so I don't even understand why she's having auditions. I can think of five bass players she should hire right now, and none of them equal me, right?
ELIZABETH: Wrong, as it turns out. Divinity auditioned in Atlanta, then again in New York. Beyoncé's father, Mathew Knowles, was the one to announce that she and the other musicians were hired.

Elizabeth plays off the self-effacing nature of the actuality, and uses negation to connect it to what happened next.

"Like an Old Married Couple"

Even though tape should be the star of audio stories, if you write well to it, your words and the actualities will become "equal storytelling partners," says Barry Gordemer. The result is like two people who have known each other for so long they could each tell the whole story themselves.

"It should be like an old married couple, where each one is sort of completing the other one's sentences or the other one's thoughts," he says. In fact, he trains people on writing to tape by playing a clip from the 1989 film *When Harry Met Sally*, where an elderly couple harks back to their courtship.

> MAN: You know where we met? In an elevator . . .
> WOMAN: I was visiting family—
> MAN: —at the Ambassador Hotel in Chicago.
> WOMAN: He was on the third floor, I was on the 12th.
> MAN: I rode up nine extra floors just to keep talking to her!
> WOMAN: Nine. Extra. Floors.

This technique is particularly effective when you have a series of actualities from a single person. On the 15th anniversary of the September 11, 2001, attacks, *Morning Edition* aired the story of Vaughn Allex, an American Airlines agent who checked in passengers on Flight 77, including two of the hijackers:[22]

> ALLEX: I didn't know what I had done. It wasn't until the next day, September 12, that I started finding out what happened. I came to work, and people wouldn't look at me in the eye. And they handed me the manifest for the flight. I just stared at it for a second. And then I looked up. I go, "I did it, didn't I?" I checked in a family. It was a retiree and his wife. I had time to talk to them. There was a student group, and I checked in a lot of those kids and parents, teachers. And they were gone. They were just all gone.

Producers at *All Things Considered* took that tape five years later, on the 20th anniversary, and wove it with copy read by host Audie Cornish at the beginning of a new retrospective segment about 9/11:[23]

ALLEX: I didn't know what I had done. It wasn't until the next day, September 12, that I started finding out what happened.

AUDIE: Back at the airport on September 12, someone handed Allex the flight manifest.

ALLEX: I just stared at it for a second. And then I looked up. I go, "I did it, didn't I?"

AUDIE: As Allex would come to learn, about half an hour after takeoff in the air over southeastern Ohio, Flight 77 had turned around.

ALLEX: I checked in a family. It was a retiree and his wife. I had time to talk to them.

AUDIE: By 9 a.m., the plane was heading back east, descending from 35,000 feet.

ALLEX: There was a student group, and I checked in a lot of those kids and parents, teachers.

AUDIE: At 9:34 the plane made a looping descent, miles from Washington, D.C. And minutes later, nose-down, traveling 530 miles per hour, Flight 77 crashed into the side of the Pentagon.

ALLEX: And they were gone. They were just all gone.

AUDIE: A hundred and eighty-nine people were killed, including the two men Vaughn Allex had checked in to the flight that day.

The host's comments are scripted in a way to help tell the story. They are tight, they are descriptive, they add context, but they don't get in the way of the ticketing agent's story. On the contrary, they make it stronger.

Clear, Simple and Short

Before we move on, here are a few more suggestions on writing to tape.

THE WHO AND THE WHY SHOULD BE CLEAR. "Listeners need to know who is talking and why the person is making the comments," says Southern bureau chief Susanna Capelouto. But don't try to throw in too much before the actuality. Often the explanation can wait until afterward.

JUST USE "SAYS." Avoid pretentious synonyms: "states," "declares," "announces," "retorts," "remarks," you get the picture. Susanna says she particularly dislikes "explains," even when it's an expert talking. "It only works if they really explain something complicated," she says.

KEEP ACTUALITIES SIMPLE. Most of the time, a long one can be split into two or trimmed down to its essence. "A bite has to really be great to hold a listener for more than 20 seconds," says Northeast bureau chief

Andrea de Leon. "One idea expressed per actuality is best," says Susanna. "Don't let people ramble on."

WRITE SHORT. "When in doubt, write less," says *Morning Edition*'s Phil Harrell, another producer who says you should "let the tape do the talking." Part of having a good interplay between tape and copy is concise prose, Andrea says. In a show-length segment, "a track shouldn't generally be longer than 40 seconds."

What About Tape That's Not in English?

I once did a training session for locally hired producers at NPR bureaus overseas, who do a lot of the translating for our correspondents, and went over the aforementioned techniques for writing into and out of tape. One of them asked me, "What if you're working with tape that's not in English?"

That's when it hit me. I'd occasionally used these techniques when I reported from abroad, when interviewing English speakers. But most of the time, I was also interviewing people in Spanish and other languages.

So here are two answers to the producer's question:

FOREIGN-LANGUAGE TAPE IS LIKE AMBI. And by that, I mean only to the listeners who don't speak the language (more on that in a second). This doesn't mean it's not important. But it's the emotional quality of the voice that matters, more so than with an English-language actuality. And since it's followed by a translation, you will still want to write into the tape using the methods discussed above.

NOT ALL "FOREIGN" LANGUAGE IS FOREIGN. Many Americans speak more than one language, particularly Spanish.[24] For them, that piece of tape is not only ambience. It has meaning. So try to get whoever mixes your story to let the tape breathe. And make sure the words that are posted in the clear match the translation.

Also, try to avoid voiceovers. They break the flow of the story, and the change of voice can be confusing. For them to work well you really need a professional voice actor, whereas the reality is that the translation is often read by a bleary-eyed production assistant at 3 a.m. "I was burned too many times," says Lauren Frayer, who was also a Madrid-based reporter. She might have interviewed "an elderly Spanish woman," and the translation was read by a 22-year old producer from one of the shows. "It just didn't fit."

So when possible, write out of non-English actualities by voicing the translation yourself, either as a quote or by paraphrasing. You can add a "he says, or she says" at the end. This is easier and it's faster, because you save the producer who is mixing your piece an extra step. When you do the translation, listen to the original and try to match the inflection and intonation, and then switch back to your own voice for your tracks.* That's how Lauren came to do it. "At first I thought that was too much of me," she says. "But it's always better. It's me telling the story. I can't be afraid of my voice."

There may be times when it makes sense to leave the full actuality up in the other language. After the 2016 Pulse Nightclub shooting in Orlando, one actuality ran for a full 20 seconds in Spanish.[25] The speaker choked up as he spoke about the moment he realized his best friends were among the dead. Adrian Florido was the reporter, and he and *All Things Considered* producer Christina Cala thought it better reflected how distraught the man was. "Sometimes you don't need to know what has been said to understand the feeling that has been conveyed," Adrian says.

Signposting

Time moves in one direction. If a listener misses something while listening to a story on the radio, they can't go back and listen again. On a podcast, they can scroll back. But why would you make your busy listeners do that? And even when you've done everything right—with clear writing that supports a strong selection of bites—people still get distracted.

Here's where signposting comes to the rescue. What is a signpost? This is the definition given in the *Rough Translation* glossary:

> Breaking the third wall (either explicitly or subtly) to tell the listener to lean in—that this information is important, or you've heard this before, or to remember what this is, or to foreshadow, or to bring them in after a break, etc.

---

* I always did the voiceover myself when I tracked the story, and let the show editors decide if they wanted to use my voice or substitute it with a voiceover. Even when they went with a voiceover, they could first listen to me doing the intonation and mimic it.

It's essentially holding the listener's hand, and zooming out a bit with comments that don't necessarily contain news or information, but help the listener follow or reconnect with the story.

Alison MacAdam, my predecessor as audio journalism trainer at NPR, identified four types of signposts:[26]

- An explicit reference to tape, such as, "Listen to this . . ." or "This is the sound of . . ."
- A step-back-and-learn moment, such as, "In order to understand *x*, we need to talk about *y* . . ."
- A transition, such as, "Which brings us to . . ."
- A moment of review or summary

Signposting can take many forms, and it doesn't have to be in the body of a piece. When *Weekend Edition Saturday* host Scott Simon announces, "It's time for sports!" that's signposting. When Don Gonyea, in a two-way with a host, inserts "But here's the thing . . ." into his electoral analysis, that's also signposting.

Signposts tend to be less explicit in podcasting. "In other words, we often try not to announce their presence," says Adelina Lancianese, senior producer of NPR's *Embedded* documentary podcast. "We think about these moments of recap or clarity structurally, so a signpost could be replaying a distinct ambi sound you heard at the top of the episode, repeating a piece of scoring for thematic reasons, or a large graf of host analysis."

They also tend to be more literary, more about themes than indicating the sound you're about to hear. For example: "And this was the moment Raquel realized she couldn't wait around for her favorite superhero—no one was going to save her."

Here's an example of a signpost in a piece by former *Invisibilia* host Alix Spiegel, who interviews a man named Jason Comely about handling rejection:[27]

ALIX: Jason's heard from a teacher in Colorado, a massage therapist in Budapest, a computer programmer in Japan and even a widowed Russian grandmother. She's using rejection therapy to pick up men.

COMELY: It's really cool. So there's an 80-year-old babushka playing rejection therapy.

ALIX: *So what has Jason learned from all of this?* That your fears, most of them anyway, aren't grounded in reality in the way that you think that they are. They're just a story that you tell yourself, and you can choose to stop repeating it. You can choose to stop listening.

The italicized sentence gives the listener a chance to catch up and signals something important to come. Again, there's no new information. A good editor would probably cut it from a story written for the web. But for audio, it may be the difference between keeping and losing listeners.

Getting Creative

As a journalist, you report facts. But the last thing you want is for your stories to sound like a lecture or a dry recitation. That's when borrowing narrative techniques from fiction can help. They can inspire you to try something different—or teach you what not to do. "It's been very helpful to me, actually, to watch good television and bad television," says Jasmine Garsd, host of *La Última Copa/The Last Cup*, a bilingual eight-episode podcast about soccer star Lionel Messi's immigrant story. With each episode, she wanted to create a sense of anticipation, with the listener saying, "Oh, I can't wait until the next episode!" And she knew how it felt not to want more. "I've seen TV where I just I can't get past episode 1."

Especially on podcasts, producers and editors sometimes find a term of art to frame a particular episode. Adelina, who has also produced the podcasts *On Our Watch*, *Rough Translation* and *Louder Than a Riot*, says she likes to "fantasize about a dream structure" for an episode.

> If this were a novel, how would I write it? What would be the opening scene? What would be the flashback sequence? The denouement? And then understand the ways that the facts either make or break that structure. Breaking it can be a really good sign, because it means you're letting the facts lead you, not the other way around. After all, it's often true that reality is stranger than fiction. But that doesn't mean using fiction as inspiration is fruitless. It gives you the license to right-size your creativity to your journalism.

The *Rough Translation* podcast team, including host Gregory Warner, reporter Justine Yan and editor Luis Trelles, did something along those lines with an episode Adelina produced about the Lindy Hop, a jazz dance that originated in Harlem in the early 20th century.[28] The first person the listener meets is an African American woman who learned the dance from her great-grandmother. Then the listener meets a white man from rural Sweden who saw it performed on national TV and went on to become a professional Lindy Hop dancer. And then, the listener hears how their worlds and their perspectives on the dance, and its cultural resonance, collide.

The idea of meeting one person and then another is also a technique Adelina sees in theater, where the audience gets to know one character at a time. "For the most part, if we're in the thick of a story with person A, we want them to either exit the stage before person B comes on, or if the two of them do interact on that stage, think critically about why and how," she says.

The cultural chasm the two experience over the role race plays at a Swedish Lindy Hop dance camp is loosely inspired by the Rashomon effect, named after a 1950 Akira Kurosawa film by that name, where a murder is presented from different angles. The movie has inspired all kinds of storytellers who have used it to contrast divergent understandings of the same thing.

Adelina says she's always careful not to let fictional devices influence reality, and that where they are most helpful in telling journalistic stories is in allowing the story to unfold in an inventive way. "As long as you present the fullest picture of the truth, the order in which that plays out is something that you can have creative license over." So she might not disclose a key fact about someone in episode 1, but save it for a reveal later on. "We're not withholding information," she says. "We're just presenting it in an order that is very compelling and very narratively exciting for the listener."

The order in which you present information to the listener can also help them understand and retain that information. "Sometimes," Adelina says, "the listener has to understand one thing before understanding another. You might not want to take a 102 class before taking 101, just like you wouldn't watch a movie's sequel before the original."

What you don't want to do is manipulate someone's words or misrepresent their story. And to guard against that, Adelina says she'll routinely have fact-checkers look at scripts and send them out to a wide range of people, including those inside the community, for "friendly ears listens or sensitivity listens."

On *Planet Money*, episodes often center on an individual who goes through some kind of transformation. "I want a character who faces a choice," says Jess Jiang, senior supervising editor at *Planet Money*, "and they make a decision and it leads them down a road that they didn't expect." That kind of story structure is a form of what's known as the hero's journey, an age-old narrative template in which a main character leaves home, undergoes a crisis and emerges a better person.

Planet Money tells the story of Charles Ponzi in an episode hosted by Nick Fountain.[29] Ponzi may be one of the best-known con artists, but the main character in the story is a former *Boston Globe* banking reporter named Mitch Zuckoff, who became obsessed with Ponzi and went to Italy to trace his roots. Zuckoff eventually wrote a book called *Ponzi's Scheme: The True Story of a Financial Legend.*

In doing his research, Zuckoff discovered that Ponzi, upon hearing that a nurse he barely knew had been burned in a gas stove explosion, offered without hesitation to donate 72 square inches of his own skin, and ended up spending months recovering in the hospital afterward. Zuckoff tells Nick how that discovery changed his view of the man history regards as a scoundrel:

> ZUCKOFF: I thought I was going to write about a guy who was just a schemer, who was just a snake. And the deeper I went, the more I felt sort of the stirrings of sympathy, the stirrings of empathy. And, you know, what he did was wrong. And what—you know, and I'm not confused by that. But I kind of understood the motivation of a guy who couldn't stop, as he put it, you know, the snowball once it started rolling downhill and not get rolled over by it.

Nick could have told the Ponzi story chronologically, or explained how the pyramid scheme worked. Instead, it's much more captivating to see Ponzi through the eyes of someone whose impression of him changed in the course of investigating his past.

So, next time you read a good book or watch a spellbinding film, let it inspire the storyteller in you.

Endings

Or you can search online. That's what Thomas Lu did once as a producer at *Hidden Brain*. He and host Shankar Vedantam were looking at different ways to end an episode on AI-generated voices.[30] "Does Shankar come back in and do a coda, and then end with music? Or do we have a guest end with a snappy sound bite? And we tried different things, and nothing quite worked." So Thomas decided to see how great movies end. "I went onto YouTube and typed 'movie flashback scenes.'" And he came across

The Theory of Everything, the 2014 movie that ends with a retrospective montage of scenes of the life of the British physicist Stephen Hawking, who had a debilitating neurological disease. Although the movie does it with images, Thomas says, "I really tried to emulate the feeling that this ending scene gave me." He took the tape of the voices that had been featured at various points in the story, and replayed them in a beautifully layered montage at the end.

MAKE IT MEMORABLE

Getting inspiration for an ending can add a little zing to a podcast episode. For a short-form news story, you're more limited in what you can do. That doesn't mean the ending is unimportant—on the contrary, it's often the one thing that listeners take away from your story.

An ending can do one of the following:

- Recap
- Leave the listener thinking
- Look ahead (but with substance, not "only time will tell")

Ideally, you want to aim for two out of the three, or even all three. NPR Midwest bureau chief Cheryl Corley covered the 2022 racist shooting at a grocery store in Buffalo, New York, back when she was a correspondent. She concluded her story this way, after an actuality about the security guard who was killed trying to confront the gunman:[31]

> CHERYL: One of the 10 lives lost in this community, as people sort through the emotions of what happened a week ago and figure out what to do next.
>
> Cheryl Corley, NPR News, Buffalo, New York

There's no new information in her ending, which is the point. It recaps the story, lets listeners sit with the gravity of it, and hints at a future of grieving and processing the loss. When writing an ending, Cheryl suggests thinking about "what the story is and how you want listeners to feel as you end the story."

But don't sweat it. If you've done a good job with the story, the ending will come naturally. Often, you may already have it somewhere in the story, and moving it to the end will make the middle flow better too.

Actually, there's also a fourth option. Have fun. Cheryl did when she reported on the pink Cadillacs lining up for Aretha Franklin's funeral.[32] After riding in one, we hear her getting back into her own vehicle.

CHERYL: Now, my rental car is no pink Cadillac—
(*Car door closes.*)
CHERYL: —but I can still pay tribute to Aretha Franklin by taking a drive along the freeway.
(*Ambi of Aretha Franklin singing "Freeway of Love."*)
CHERYL: Cheryl Corley, NPR News, Detroit

AVOID ENDING WITH AN ACTUALITY

It's OK to end with ambi, as Cheryl did, or even with a piece of hot tape. But only in rare cases does an explanatory actuality work. As the reporter, you owe it to the listener to bring the story to a close. "The reporter is in control of the flow and the thought process you want your audience to be going through," says Northeast bureau chief Andrea de Leon. "You told them this was the beginning, you told them this was the middle, and you should have the last word." In reports on controversial subjects, ending the piece with an actuality may give the impression that the reporter sides with the speaker; it's as if the reporter were saying, "What can I possibly add to that?"

But I did say it's OK in rare cases. How rare? In my nine years as a radio correspondent, I can only remember doing it once. It was the other story I did about American veterans of the Spanish Civil War. They were getting together for what was likely to be their final reunion.[33] And I concluded with one of them, 85-year-old Jack Shafran, talking about how he went with his son Seth to the tomb of Gen. Francisco Franco many years after that epic fight against fascism.

JEROME: But first, he stopped at a drugstore and bought a vial—
SHAFRAN: —and took it back to the hotel, filled it with urine. Seth and I went to the cathedral, the Valley of the Fallen, and, standing over Franco's grave, I poured it out.
JEROME: For NPR News, this is Jerome Socolovsky, in the Ebro River Valley, Spain

So that's my standard for ending with an actuality: when the answer to the question, What can I possibly add to that? is, nothing.

CHAPTER 10

Editing

By now we know how producers and reporters work: They pitch ideas, do interviews, gather tape and information and write copy, and—at NPR, unlike many member stations—the producers mix the audio. At first glance, that may seem all you need to put together a story. But that would be neglecting one hugely important role—that of editors.

It's an odd role. The editor is a kind of friendly opponent, putting up hurdles each step of the way. Not for the hell of it, or from a desire to be a naysayer. On the contrary, when done with tact and deliberation, the editor's interventions make for more listenable, more accurate stories. The editor brings a critical ear and expertise in storytelling with the aim of pinpointing anything that might fall short of solid journalism.

At NPR, the editor serves the following functions:

AUDIENCE PROXY. The editor is the first person who hears the story fresh, essentially as a listener does. The editor is also an expert on the kinds of people who listen to NPR shows or podcasts or read our website, recognizing that each of these audiences is diverse.

STORYTELLING EXPERT. Largely as a result of being the audience's proxy—but also through training and most likely having worked as a reporter or producer—the capable editor amasses a wealth of expertise in journalism and storytelling and brings it to each edit.

QUALITY CONTROL. The editor plays a prosecutorial role in ensuring everything in the story has been fact-checked and that the sourcing and analysis are sound. The editor listens for tone, language, fairness and balance. The editor ensures diversity and proper representation of the stakeholders and groups involved in or affected by the issues and actions covered in the story. The editor also makes sure that the story is free of prejudice, bias and misconceptions about the people and topics being covered.

PARTNER. The editor may be in charge, but also acts as a partner in the creation and execution of the story, serving as a sounding board for the reporter, producer or host to bounce ideas off of and as a source of wisdom and perspective who can help keep the newsgathering effort focused.

SUPPORTER. The editor is an emotional and journalistic backstop, listening closely and making sure the reporter, producer or host have what they need to do their job and that they are physically and emotionally in good shape. The editor checks in regularly with anyone who is out on a potentially dangerous assignment. When working with a reporter in the field, the editor protects them from the back-and-forth of newsroom coverage discussions so that they can focus on getting the story.

NEWS MANAGER. The editor assumes responsibility for the coverage, deciding whether to do a story or not, to prioritize one story over another, or to do more than one story. The editor also plays traffic cop, arbitrating between reporters in the field and producers on the shows and desks, making judgments on story requests from the latter and story ideas from the former. In performing these duties, the editor decides not merely what's news and what isn't, but what is coverable given priorities, staffing and audience interest.

Some of these roles may not sound all that different from those of a newspaper editor. But one key difference at NPR is that for stories that go to air, editors have to edit with their ears. We'll talk about the "ear edit" in this chapter in addition to expanding on the roles above.

But first, let's make a distinction. NPR editors' jobs are differentiated primarily by where they work and whether they are editing reporters, or hosts and producers:

DESK EDITORS. Mainly work with reporters on beats and are often responsible for areas of coverage that include those beats. On the national and international desks, each editor may be responsible for a geographic region, or a topic such as immigration or diplomacy. On the science desk, an editor may have a cluster of reporters covering health care or the space program, while on the Washington desk, there are different editors for the White House, Congress, justice and national security beats.

SHOW EDITORS. Mainly work with hosts and producers, editing on-air interviews and stories reported by show staff. On every show there's a line editor who has a last look at the scripts and is the overall content editor as the show goes to air. On podcasts, the editors' roles vary depending on the type of podcast: editors of roundtables are like show editors, whereas those on narrative podcasts are more like desk editors in that they work with podcast hosts and producers who are reporting the stories that make up the episodes.

"Be the Editor You Always Wish You Had"

Editing is rarely an entry-level job. At NPR, the people who become editors have worked as producers or reporters; they have a good sense of the kind of editing that made their stories better, and the kind that didn't.

"Coming up as a reporter, I was very, very lucky," says newscast editor Elizabeth Wynne Johnson, who spent 20 years as a reporter. "I had really smart editors who took the time to tell me why they thought I should do *x* instead of *y*." She tries to do the same. Which is not easy in the Newscast unit, where editing is fast paced. Sometimes there's just a half hour between broadcasts, and edits have to be done in a matter of minutes. Still, she tries. "When I make changes, I try to tell the reporter why I've done it."

Sadly, some reporters are not so lucky. Before coming to NPR, climate editor Neela Banerjee's long reporting career included stints at some of the biggest newspapers in the country, and occasionally an editor took her story and "just changed it to make it sound like their voice." NPR's Northeast bureau chief Andrea de Leon, another former reporter, says: "I had some excruciating edits that I didn't learn a lot from beyond, 'I'm never working with so-and-so again.'"

Still, even bad experiences can make you a better editor. "It's kind of like hazing," Neela says, only half-joking. Once you've had that experience, you either decide to do the hazing yourself or vow that you're "never going to do that to anybody."

Neela is firmly in the latter camp. Whenever she thinks that a reporter's story needs retooling, she always stops and asks herself, "Why am I making that change?" Neela reminds herself that although *she* might express a point differently, it doesn't mean her way is better.

So, do the things your editors did right, and don't do the things they did wrong. "That little editor on my shoulder that yells in my ear," says Elizabeth, "that's the collective voice of all the really good editors that I was lucky enough to have." Or as Neela puts it, "Be the editor that you always wish you had."

HONOR THE VOICE

The best editors let the reporter write the story. Especially in audio, where there's so much emphasis on the voice, the words in the script should come from the reporter. That's why I recommended earlier in the book to "mouth edit" the story: say the sentences out loud before writing them, and only use words that naturally come out of your mouth. Extending that idea to editing means that while an editor identifies the problems in a story, the reporter should do the fixing. "I don't want them to write my story for me," national political correspondent Don Gonyea says of his editors.

He also doesn't want to be told what the story is. "I want an editor who gives me the freedom to find the story that I find," says Don. Before he goes out reporting, he and his editor will discuss what they think the story is, "and then maybe I find that, or maybe I don't," he says. "They have to trust me. If I find something else, that's where the story is going to go."

That doesn't mean the editor shouldn't challenge the reporter. "Nothing makes me happier than being able to sign off on something and say, 'This looks good, no changes, thank you very much,' " says Elizabeth. "But I don't rubber stamp. If I'm suggesting a change, it's because I know and or believe that that change is for accuracy or clarity or it's going to be a better experience for the listener."

Still, editing is a collaborative process. The idea is to make the story better together. Sometimes, even the best reporters can get stuck in the weeds and miss important elements of the story, so it's always advisable to have two people brainstorming and looking it over together. "It's a conversation that takes place over the life of the piece or the spot," Elizabeth says. Most of the time, when she's editing a reporter, she's "nudging them in a direction, so it's still their voice, their words, their reporting." Sometimes she'll make suggestions for specific changes, and the reporter will say, " 'How about this instead?' And invariably, what they come up with is the best. And I've done my job because I've gotten them from point A to point C, by offering a suggestion to point B."

On shows and podcasts, editors also work to help the hosts sound like themselves and catch errors as they distill an extraordinary amount of information and talk about it live in front of an enormous audience. "They're out there alone with the listener," says *Morning Edition* senior editor Jacob Conrad. "And my job is to make sure that they have everything they need for that encounter. I might not always agree with them with a choice they make, but there's a reason they're the host."

EDITOR AS COACH

When a script has a problem, the editor should not tell the reporter, "This isn't working, you need to do this." Neela favors an approach that brings psychotherapy to mind: "'I don't feel like this is working here. What is it that you're trying to do?' You're trying to tease out from people, 'Well, what are you trying to accomplish?'"

When assigned to work on a segment, Ashley Brown, editor on *All Things Considered*, says she tries to get at "what motivated the producer or the host to pitch the story." She tries to find out what's most interesting to them about the story or the particular guest, "so that we make sure that that is in the central question or that it shines through in the intro."

Middle East editor Larry Kaplow says a desk editor should, like a therapist, pay attention to the words the reporter uses. "Listen to what they're saying. Listen to what they're not saying," he says. Occasionally he'll notice something missing or written awkwardly in their piece. "They'll say something in a passive voice, and I'll say, 'Well, how do we know this?' And they'll say, 'You know, actually, I'm not sure about that. Maybe we shouldn't have it in the story.'"

Being a good listener also means being attuned to the reporter's well-being. A few days before Larry talked to me about editing, he was working with a reporter who was "very slow to respond" to a question he had asked. "And I said, 'Why the long pause?' And they said, 'It's just been a really long day, and I'm exhausted.'" When that happens, Larry says, he tries to scale back what's expected of them.

Giving the reason something needs to be changed helps the reporter develop as a journalist. "I really see my job as helping build reporters as well as getting stories on the air," says Andrea, the Northeast bureau chief. She edits mainly member-station reporters for the national desk, taking stories from their region and helping turn them into national stories. "What I hope I'm doing is giving them some things that they will remember and think about the next time they go out to report or write."

As NPR and many member stations have increasingly posted text versions of their stories online, audio reporters have had to learn to write for digital platforms. Neela trained some of her reporters on writing ambitious features for the web by taking stories from other publications and annotating them for her reporters. "You don't have to write it in the same style," she told them, "but I want you to understand what the writers are doing here."

The Sword of Judgment

It may sound harsh, but one of the most important tools the editor has is the power to kill a story. Why do that? Consider these scenarios.

THE STORY HAS FATAL ERRORS. There may be inaccuracies that can't be corrected or facts that can't be confirmed, either because there is no way of discerning the truth or because there isn't enough time to do it before the show. Remember, a story should only be killed or delayed if the fact in question is central to the piece. If it's not, try cutting that part out and seeing if the story still holds.

THE STORY IS INCOMPLETE OR BIASED. Every story must be complete, in that any assertions or claims that call into question the story's fairness should be given a counterpoint. Tacking another voice onto the story is one option. But if that gets in the way of the narrative, it might be better to incorporate the countervailing viewpoint into the reporter's tracks. With a host interview it's trickier, because the conversation is usually between a host and guest exclusively. One solution is to write the opposing viewpoint into a host question. Or, if you have tape expressing that viewpoint, you can play it in the intro. This method works best if the tape is of a well-known figure and the host interview itself serves as a counterpoint. Another way of using that tape is to play it during a host question and ask the guest to respond specifically to the statement they just heard.

THE STORY WON'T INTEREST OUR AUDIENCE. If it doesn't meet enough of the basic criteria of an interesting story that I talked about in chapter 2, and also isn't written and structured in accordance with some of the basic principles of writing and storytelling, then it shouldn't go on air. While this may be the most subjective of these three criteria, it doesn't necessarily have to lead to an entire story being spiked. If you accepted the pitch, there's probably a story there: you and the reporter just have to figure out a compelling way to tell it.

As a show editor, your responsibility is to make sure that no story that fits any of these descriptions goes to air. As a desk editor, if a story you've commissioned still suffers from one of these problems despite your best efforts, you need to keep it from being broadcast. Whatever the reason, the decision shouldn't be taken lightly: It may undermine days or even weeks of work by a reporter or producer. "Sometimes it feels like it's a mean thing to do, to say this doesn't look or sound good, but you just have to break through that feeling," says Ashley. "And it doesn't mean you're a jerk or that you can't be kind in that process. But you have to be very unafraid to pull a story entirely if it doesn't feel right." And even then, first have a conversation with the team to see if they can make it better, if time permits.

Once, *All Things Considered* was going to run an obituary about a well-known sports fan who had been unapologetic about mocking Native American traditions even after his team changed its name. "Tonally, I thought, this is a weird thing—to celebrate a guy who was openly disrespectful to an entire marginalized group of people and who, when they said to his face, 'Please stop,' he said no." Ashley told executive producer Sami Yenigun she didn't think the story should run. He agreed and pulled the story from the show rundown after letting the desk editor and the newsroom leadership know about their concerns. That all took place just a few hours before the piece was slated to air. The move had consequences. They had to look on the shelf for another story of the same length to replace it, and race to get it ready for air.

While the first two scenarios for spiking a story are clear, the third, that the story won't interest the audience, requires a sophisticated appreciation of what works and what doesn't in storytelling. It can take years of editorial experience to develop that. Essentially, the editor has a duty to guarantee NPR's large and diverse audiences that the stories we give them are worth their time. "Everyone is very busy, and it's our challenge to keep people interested," says national desk correspondent Jasmine Garsd. "And that's where a great editor can come in and say, 'Sorry to hurt your ego, but this isn't interesting.' I'd rather an editor tell me that than thirty-three million people tune out."

It's because of our experienced editors, as well as our diligent reporters, that killing stories at the point where the reporter hands in a script is something "we almost never do," says Larry. The editing process, starting with the consideration of the pitch, is designed to make sure reporters are headed in the right direction, to weed out any errors or biases and, if need be, talk through the story early in the reporting phase before reporters "go too far down the pike," he says.

But while killing a story is rare, killing parts of it—in other words, editing stuff out—is not. Most edits result in sections either being excised or reworked. It's one of the editor's primary roles: to identify elements that don't work, or that might have worked at a longer length but not for the time allotted on the show.

Speaking of time, pieces should always be edited with a stopwatch. If the piece is too long, the show won't be able to use it. Failing to edit to time can result in an airworthy piece being spiked, and it throws a wrench into the show planning. Even if you're editing for a podcast, it's good practice to have a target length. Though an episode can in theory go on forever, the stopwatch will help you keep the episode focused and limit the post-production work.

THE NUT GRAF FOR AUDIO SCRIPTS

As an editor, what can you do if you get a story that's accurate and fair but still may not interest the audience? When that happens, Larry tells the reporter to write a line of copy that communicates to the audience why it's relevant. Essentially, you're asking them to write what in print journalism is called the nut graf.

Larry recalls a story by reporter Emily Harris when she was NPR's Jerusalem correspondent. A mother she met in Gaza, who worked for a development aid organization, didn't want to raise her daughter the way she was raised in conservative Palestinian society, where boys are given more freedom. But she was upset after the 3-year-old's preschool teacher made her feel that she had a "bad daughter" because the girl, at the instigation of a couple of boys, had taken her clothes off in the classroom. To Larry, the story at first didn't seem that compelling, at least compared to everything else going on in the world at that moment. "It was just a kid doing a weird thing. Why is this worth four minutes of a listener's time?" Larry asked Emily. "What you're saying," Emily responded, "is I need to reach out to the audience." So, she wrote a line for the intro saying it was a story about "modern parenting in a very traditional society," which convinced Larry, and it ended up running as a Mother's Day story.[1] And, of course, it was a hit. "People said, 'Oh, I love Emily's story,'" Larry says, adding with a self-deprecating laugh, "And I didn't want her to do it!"

Larry now asks for a nut-graf-like statement for every story. It helps keep the reporter focused, and it can also prevent reporters from going down rabbit holes. "A lot of times the debate you'll get into with a reporter is whether something fits in a story." It may be an interview with an interesting individual that's not central to the story, or something that may be funny or intriguing but sidetracks the story. Larry tells the reporter: "If you can write one line in a track that tells the audience why this fits in the story, then it probably works. But if you have trouble writing that line, then you probably shouldn't have it in the story."

This approach in audio storytelling, though, has its limits. There are moments that can only be justified as being, in that time-honored term, "great radio." A person laughing infectiously or crying, or singing, or the sound of children playing. Or anything that adds personality to an individual and helps us relate to them as a human being. For more on such moments, go back to "Making Them Human" in chapter 9.

PULLING REPORTERS OUT OF THEIR BUBBLE

Editors who have designated reporters try to check in with them on a regular basis, sometimes once a day, to find out what they are up to, and what's new on their beat. Neela says she asks about the stories they're working on, the stories they're looking into, and helps them decide which ones are not worth pursuing. She also makes sure they budget enough time and plan to collect the right material, when required, for both radio and web versions. "You're constantly helping them manage their time and their priorities and helping them focus."

When chasing a story, it's easy for reporters to lose perspective. Or to forget the thing that made their pitch compelling. I should know. When I reported the story from Spain on Iberian ham that I mentioned in chapter 2, I visited the first slaughterhouse in the country certified to export it to the U.S.[2] I'd never been inside a slaughterhouse, and the scene rattled me. But as a good reporter, I sublimated my distress into my work and did a standup while interviewing the quality manager, Carlos Davila, where I described the pigs being killed before my eyes.

> (*Ambi of machinery clanking.*)
>
> JEROME (*standup*): Two large electrodes come down from the ceiling, grab them on both sides of the head, and give them an electric shock.
>
> (*Ambi of squealing pigs posted, then faded down.*)
>
> JEROME (*standup*): Cuanto voltage es?
>
> DAVILA: Seiscientos cinquenta voltios.

JEROME (*standup*): They get a shot of 650 volts, which kills most of them on the spot.
(*Ambi of squealing pigs, faded up.*)

Fortunately for me, and for NPR's listeners, that part didn't make the story. My editor at the time, Kevin Beesley, felt it didn't belong. "Jerome, this is supposed to be a story about a delicacy," he told me. "What I just heard is not very appetizing."

Most of the time, what stands out to the reporter will interest the audience. That's why reporters should always note down three or four things that struck them after each interview. But that instinct can fool them, as it fooled me. If they're too close to a story, small things can look big. If they're following developments daily, they may also get excited about incremental changes that don't mean much to outsiders. Or they may overly report the procedural stuff, which they may need to follow to know what's happening. One of the most valuable services an editor can provide is to pull the reporter out of their bubble.

In podcasting, the reporter can become so inundated with information that they have trouble stringing it together. "The thing that I experience editing on *Life Kit* is people getting really lost in their own piece," says the podcast's managing producer, Meghan Keane, who also serves as its editor. "And that's where you as an editor have to hold their hand and guide them out of the dark forest into the light and show them that there is a path. You're responsible as the editor to see the big picture."

The nut-graf exercise can also force the reporter to zoom out and consider the question why, as in, "Why do we care?" or "Why is this important?" or "Why is this happening?" In chapter 6, I talked about "why" being the crucial question reporters must ask interviewees. It's also the crucial question editors should ask reporters. "They'll write a whole piece about a hospital ward where people are malnourished. And I'll say, 'Wait, why is this happening?' " says Larry. " 'Why' is always the most interesting question for the audience."

It's not that the reporter doesn't know the why of the story. They may have expressed it when they pitched it, but by the time they wrote the story, they assumed it was common knowledge. Or, they think the why is not worthy of their audience. As much as podcasting thrives on informality, says Jasmine, host of *La Última Copa/The Last Cup*, there's also a tendency for podcasting culture to be "rarefied," as though people "should just be interested." But even though they might be interested, listeners still need to be able to follow the story, whether they know the background or not.

Audience Expert

We've discussed the role of the editor as therapist, trainer and judge. There's another role that's unique to audio stories: first listener. The editor hears the reporter's story before anyone in the audience does, and should vet it for anything that a critical but smart and sincere NPR listener would experience as problematic.

Andrea often tells a reporter that they are the expert on the story and she is the expert on the audience. "They have done the reporting, and I don't want to run them over with what I think I know about something," she says. "But what I do know is what's going to work for the national audience." Member-station reporters in her region are "really good at telling stories for a local audience," but her job is to focus on how their reports will be heard by a listener outside her region: "Why am I listening to this in Idaho and still paying attention at minute 3?"

Andrea's point touches on one of the main things reporters do: take information that one group of people knows and share it with a larger group of people. As journalists, part of our public service is to make the news relevant to as many listeners and readers as possible. But there are two traps reporters sometimes fall into that editors can correct:

HAVING THE INSIDER PERSPECTIVE. The reporter is too involved. They may forget the context that people outside their geographical area or community need. The editor can help by taking advantage of their position on the outside and pull the reporter out of their bubble.

HAVING THE OUTSIDER PERSPECTIVE. In trying to make the story appeal to outsiders, the reporter makes generalizations or relies on tropes or stereotypes that are unfair and offend people inside the group. Or the mere act of explanation marginalizes the group by treating their customs or traditions as outside the mainstream. While I call this the "outsider perspective," it's not always the result of an outsider covering a story; it can also be the result of an insider making too much of an effort to make a story relevant for outsiders.

I don't want to suggest that either of these perspectives is inherently a problem. On the contrary, being an outsider can help you identify context the audience needs to understand a story. And being an insider—belonging to a community or hailing from a place and knowing the ins and outs of the culture and the traditions—can be a boon for the journalistic effort. *All Things Considered* producer Christina Cala went on a host trip to the Colombian-Venezuelan border to cover a refugee crisis in 2019. Her fluency in Spanish and the fact that her family was from there made the reporting better.

"The last time I was here," she remembers thinking, "there were not a million people in the park camping." She also knew how treacherous the mountain roads were, even for people who weren't carrying suitcases the way the refugees were. Less obvious was the change she sensed in Colombians' perception of Venezuelans because of the refugee crisis. That kind of cultural knowledge, which "would take much more steeping into culture or expertise or living in a place" for someone else to gain, she says, was "automatically there."

Insiders can also catch problems in reporting that others might miss. The team at *Rough Translation*, a podcast about how familiar things and ideas are translated across cultures, routinely asked people from those cultures to audition an episode. "It's really important that if we're writing about any place, that the people who identify with that community or that place are listening" as a check, says Gregory Warner, who created the podcast. Such vetting is essentially a backread or listen by someone working for NPR who is not on the episode's editorial team but who has experience or expertise in the subject matter. If no such person is available, NPR may sometimes consult an expert outside the newsroom to get their input, though without ceding editorial control.

In podcasting, this step is sometimes called a sensitivity listen, though Greg dislikes that term. "I'm not a big fan of that phrase, because it sounds like what we're asking them to do is just iron out stuff that may be offensive," which shouldn't be the focus, if you've asked the right questions from the start, he says. Instead, what the insider can offer is "a more nuanced analysis, where you're asking the right question, but what you're missing is this idea, or its implication." Greg often thinks about skipping that step for practical reasons, "because we're on deadline and we have no time, and I don't want to be making last-minute changes." And there is an added temptation. "We have a Portuguese reporter anyway. Why do we need a Portuguese listener?" he says, referring to a *Rough Translation* episode on work-life balance in Portugal.[3] "And every time, I'm glad that I did it, because we're all limited by our own sensibility."

The question of offending a community brings us to the second trap. Most serious journalists take pride in reporting the truth. Sometimes, though, an explanation can be technically correct yet make you "come across as potentially offensive or just off," says podcast editor Luis Trelles, who lives in Puerto Rico. "This is one island that has been studied to death by anthropologists, ethnographers, social scientists, etc.," he says, and all that outsider-explaining made him resolve to be different. "I always try to resist the impulse to go to a community and say, 'I know you better than you know yourself.' And so that's the rule of thumb, not coming across as sounding like you get the full experience, because it's impossible for you to get the full experience if you haven't lived it."

That kind of outsider-explaining is also found in religious references. Stories about Christmas routinely assume listeners know what the holiday is about. But references to non-Christian holidays such as Eid al-Adha and Yom Kippur and observances such as Ramadan and the Hajj often come with explanations for those who don't practice them. That's not to say the explanations are unnecessary, but they may serve to place Muslims and Jews outside the realm of everyday American life.

It's also fraught to explain one culture in terms of another. Continuing on the theme of belief, the Bible is a collection of scriptures that are sacred to Christians. So it's wrong to call the Quran "the Muslim bible." Better to call it the Muslim holy book or holy text. And while the texts in the Christian Bible that are also sacred to Jews were written mostly in Hebrew (hence the term "Hebrew bible"), using the Christian term 'Old Testament' to refer to them collectively when not talking expressly about Christianity is both nonsensical and offensive to Jews. Christians call it "old" because they believe its teachings were superseded by the New Testament—a notion, as it happens, that justified centuries of persecution against Jews.*

Anytime there's a cultural reference of any kind in a story, there's a risk of excluding people. An episode of NPR's *Code Switch* podcast, called "Hold Up! Time for an Explanatory Comma," explores explanations that journalists often present in the form of dependent clauses set off by commas. To illustrate their point, the hosts cited criticism they received about an earlier episode where they did not explain who the rapper Tupac

---

* The actual Hebrew term for the Hebrew bible, which contains the Torah and other prophetic and scriptural writings, is *Tanakh*.

Shakur was, and argued that people of color are often expected to explain themselves. Noted the Poynter Institute's Kelly McBride, who also served as NPR's public editor, "if you have to explain the significance of Tupac Shakur, you are probably writing for a white and/or older audience. When journalists do that, it's a signal to everyone in that culture that this journalism wasn't created for them."[4]

The explanatory conundrum is particularly difficult for broadcast shows, which, as the term implies, are made to reach a broad audience. By contrast, podcasts are made for listeners who have an interest in a particular subject matter and come to the podcast to hear about it. Still, any explanation has the potential to alienate, and editors need to be vigilant.

How? Is there a way to explain something that will enlighten those outside a group while not making those on the inside feel excluded? It's not always possible. But sometimes it is: Write in a way that makes the explanation as unobtrusive as possible. Because I'm Jewish, I was asked to do a sensitivity read on a piece about cantorial music that began this way: "Cantors lead Jewish congregations in prayer and song . . ." I noted that this explanation made me feel the story was not for me, and suggested simply adding the word "when" to the beginning of that sentence, to present it as descriptive rather than explanatory:

> When cantors lead Jewish congregations in prayer and song, they hope to inspire people spiritually . . .

If it were my story, I would have also dropped in the Hebrew word for cantor, *hazzan*, somewhere in the script. That would have lent authenticity and also signaled to Jewish listeners, "This story is also for you." Sometimes, a discreet insider reference is like a small gift to people in the community, and it might just make up for the explanations you have to give for everyone else's benefit.

Editor as News Manager

Reporting is hard work. It involves a lot of researching and fact-checking, going to press conferences or trying to get people to agree to be interviewed. But through it all, you pretty much have control over your project.

As an editor, it's not like that. You are juggling a multitude of things that are out of your control: "the request from Newscast, the request from

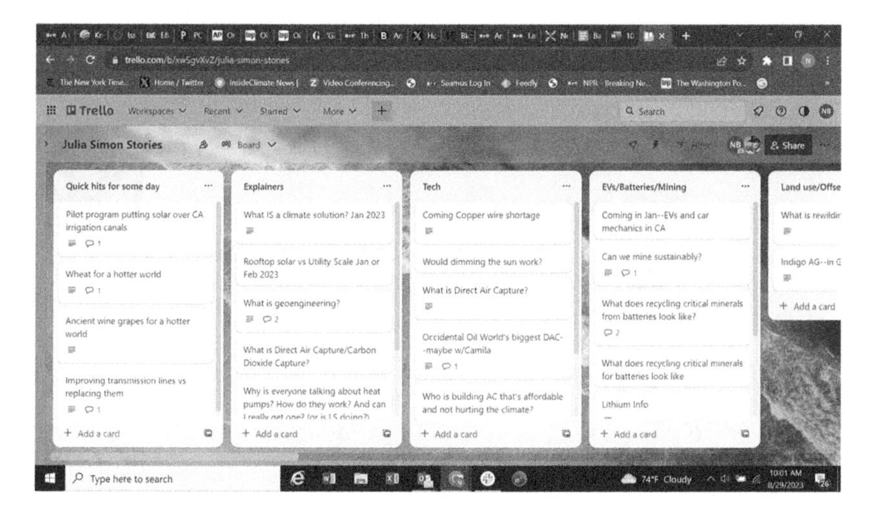

Climate editor Neela Banerjee uses Trello, a card-based scheduling system, to keep track of her reporters' stories. This one shows the different story ideas Neela brainstormed with Julia Simon, the climate desk's solutions reporter.

Newshub [the NPR website news desk], the requests from the shows, the editing of all the stories that go back-and-forth," says Gisele Grayson, who has worked as both a producer and editor on the science desk. "A good editor is triaging everywhere."

"Being an editor is like being a waiter with 12 tables," says Larry, quoting wisdom that was passed on to him from one of his editors. He has four staff reporters in his region, and sometimes also edits diplomatic correspondent Michele Kelemen and international affairs correspondent Jackie Northam. He's fielding requests for two-ways, newscast spots and pieces, and also gets asked for feedback on podcast episodes or show-guest ideas. "You have to know when to go back to this person and see what they need. You have to know that this person's finished and they need an edit. You know that this person needs a decision from the deputy managing editor," he says. "It takes organization."

As editor, you are also the conduit for messaging coming from the newsroom. Coverage priorities are set at the daily editorial meetings and passed along by the editor, but there may also be training sessions the reporters need to know about, or software updates and other administrative necessities.

Some editors use software to keep track of their responsibilities. Climate editor Neela Banerjee, whose six reporters often work on enterprise

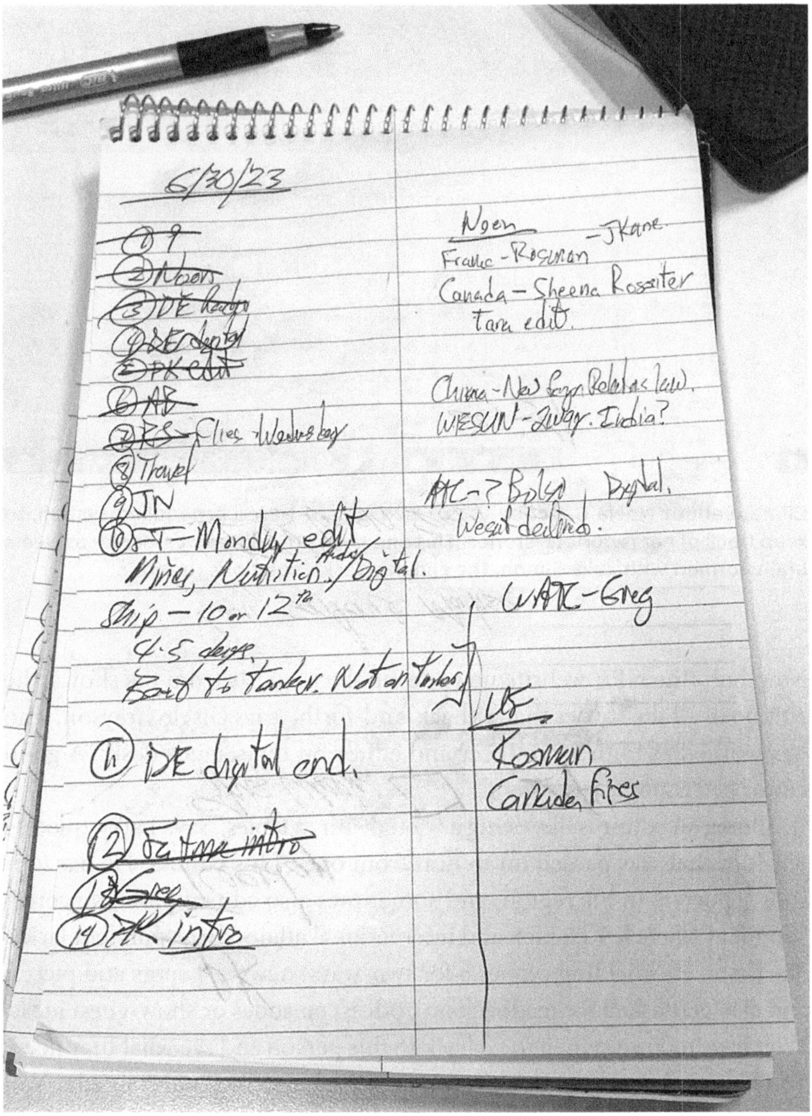

Middle East editor Larry Kaplow uses a paper notebook to keep track of his reporters and other responsibilities. The numbered list signifies the reporters (identified by their initials) who have a story that needs editing or who he just needs to check in with. On the right are stories in the works for the entire international desk that are coming up on *All Things Considered* and the weekend shows, which he will talk about at the Friday noon newsroomwide editorial meeting.

reporting projects that go much deeper than breaking news stories, keeps story lists in Trello, a card-based scheduling system. Larry prefers an old-fashioned paper notebook. "I make a list every morning. And then I work on it during the day and I circle the stuff that didn't get done. At the end of the day, I make the list for the next day and put those things on it."

There's also a lot of triage involved in editing newscasts. Editors on this team don't have reporters assigned to them; they take spots from reporters all over NPR, member stations and freelancers. Newscasts are broadcast on the hour throughout the day, and on the half hour for parts of the morning and evening; each broadcast typically has three reporter spots, as well as host copy, that need to be edited. At the same time, there's a constant stream of copy coming in for editing. Elizabeth Wynne Johnson edits around 30 spots a day as well as anchor copy. "I try to take things in the order first come, first served." But she'll make an exception if there's a late-breaking story and a spot is needed for the top of the hour. "Or, if it's 5 p.m. here [in Washington, D.C.,] and somebody is filing from Ukraine [where it is midnight], I'm going to bump that to the top of my list."

The fact that she doesn't have "complete control" may be the most important thing a reporter working with her should know about her job, Andrea says. As NPR's Northeast bureau chief, she edits dozens of reporters, producers and news directors at member stations around her region and also handles the addiction beat with NPR correspondent Brian Mann. While working with reporters on their stories, she negotiates with show staff, who have their own demands and constraints. If she feels a story deserves more time than show producers are offering, she might argue for a longer piece. But, she adds, when a reporter wants more time, "I can't wave a magic wand and say, 'Yes, you can have eight-and-a-half minutes' just because."

Editor as Talent Agent

What editors can and should do is shield their reporters. In chapter 5, I wrote about how, just after the 2023 earthquake hit Turkey, Ruth Sherlock was initially NPR's only person on the scene. Everybody wanted something from her—the daily shows, the weekend shows, NPR.org, the video unit producing stories for TikTok and Instagram. But everyone knew to funnel their requests through Larry, her editor. "Nobody at NPR reaches out to me directly," says Ruth. And that's something she and many other reporters deeply appreciate, because they already have their hands full chasing the story. "What I've liked about some editors is that, when shows are clamoring for the piece to come in, they can pull that energy back from you, so that you as a reporter can go ahead and do what you need to do," says Cheryl Corley, who was a longtime national correspondent before she became Midwest bureau chief.

The shows are clamoring because there's a tremendous amount of programming to fill. The weekday shows—*Morning Edition*, *Here and Now* (an NPR/WBUR coproduction) and *All Things Considered*—each have two hours of airtime every day, plus more when news happens and stories need to be replaced or updated. And the Newscast unit has 24 top-of-the-hour broadcasts a day, plus another 12 that run at the bottom of the hour when *Morning Edition* and *All Things Considered* are on the air. So it's no surprise that there's a voracious appetite for content. But not every story idea pitched to an editor needs to be passed on to a reporter.

"You can be honest" if your reporter has too much on their plate, says Larry. You can say, "This is a good idea, but they don't have time." Sometimes just suggesting a few possibilities for an outside guest will go a long way toward helping a show get the coverage it desires.

People are generally understanding and willing to work with the editor in brainstorming other ways of covering a story. "We know that everyone's resources are finite," says Newscast senior producer Nathan Thompson, who regularly sends out requests for spots. Most people on the shows also understand that desk editors are doing everything they can to provide coverage. Nathan always tries to be judicious about his requests, because he knows he will rely on the same people "tomorrow and the next day and next week." He also doesn't want to get a bad rap, with editors thinking, "That Newscast shift is really tough. They're asking more than we can give, constantly."

On the other hand, providing the Newscast unit and the rest of the shows with journalism is the reason you're there. And reporters have their own ambitions. Sometimes, Larry says, he used to decline requests on their behalf, but they would say, "'I want to do that!' Which is great." So now he handles it differently. "I'll say, '*Morning Edition* is asking for this. I don't think we *need* to do it. But if you want to, let me know.' It's all about communication."

As I mentioned in chapter 2, one of NPR's strengths is that anyone in the newsroom can suggest a story idea. In fact, plenty of reporters' stories originated with producers, assistant producers and even interns on shows who read an article on another news site, or saw an interesting development mentioned on social media or simply had an idea or question about something in the news. There's an advantage to having a fresh eye, or seeing things from a distance, that comes from show staff looking for stories, but editors should keep in mind that their reporters are the experts in the field, the ones who are following the beat closely. When he gets a show pitch, Larry tells his reporters, "Let's try to do our own thing." Often the result is better and more nuanced than the other outlet's story that may have inspired the pitch. Or it covers new ground. And sometimes a correspondent's reporting debunks the story. Larry recalls one such case, which came from posts someone saw on Twitter about new laws banning dangerous breeds of dogs in Egypt. When Middle East reporter Aya Batrawi did a little digging, what she found raised doubts about whether the laws were being enforced and whether they really constituted something new. "It would be a clickbait-y thing—'Egypt Bans Dogs,'" Larry says, but it wouldn't reflect reality.

The Story-Editing Process

When possible, the editor and reporter should collaborate from the start—before the reporting has begun. One of the editor's main aims should be to help the reporter (and, in some cases, the reporter and producer) give the story focus; a reporter-editor conversation early on can save time when it comes to doing interviews and tracking down facts.

This stage is sometimes called the front-end edit. For Andrea, it involves a conversation with the reporter about the story's focus and length. She'll also try to get a "shopping list" of "the ingredients" of the story, including possible scenes and interviews. Andrea considers the front-end edit the most important part of the process. "Usually, if I get a story that I think is really flawed, when I get the script I realize that it's because I skipped that conversation."

After starting an assignment, a reporter usually checks in regularly with the editor. For Neela, that "constant communication" enables her to make sure her reporters are staying focused and helps her find out if a story is changing as a result of their reporting. "The worst thing, for both reporters and editors," she says, is for the story to be "a surprise to the editor."

If a story doesn't pan out, the editor wants to know as soon as possible. Andrea recalls one time when a reporter went on assignment to a museum in New York where immigrant women were being offered a course in citizenship. Only one person showed up to the class, but Andrea found that out only after she was given a story featuring a single student. Inexperienced reporters tend to blame themselves when stories fall apart, but it happens even to the most seasoned correspondents. "You should have called me," Andrea told the reporter. "I wouldn't have bitten you."

"With my best editors, it's a partnership," says political reporter Don Gonyea, commenting on their regular check-ins. "I don't want to feel like I've surprised my editor with the story that I do. I like them to be surprised by how good it is." As he does his reporting, he'll bounce ideas off the editor he's working with, whom he uses as a sounding board and a reality check. "I might call him from the road as I'm driving back to the hotel to tell him about that conversation with the lady who said she should be afraid of me because I was vaccinated," Don says. That call puts a marker on a moment that the editor might feel would be good to have in the story. "So if in three days we're editing the story, he might say, 'What happened to the lady? What about the vaccine?' "

Not every editor-reporter relationship spans the life of a story. For newscast spots, reporters are also working with other editors, often as part of a bigger coverage effort that another editor, usually working on a desk, supervises. For the daily newsmagazines, a story may have been pitched to one editor and then handed off to another, or to several. When that happens, the editor should not try to redirect the reporter but focus instead on the editing tasks involved in whatever stage the editor is responsible for.

"Generally, my role starts when the spot comes to us," says Elizabeth, the newscast editor. "It's been reported. It's been written. And submitted for edit. So essentially it's a done deal." She may make no changes if it's in great shape, or ask for significant changes if warranted. But her mandate is clear: "Number one, accuracy. Number two, clarity. Number three, context."

Morning Edition has daytime ("dayside") editing shifts, a swing shift and an overnight shift. Each one is handing off edits to the next. Jacob Conrad says that when he does the overnight editing shift, he always keeps in mind the "very thoughtful and editorially informed process that my predecessors on the dayside and swing shifts have been through in creating the two hours that they've delivered me." News can develop or change during his shift, especially with overseas stories that are in different time zones, so there's always a possibility of having to bump a story. "If a story emerges overnight, we're going to probably have to sacrifice some element that had been planned," he says. Because of that, *Morning Edition* has a priority system that categorizes stories on the board by time sensitivity, as "must run," "should run," or "can hold." When stories need to be bumped, editors try to find one in the last category that can be moved to a later day.

Editing by Ear

Editing audio stories is about much more than merely revising text. The editor serves as a surrogate for the listener—a one-person focus group who should determine whether a report is suitable for radio or podcast.

Unlike in text-based news, where the editor reads the written story, the radio reporter reads their story out loud to the editor. This process is called the ear edit, because the editor is focused on what they are hearing, not what they are reading in the script. Some editors won't look at a script ahead of time; they want to hear it fresh, as if they had just turned on the radio in time to catch this story at the beginning. Other editors ask for a script but use it mainly as a notepad; they just glance down at the copy as they're listening, and circle sections they have questions about or think need extra work.

A few editors do like to see the script beforehand, especially if they are concerned about the way the reporter has structured the story and think it might need major surgery. But most, especially those editing feature stories, will end up either sitting with the reporter or listening to them on a call. They will edit by ear to determine whether the story is intelligible, accessible, interesting, paced correctly and generally ready for broadcast. For Andrea, it doesn't matter how a script reads; she only cares how it sounds. "Sometimes I'll hear a paragraph or hear a sound bite, and I'll think, 'Huh?' Then I'll look at the script, and say, 'That makes perfect sense.' But I didn't get it the first time." For radio listeners, there is no second time. They get one chance to make sense of a story. Even on podcasts, you will lose listeners who don't feel like scrolling back. So if a sentence doesn't work for the ear, it doesn't work, period.

In an ear edit, the editor should still listen to all the sound that will be mixed into a story and judge whether it's airworthy—that is, whether it will be easily intelligible to someone listening on their car radio while driving, or on their smartphone in the bathroom while they're brushing their teeth. Have the reporter send the audio files. Because they can become overly familiar with and almost emotionally attached to their interviews, reporters can sometimes recite an actuality word for word. The editor should provide a reality test, to avoid ending up with a report built on a foundation of incomprehensible actualities.

The ear edit is the time to focus on the overall feel of the piece. Does it contain all the necessary elements? Does the structure make sense? It's best not to get too focused on the detail, because you may need to send the reporter back to rework entire sections. Then, when the reporter comes back for subsequent edits, that's the time to go over the text line-by-line and give more detailed feedback.

Developing Your Editorial Voice

Back when Meghan Keane was a show producer, she knew the medium well. She was a creative storyteller and adept at mixing. But her interactions with reporters were limited to requesting minor fixes: "Can you tape that again? You misspoke this word." When she started editing, her feedback had to rise to a whole other level, as in, "Here's how you should frame this story differently." At first, she says, "I felt really shaky and insecure about giving feedback to other people." So she had to develop her editorial voice.

Part of the problem was gauging the reactions in her mind prompted by listening to a draft story. Fortunately, at the time, she was working at the NPR podcast *Invisibilia*, which did group edits, and she noticed that other people were articulating some of the same thoughts she'd had but was afraid to express out loud. She realized that she didn't have to wait for someone else to offer a critique first for it to be valid. "So I made a deal with myself to go first in the group edits as much as I could. And even if I didn't have the smartest feedback in the room, or the biggest rethink of how the piece should actually go, I started to gain a bit of confidence."

That newfound self-assurance helped her later when she became managing producer at *Life Kit*, and suddenly, she was "the person everyone was looking at" for judgments on stories. That's because an editor is expected to critique other people's work, though it's best to do it with care and respect.

One of the surprises for me when I started editing was that journalists who sound so composed and assured on air can be humble and sometimes even insecure when they are being edited. "The moment at which I have the least confidence in my story is always when it's done and I'm about to be edited," says Shannon Bond. As the editor, at that moment, you should deliver your feedback gently and carefully.

Of course, Shannon knows deep inside that her reporting is sound. But like many reporters, she still tries to anticipate the editor's feedback and head it off before it comes. "I try to restrain myself, before I do the read, from undercutting it and telling them all the things I think are wrong with it," she says. Some editors might stop the reporter and say, "Let's listen to your story first." Having been both a reporter and editor, I personally think it can cut both ways. Some reporters feel better getting it off their chest before the read. With others, it might be better to wait. If the reporter seems to really need to talk before the edit, let them. But store that information in the back of your mind somewhere and address it after the read. Don't let it prejudice your first experience of the story.

When the reporter finishes reading, that first moment of silence is a sensitive one. Restrain yourself. Instead of launching in with all that you think is wrong with their piece, let them have the first word. "How do you feel about the story, having just read it to someone else for the first time?" is something I used to say.

This kind of editing not only makes for a better story, it also makes for a more fruitful editor-reporter relationship. Instead of just rewriting, steer the reporter in the right direction by encouraging them to use words they're comfortable saying. One way of doing that is to ask them to set aside the script and tell the story, especially the parts that need work. Then you can note whether the phrases used in speech are better than the ones in the first draft of the piece. For example, after hearing a reporter's story about farming and climate change, you might tell the reporter something like this: "See, you didn't call her the U.S. Department of Agriculture Under Secretary for Research, Education, and Economics and Chief Scientist, you just called her a USDA scientist. And it was much easier to follow you when you talked about how she thinks agriculture needs to adapt to climate change. All the detail in the script had me confused. Now go back and try writing it the way you just told it to me."

This approach will also mean less work for you. A reporter shouldn't expect an editor to reduce an eight-minute piece to four. If a story is twice as long as it ought to be, offer general guidance about what doesn't need to be there, but leave it to the reporter to do most of the cutting. Chances are, that small amount of feedback will result in better cuts, since the reporter knows their material better than you do.

And since you can't know everything about every subject you're editing, focus on identifying the facts that need to be verified. Make sure that the reporter has checked and double-checked all the names, dates and places in the script (and don't forget that someone who was 28 when she was interviewed last summer may be 29 now). Get in the habit of taking a hard look at all numbers ("half of all 14-year-olds have tried vaping"), at assertions from individuals or groups with political agendas ("the state government has made no effort to help unemployed workers find jobs"), at historical interpretations that may require attribution ("the Japanese invasion of Manchuria unleashed forces that led ultimately to the attack on Pearl Harbor"), at anyone characterizing another person's motives ("Republican Congressman Jim Smith says Democrats are introducing the bill to embarrass the president"), and similar statements. They may not be wrong, but you need to know why the reporter felt comfortable making them.

And don't forget to praise the parts of the story that deserve it. The reporter may have been living and breathing this story for weeks and given thought to every word in there. Even though you hold the power to change the story, don't be that caricature of the editor "who is really hard to please and never says anything that the reporter wants," Meghan says. "A good editor is a collaborator, who puts up good boundaries. They're not giving a host or reporter everything they want. They're having a conversation to ultimately make the work better."

The Gratified Editor

When my son was in college, he called me once to tell me that one of his professors played an NPR interview in class with someone who had just completed a 3,000-word translation of the Hebrew bible.[5] My son had no idea I was the editor on that interview. Neither, for that matter, did the millions of others who heard it when it aired on *Morning Edition*. They heard Rachel Martin doing the interview, but not how I'd briefed her on the project. Nor did they know that I had enlightened her on the linguistic differences between modern English and biblical Hebrew, informing a question that prompted one of the interview's best moments. None of that mattered, though. The mere knowledge that I made the interview just a little more interesting than it might have been otherwise, and that it left an impression on at least one—admittedly rather important—person, filled me with pride.

"Every now and then, somebody will say 'Oh, I remember that story,'" about a segment she edited, says Ashley Brown. And it takes her by surprise every time. "Are you bullshitting me?!" Because, she says, there's something very gratifying about creating a memorable piece of journalism.

And then there is the satisfaction editors get when reporters feel an edit made their story better. "When we break through and take their work to the next level," says Elizabeth, "when two plus two—two being the editor and two being the reporter—equals five, that's the best part of my job."

CHAPTER 11

Voicing

NPR vowed at its founding to "speak with many voices and dialects," and in many ways it did.[1] NPR had a cast of women anchors and reporters, including "founding mothers" Linda Wertheimer, Susan Stamberg, Nina Totenberg and Cokie Roberts, at a time when the major TV network's evening news anchors were mostly men. Vertamae Smart-Grosvenor, the griot who hailed from the Gullah community of North Carolina, was a frequent commentator, as was Romanian American writer Andrei Codrescu, who delivered trenchant reflections in thickly accented English, and the late Wade Goodwyn, whose Texas-inflected drawl was once compared to "warm butter melting over barbecued sweet corn."[2]

In the days of AM radio's dominance, a resonant baritone may have compensated for a weak signal. But that's no longer the case with the superior quality that podcasts and FM radio deliver. And there's been a turnabout in thinking across American public radio, in no small part thanks to people like podcast host and professor Chenjerai Kumanika. In a 2015 article for Transom, he pointed out the pressure faced by journalists "of various ethnicities, genders and other identity categories" to imitate what he described as "culturally dominant 'white' styles of speech and narration."[3] A distinctive "NPR voice" did develop over the years that seemed free of regional and other dialects, making it easy when you turned the FM dial to tell when you'd landed on an NPR station. But in recent years, the network has made a concerted effort to bring the full range of speech and dialects that reflect who our journalists really are, and live up to that original mission statement in the fullest sense.

So listen to NPR and learn from our journalists' on-air delivery, but cultivate your own voice. "Don't try to change the way that you sound," says *Weekend Edition Sunday* host Ayesha Rascoe, whose unmistakably warm and vivacious personality comes through on air, which she says has a lot to do with her coming from Durham, North Carolina. "If you grew up in Boston, you're not going to sound like me. If you grew up in Oakland, you're not going to sound like I do."

But, she adds, there's a difference between changing your voice "and trying to figure out, 'How can I get these words out without stumbling?'" she says. No matter what your voice is like, sounding natural, and not like you're reading, "is the hardest thing in the world."

In this chapter, I'll help you tackle that "hardest thing in the world" by sharing what has worked for me and for my NPR colleagues. We'll also get some tips from voice coaches who've worked with NPR folks.

"Everybody has a voice for radio," says Rachel Martin, who hosted *Morning Edition* and *Weekend Edition Sunday* and was former national security and international correspondent. "You just have to own your own voice and talk like you actually talk in real life." And while a certain amount of theater is involved, it's something anyone can learn. "It takes a bit of work to sound like yourself," says *Life Kit* managing producer Meghan Keane. "But when you get there, it feels awesome."

Perform the Story

Let's start with a couple of don'ts. Don't just read your script aloud. It will sound like you're bored, and your audience will hear it. Don't exaggerate as though you're reading a children's story: You will sound patronizing. And while you should sound natural, "sounding like yourself means not that you're at a bar talking to your buddy over a beer," says Jessica Hansen, who was NPR's voice coach for eight years. "You have to perform the story." The performance trips up many journalists who come to audio after a career of letting their words speak for themselves. "I am not comfortable with the performance part of our job," says Eyder Peralta, a former newspaper reporter. "But it is a big part of our job."

Why? Because the listener can't see you. They can't follow your lips moving, your hands gesturing. They are getting none of the visual cues that animate speech. So you have to make up for it by emphasizing important words, varying your cadence and timbre. It's the same as with the people you interview: How a voice sounds sends signals that help the listener absorb the information they're getting.

That's something people do naturally in conversation. "When somebody tells you something surprising," says Eyder, "you don't think, 'I need to act surprised.' No, you act surprised. Or you smile when something's funny. People can hear you talking through a smile. If you're in the moment, you'll do that naturally."

"Live the script," adds Eyder, and the performance will happen by itself. Whether you're tracking in your hotel room, in your bedroom closet, or in the studio, imagine you're still on the scene, surrounded by all the sights and sounds and personalities. "That's a huge part of the performance. It's being in the moment, being in that world when you're actually reading."

Let's say you were talking to firefighters battling a wildfire, and you've written evocatively about their faces being covered with soot and the trees behind them charred to a crisp. Take yourself back there as you track. If you're talking about someone whose home burned down, remember what they looked like, and how you felt, as they told you about it. Don't do it just for emotional stories. If you're reporting on a new kind of battery technology, think about what the inventor looked like when they described it to you.

Admittedly, it's harder to do that if you didn't report from the scene, or if you're reading a script someone else wrote. But it can be done. Just listen to how NPR hosts read intros, putting their whole selves into every word. What they're doing is thinking about the meaning of the words coming out of their mouth. Which is what you should do too. If you're voicing a spot about a plane crash that killed 73 people, don't just read "73 people died." Think about what that loss means to the bereaved families. Don't rush. Don't be boring. Give it the inflection that it deserves.

LISTEN TO THE CLIPS

And match the mood of your piece. If it's a fun story, smile as you speak. What Eyder said is true: people can hear you smiling! If it's a sad story, your voice should reflect it.

Sometimes reporters just want to get the delivery over with. So they read the tracks, and ignore the sound and the actualities. That's a mistake. Your tone should do justice to the emotional quality of clips. If someone tells a joke in an actuality, come out of it with a little chuckle. If they just spoke through tears, you should talk like it moved you. So listen to the actualities as you track. If you're tracking in a studio, ask a producer to play them for you when they appear in the script.

Rachel Martin and David Greene did that with a humorous tape and copy about a congressional session in which lawmakers quoted songs by the singer Meat Loaf.[4] You can hear the hosts laughing as they tell it. "If I hadn't played the tape for them as they were reading it," says *Morning Edition* producer Phil Harrell, "and if I hadn't played the music, it would have sounded completely different. But the way it came together, it was a sparkling moment."

If you're not in a studio, you can still play the acts back to yourself on your computer as you read. If even that's not possible, listen to the tape so many times that you can play them in your head as you track. No matter how well your piece is written, and how good your tape is, it's a waste if you sound like you're telling a completely different story.

BE EMPHATIC

When she began hosting *Weekend Edition Sunday*, Ayesha would lower the energy in her voice when delivering sad news. She thought it would be an appropriate tone, but she was told that her voice actually sounded flat. The lesson: The delivery must be robust, no matter the tone. "Whatever I'm reading, even if it's sad, you still have to bring energy to it. It's not a happy energy," she says, but it is a forcefulness that matches the impact of what you're saying. A script might read, " '17 people died in Ukraine today,' and I think in my head, 'This is important, this is urgent.' "

"The exclamation point is in your voice," says longtime correspondent John Burnett. "You have to get emphatic about things. And you can sound outraged. It's not editorializing as long as you're not giving your opinion. But the voice says that 'this is fucking important.' And listeners want that because that's the boldface, that's the lede, that's telling people what the takeaway is."

A trick that Meghan picked up from Jessica's training was to take her script and read it with different emotions: proud, loving, angry, bored, jealous and so on. "It's really interesting to see which emotions actually make you sound the best. And doing that over and over again, you have a shortcut word if you're slipping into sounding like you're stale or you're rushing."

You can use emotion in your voice to signpost stages in the story. "Whether it's a seven-minute feature or a newscast spot, there's a beginning and a middle and an end," says Jessica. "What's the journey? What's the good news? What's the bad news? What's the high point? Where are you comforting? Where are you welcoming? Where are you breaking it gently?"

Pretend You're Talking to One Person

When you're on the air, the listener experiences it as a one-on-one conversation. It may be a performance, but you're not an actor in front of an audience. You're not a professor lecturing a class. As far as that listener is concerned, you're not reading a script. You're just talking to them. That's why voice coaches suggest picturing a specific person when you're tracking: It can be a relative, friend or colleague. Imagine where that person is, what they're doing and even what they're wearing. Tape a photograph to the wall, or to your computer screen, if that helps.

"When I started in radio," says Don Gonyea, "I always pictured my mom sitting at her kitchen table listening to the radio. And my mom was wicked smart. But I also knew she got distracted, by the cat wanting to be let in the back door. Or, my dad would make some noise in the other room. So I would picture all of those things that might distract her, and I had to pull her along through this story." Mothers do so much for us, they even improve our on-air delivery! Ayesha also imagines her mother listening and thinks of her when she signs off the show. "You're listening to NPR News," Ayesha says as though she's on the air, and then with a big smile adds, "Goodbye! See ya later!!"

No shade on our dear mothers, but it can be useful to have a cast of individuals at hand, depending on the story. "When I coach people, I help them find four or five people," says Jessica. "Maybe for sad stories they need somebody like a grandmother or a niece or somebody that they need to be gentle with, and maybe for a silly, lighthearted story they tell their childhood best friend."

Who we imagine we're talking to really does influence our delivery. "We all code switch a little," says NPR music writer and podcast host Rodney Carmichael. When he tracked the pilot for the *Louder Than a Riot* podcast, he couldn't help but think about the managers and executives who would be deciding whether to greenlight the podcast. But when he did the podcast, he imagined a different kind of listener. "We all kind of have a work voice, and we have an 'I'm not at work' voice," he says. "And I just wanted people in my life outside of my work, and outside of even knowing my job and what I do, to totally recognize me if they heard this thing."

Mary Louise Kelly on Talking to One Person

I used to work at the BBC, where it's more of a "Voice of God" telling you the news. And I had remedial voice coaching when I came to NPR, both as a reporter and as a host. It sounds strange, that just speaking informally in conversation would be a skill you would need to practice. But you get in front of a microphone, and you're talking to important people doing important things. And my tendency is to tighten up. I often think when I'm in front of a mic and it goes live for *All Things Considered*, there's probably a few million people listening. But you can't think of that. You have to think about just one. So I talk at the volume and pacing that I would if I were just talking to one person. And that helps me stop projecting and talking at you, because I'm just talking with you. Because it's just one of you, and I'm imagining you sitting across from me.

Mark Up Your Script

Encouraged to put a little life into their words, some newcomers to audio overdo it. They stress too many words. Perhaps because they want to sound authoritative, or they feel so much of what they've written merits attention, they accentuate nearly every other word, or the last word in every sentence, or all of the above. As a result, their pitch falls into an unnaturally regular pattern.

Listen to the way people really talk when they're not on the air. You'll hear that most of the time, they vary their pitch. They emphasize just one or two key words in each sentence.

Wait. Let's stop for a second and consider how we might say what I just wrote. Here's how I might do it:

> Listen to the way people <u>really</u> talk when they're <u>not</u> on the air. You'll hear that <u>most</u> of the time, they vary their pitch. They emphasize just one or two <u>key words</u> in each sentence.

You might do it differently; the point is that in natural speech the stressed words don't fall into any predictable pattern. In contrast, someone who is emphasizing too much might sound like this:

> <u>Listen</u> to the way people <u>really talk</u> when they're <u>not</u> on the <u>air</u>. You'll hear that <u>most</u> of the <u>time</u>, they <u>vary</u> their <u>pitch</u>. They <u>emphasize</u> one or two <u>key words</u> in each <u>sentence</u>.

The problem with reading this way is that it doesn't help the listener know what's important and what's not. The abundance of stresses, and their regularity, sounds a little like those horrible voices on robocalls.

EMPHASIZE THE NEW STUFF

Read your story out loud before you mark it up, and listen to yourself. You'll probably find yourself underlining one or two words per sentence. And those words will most likely represent names or ideas you're saying for the first time, because it's natural to stress what's new. For example, in the following sentence, the emphasis might look something like this:

> <u>President Biden</u> is in <u>Japan</u> for a meeting of the world's wealthiest nations. High on the agenda is how to tackle <u>China's</u> economic <u>retaliation</u> against countries that challenge its policies.

Remember, the underlining is only a mnemonic. The goal of marking up a script is to remind yourself how you would say the sentence if it were not written down at all, not to force you to read it with artificial emphases.

There may be times when you want to remind yourself what word *not* to stress. For instance, sometimes reporters incongruously emphasize the word "people," as in, "a demonstration by about two thousand <u>people</u>." Did the speaker worry we might think the demonstration was staged by armadillos? If you have a tendency to misemphasize a word, find a way to remind yourself not to stress it—maybe by putting a squiggly line under it.

PAUSE, PAUSE, PAUSE

Speaking of lines, some people like to put slashes or vertical lines where they want to pause. However you mark it, don't forget to do it. "In some ways, pausing is more critical than emphasizing a particular word," says

Sora Newman, another former NPR voice coach. "Pausing lets the listener have a second to take in what somebody's saying, particularly if it's complicated, and it lets you, the reader, breathe and stay aware of what you're talking about."

The advantage of vertical lines or slashes is that you can add them after you write your script. Some reporters draw little yield signs. Others anticipate the pauses and put ellipses . . . as they are writing the script. At the very least, heed your own punctuation. Many people use commas or dashes when they write, and then ignore them when they read—plowing through their scripts headlong, with scarcely a break. Here's how I might mark the pauses in the sentence above.

> President Biden is in Japan . . . for a meeting of the world's wealthiest nations. High on the agenda . . . is how to tackle China's economic retaliation . . . against countries that challenge its policies.

Another way to mark up a script, especially a longer one, is to put notes in the margins as guideposts to your desired tone: Writing "suspenseful" next to a section, or a reminder that you should be "soft, tender" or that this part is "joyful," says Jessica, "takes some of the weight off you, so you can really just live in the moment."

As with everything, relying too much on marking up a script can get in the way of a natural delivery. It can also lead you to focus too much on how you're saying the words, rather than thinking about the meaning of the words. So use it as a crutch. And as you become more practiced in your delivery, rely on it less and less. "When I first got here," says correspondent Frank Langfitt, "I would highlight words and bold them. I don't really do that anymore. I try to read what I'm going to say and then look away from the screen as I'm talking."

Meghan Keane on Tracking as Music and Other Tips

Think of tracking as almost like a piece of music. Very few people talk in a staccato-like way. Sometimes you're legato, where you kind of space things out. For people who stress ev-ery oth-er syllable, try talking into it. Before you even get to the sentence you want to say, introduce it first, like, "I'm going to now talk about why the debt ceiling is about to be raised, and what that actually means," and then you kind of dip into it: "So, 'Speaker of the House Kevin McCarthy recently' blah blah." It's almost like going up a roller coaster, and then you let it go down the track and fall into the sentence.

Or just say it without thinking, three times fast, to get in the habit of saying it. Or don't even look at the script. Say it a few times and then say it a few more times without looking, especially if you have a short script.

And, honestly, don't be afraid to rewrite so long as it's factually correct. If you're tracking a sentence that doesn't feel right coming out of your mouth, then reword it.

Tracking Problems

CADENCE AND BREATHING

Delivery will vary depending on what kind of story you're voicing. A longer news feature or podcast episode naturally allows for more variation and conversational styles than a newscast spot. But even within the same story, not every sentence—or every part of a sentence—should be read at the same speed. You don't want to race through a phrase so quickly that listeners can't keep up with you. Instead, adopt the same varied pace in your on-air delivery that you have when you speak. "If you're introducing a new idea," says Meghan, "slow it down." Do the same if you have a really important idea or reveal that you don't want the audience to miss, or a study where you want them to understand the results. But if you're relating something for color or indicating motion, you can read a little faster. "The reason I think about the cadence so much," Meghan says, "is that people's natural speaking voices, when they're telling stories, goes in peaks and valleys."

Overall, you should be speaking at a much slower pace than when you read. "You're always going to sound slower in your own head than

Want to sound your best?

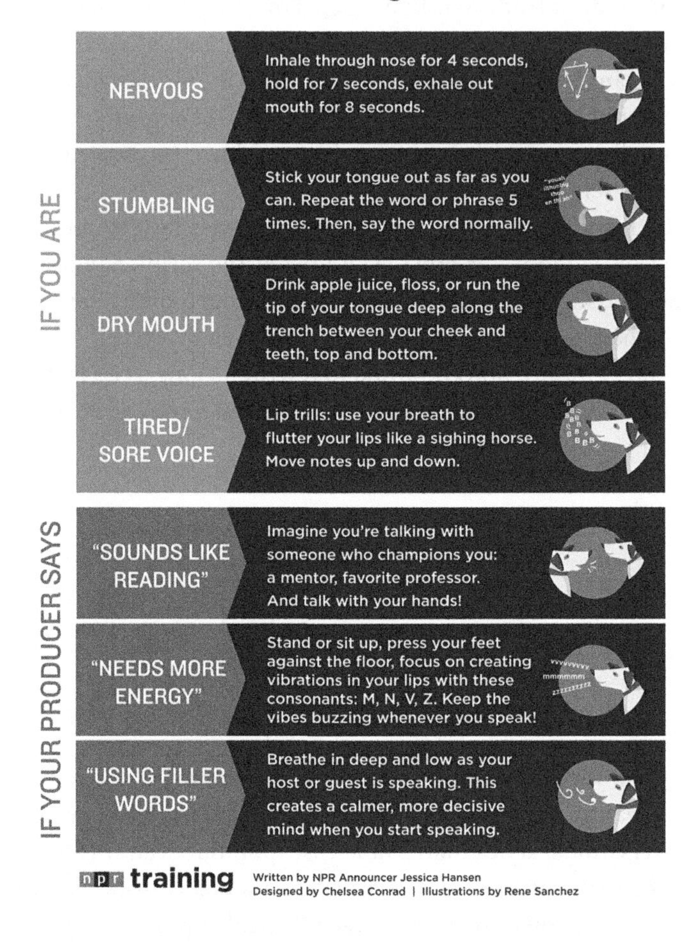

Voice coach Jessica Hansen's tips for sounding your best.

you are to other people listening," Meghan says. "But trust me, it sounds just right!"

However, telling someone to slow their delivery when they are trying to track is voice coach Jessica's "least favorite" advice. "If you're a producer or an editor and you tell somebody, 'You're talking too fast, you need to slow down,' they won't necessarily know what you're hearing and how to fix it." Do they need to pause more, breathe more, say individual words more slowly? Jessica says people usually read too fast because they're nervous, and they need to "downregulate their parasympathetic nervous system." For that, you might want to suggest pausing for some breathing exercises.

Help them sound their best

SAY, "COULD YOU TRY..."

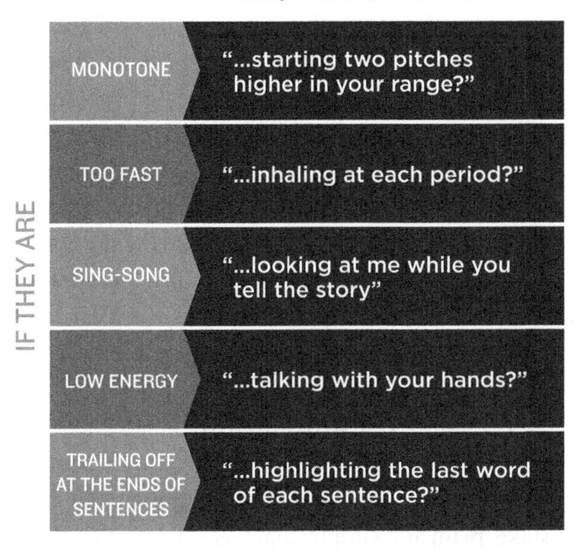

npr training
By NPR Announcer Jessica Hansen
Designed by Chelsea Conrad

Jessica's tips for helping someone else sound their best.

DO IT AGAIN AND AGAIN

Tracking takes time. Make sure you plan enough time to track and listen to yourself and do many takes. Even the best reporters get frustrated. "I track and retrack and retrack and retrack and retrack and listen," says Eyder, yet it still "sounds like shit!" It's part of the process. Once you've done the script many times, you'll get better at it. Try looking away from the script (perhaps at the picture of your mother?) and saying it. Or maybe have a real person in the room. It doesn't have to be a journalist. "Once you get behind the microphone, it's hard to hear yourself," says Meghan. "So have a buddy who can tell you, 'That doesn't sound quite right.' "

Another trick is after you've tracked the whole thing, go back to the top and do the beginning again. "I find the first page is always the hardest for people," says Meghan, especially if it's a long piece. "You will sound so much different an hour into that than you did the first 10 minutes."

READING MISTAKES ARE WRITING MISTAKES

You've got that picture of your mother taped to the wall and you've marked up your script, and you're still stumbling, especially over that one sentence. "Often tracking problems are really writing problems," says education correspondent Elissa Nadworny. She also does some voice coaching, and when someone is tripping over a sentence, she tells them to put down their script and talk to her about what are they trying to convey with that sentence. While that's happening, Elissa records what they say with her phone. "And then we basically go back to the voice memo track and I write down exactly the way they said it, and then they say that."

Don't Forget Correct Pronunciation

Mispronunciation is to audio journalism what misspelling is to the written form. It's a mistake, plain and simple, that weakens your credibility as a journalist. Finding out how to say a name, or an unfamiliar word, is a key part of your reporting. Getting the pronunciation right establishes you as an expert, evinces respect for the people you interview and saves you the embarrassment of millions of people hearing you or your host mangle it on the air.

So start every interview by asking interviewees to say their name. While you're at it, ask them to spell it, for the web version. If you are naming someone you didn't interview yourself, look for a recording of them saying it themselves. Don't rely on someone else saying the name, unless they're a close friend, colleague or family member who you're sure knows how to pronounce it.

At *Pop Culture Happy Hour*, the hosts talk about a bevy of cultural figures on nearly every episode, and they need to get all the names right. Co-host Aisha Harris says the podcast's producers "spend so much time trying to find audio of creators, actors, performers saying their own name, because sometimes you can find an interview, but they don't say their own name and another person says their name. And that doesn't necessarily mean that it's the correct way to say their name."

Once you've confirmed the pronunciation, write it phonetically in your script, next to the actual name, with the accented syllable in capital letters.

NPR Pronouncer

VOWELS	
a	apple, bat
ah	father, hot
ai	air, pair
aw	law, long
ay	ace, fate
e	bed
eh	item
ee	see, tea
ih	pin, middle
oh	go, oval
oo	food, two
ow	ouch, cow
oy	boy
u, uu	foot, put
uh	puff
ur	burden, curl
y, eye	ice, time

CONSONANTS	
g	got, beg
j	gem, job
k	cap, keep
ch	chair
hh	strongly aspirated 'h'
kh	guttural 'k'
s	see, citrus
sh	shut
z	zoo
zh	mirage

NPR's pronunciation key.

For example, I would write Jerome Socolovsky like this: jeh-ROHM sah-kuh-LAHV-skee. To do that, I followed NPR's pronunciation key.

The key also works for rendering names and words in other languages. It doesn't perfectly represent every sound that exists in the thousands of languages spoken by humans. But it does capture many common sounds, and the spellings are specific enough to avoid confusion when they are read, say, by a host on deadline who is looking at the pronouncer for the first time.

It's customary to have pronouncers for any name that might be unfamiliar to the host, or whoever is voicing the story. In my opinion, all names should have pronouncers, even supposedly familiar surnames such as Smith and Johnson. With the flourishing variety of names in the U.S., and the variations on pronunciations of traditional names, you can never really be sure about pronunciation from the way it's spelled.

And be careful with place-names, even here in the U.S. Did you know that Pago Pago is pronounced "PAN-goh PAN-goh," and that Versailles in Kentucky sounds nothing like the one in France. It's "ver-SAYLS," not "ver-SYE."

THE FRAUGHT QUESTION OF PRONOUNCING FOREIGN PLACE-NAMES

NPR gets a lot of mail, and there's a lot of internal debate, over how we should pronounce names of places beyond the borders of the 50 U.S. states. Some people think that if an anglicized version of the name exists, we should always use it. But just as the notion of the single NPR voice should be rejected, so is the idea that there is one correct way to say every place-name.

Adrian Florido was born in Southern California and spent a year on a special assignment in Puerto Rico in 2018. He says the island's name in Spanish, but he'll use the "anglicized incorrect pronunciation" of Los Angeles on air because it comes naturally to him when speaking in English. To him, listenability is what matters. "At the end of the day, we all know what we're talking about. Whether I say 'Los AN-jehl-luhs' or 'Los AHN-heh-les,' you know that I'm talking about the second largest city in the country."

For some reporters, it's about balancing what comes naturally to you with what's familiar to listeners. When I was in Spain, I almost always used the English pronunciation of Madrid, but pronounced many other place-names in Spanish because I grew up speaking the language. Some listeners wondered why I didn't pronounce the "c" in Barcelona with the *th* sound that most Spaniards make. That's because I learned Spanish from my parents, who were immigrants from Argentina, where the "c" sounds the way it does in English.

The standard of listenability does mean it's often a judgment call. If reporters who are fluent in the local languages chose to say "pah-REE," "dih-MASHQ," or "maws-KVAH," would listeners know what cities they are referring to?* Perhaps in context, but often the English name will be more widely understood.

The main objection to bilingual pronunciation is that it's not consistent. But it's the way many people speak. According to the Census Bureau, an estimated 68 million Americans speak more than one language at home.[5] I am one of them, and many of us routinely alternate languages, sometimes in the same sentence.† There's actually a name for

* Paris, Damascus and Moscow.

† My late mother, who was born in Argentina, would always tell me, "I love you *mucho mucho*!"

it—translanguaging—and it's a form of code-switching. And there's something to be said for exposing listeners to major world languages, especially when it's just a few morsels here and there.

"There's actually a lot of beauty in the inconsistency," Adrian says.

WHAT IF YOU'RE NOT BILINGUAL?

That's OK! You're in good company. Around 244 million Americans only speak English.[6] Pronouncing words in other languages is not easy. It may require facial muscles you don't normally use, at least not in the same way. Don't expect to parrot the native speaker. So if you have a non-English name in your script, do the following:

1. Listen for each syllable and transliterate them using NPR's pronunciation key.
2. Listen for which syllable is stressed and write it in capital letters. For example, the Mexican state Michoacán should be pronounced "mee-choh-ah-KAHN." Don't guess or assume it's the first syllable, as is common in English.
3. If you're stuck, talk to your editor, or your organization's standards editor, if it has one.
4. Once you have the pronouncer, put it in brackets or parentheses next to the actual spelling: Michoacán (mee-choh-ah-KAHN).

If a well-known city has an anglicized name, feel free to use it. It's perfectly acceptable to use city names such as Lisbon, Cairo, Copenhagen,* Mexico City, Geneva and Jerusalem, even though the local names of these cities are different. However, you may want to use the local name to show what its citizens call their hometowns. Same with country names.

For most smaller towns and cities, default to the local pronunciation. An exception is if it has an official English name. The German city of Köln is known in English as Cologne; Göteborg in Sweden is Gothenburg. And be grateful the seat of the Dutch government and the World Court is known in English as The Hague. That's because the city's official Dutch name is 's-Gravenhage. And it's pronounced "SKHRAH-vehn-HAH-khuh."

---

* Pronounced "KOH-pehn-hay-gehn" in English. The third syllable is not pronounced "hah," as President Barack Obama famously flubbed it when he spoke before the U.N. General Assembly in 2010.

Leila Fadel on Pronouncing Names and Places

I think we've really come a long way. When I first started and I said my name correctly, and I said other Arabic names correctly, I remember somebody telling me, "It's like you're burping in the listener's ears." That's what they said to me! I'm trying to honor the person's name, and I find lots of people really appreciate it, including people who have grown up with those pronunciations. And I'm not going to change my name's pronunciation because that's as if I'm changing who I am.

I'm also going to say the names of political leaders correctly. If it's Abdel Fattah al-Sisi, I'm going to say that because it's not far off from the English pronunciation. You're not going to lose the listener. But I don't change the English-language pronunciation of a place. I'm not going to say "MUH-ser" for Egypt or "bugh-DAD" for Baghdad, although there are constant discussions on how some places should be pronounced in English. We used to say "PA-kih-stan," now we say "PAH-kih-stahn." Should we be saying "af-GA-nih-stan" or should we be saying "ahf-GAH-nih-stahn"?

I also find that, for example in Ukraine, I had difficulty with pronunciation. I had to practice. It was not a natural thing that flowed off my tongue. So I don't fault people who don't grow up with specific letters in their language and who cannot say them properly, and do the closest approximation.

I think the more that we have listeners hear different ways that people speak, and accents—both regional and international—you're opening up the world to them. I think that's all part of not trying to sound exactly the same and honoring and celebrating the diversity of who we're talking to, who our listeners are and who we are.

CHAPTER 12

Sound Design

When Barry Gordemer started working at *All Things Considered* in 1986, multiple pieces of equipment were needed to mix audio. "You would have the reporter's voice on one tape machine, the interview cuts—the actualities—would be on another tape machine, and the ambience, all the sound that would run underneath it, would be on another tape machine," he says.

Processing those elements—cutting them into different reels and then assembling the final mix—took time and skill. "When we switched from reel-to-reel to digital, the audio editing really didn't improve that much because people here were so ridiculously good as tape editors, with razor blades and wax pencils and rocking the reels back-and-forth," says Barry. What did change was that it was much easier to learn. Whereas it used to take months to become proficient, now, "if you're working at it and you understand the NPR sound, you can get pretty good at it in two or three weeks."

And there was so much more you could do: bring in as many sound sources as you want, make sound-over-sound montages, do multiple crossfades. So the mixes became much more complex. The result was an explosion of sound design and audio storytelling techniques, and all that came at a time when the desire for long-form creativity in podcasting was booming.

Producers at NPR have a lot of latitude for creativity when they cut interviews, mix pieces and make soundscapes for podcasts. But at the same time, there are several overriding principles they are expected to uphold:

CUT ETHICALLY. Deleting words spoken in an interview must always be done in the interest of the listener's time, to deliver a story told cogently and with journalistic rigor, and never to alter the speaker's intent.

MAINTAIN TRANSPARENCY. Sound mixed into news stories should be authentic and gathered in the process of reporting. Music used to score podcasts should help tell the story, but not to manipulate the listener's emotions. We don't use sound effects, except on rare occasions, usually in long-form audio, such as when they help re-create a moment in the past. It must be clear to the listener that the sound effect is not real. And it should be discussed with the editor or a newsroom manager.

SERVE THE STORY. Sound editing should be done smoothly and skillfully and should not draw attention to itself. Production devices should not be a distraction from the journalism we are presenting.

This chapter is about using sound, not about the technical facets of processing audio, of which there are many.

Mixing Pieces

The sound-rich story that has long been emblematic of NPR usually begins life as a bunch of discrete elements—the reporter's voice tracks, actualities, and bits of sound or music. They may be transmitted in one long audio feed—"acts, tracks, and ambi"—or as a series of separate digital files. In either case, it's often up to a producer to mix the piece—to put all the parts together in just the right way.

Many producers start by going through all the files to get a sense of what's there and listen for problems like plosives—breaths sometimes emitted with certain consonants that cause mic distortion—as well as mispronunciations and deviations from the script or from NPR standards. The earlier you flag these the better, especially if the reporter needs to address them before moving on to another assignment or if they're in a faraway time zone and about to turn in for the night.

The tape filed by NPR correspondents is almost always good quality. But sometimes audio comes in from other sources with distortion or recording problems that need to be fixed with software. Producers will typically send such audio to an engineer before they begin mixing. NPR's engineers are a great resource: They are very capable at fixing bad tape. Sometimes, however, there's not much they can do to improve it, or it has too many problems and fixing them all would make the tape sound strange, so most producers minimize the amount of processing they ask for.

After a producer has established the quality of the audio, they typically "lay up" the piece. That means putting those elements in sequence according to the script.

Argin Hutchins, NPR's audio production trainer, describes the process he followed as a producer. He would start by putting the reporter's voice file in the uppermost panel of the audio editing program. He'd take the actualities and place them in the second panel. And just below that would go the ambi. He'd then space out the elements so they followed the order of the script. But up to that point he wasn't doing any editing. "I don't fix audio, I don't level, I don't do anything. I just lay it up and just see how it comes together," Argin says.

Once the "rough cut" is laid up, the editing begins—with cuts, fades, leveling, filtering and more. While producers have these and other processing tools at their disposal, here are some of the main ways they make pieces sound good:

SEQUENCING

Many producers will tell you they focus on rhythm and timing, especially in sequencing audio elements. You have to decide how much space to put between a track and an act. I have yet to meet a producer who measures the gap in seconds or fractions of a second. Instead, as with so much of the craft, it's done by feel. As he listens, there's a sweet spot Argin's looking for: "It's not too fast, where I have to go back and think, 'Wait, what did they just say?' " And it's not too slow, making him think, "Oh my goodness, this is taking forever."

"It's all about pacing," says producer Claudette Habermann, who mixes pieces on NPR's intake desk. She has a trick for getting the spacing right, which she learned from Barry Gordemer when he trained her on *Morning Edition*. At the end of each track, she snaps her fingers, "and that's when the act comes in." She does the same after the act, in judging the gap to the next track. If there's an emotional act, however, or if someone is crying, she will "give it a second to breathe before the reporter comes in."

After a reporter's standup, it's important to have a beat or two before going back to the studio track. "You want to make a clear distinction between the field and studio tracks," Claudette says. "I usually find something in the background of the standup ambi to put between the standup and the studio track to make a smooth transition."

ADJUSTING LEVELS

Everyone talks at different volumes, and natural sounds also vary in their intensity. As a producer, it's your task to mix the entire piece at more or less the same level, so no one will have to adjust the volume as they listen.

Audio editing programs give you several ways of adjusting levels. The most commonly used one is a line drawn over the waveform. It represents the fader. You can move it to adjust the entire section, or set points on the line and adjust those points upward or downward as needed to get a smooth, consistent mix.

By converting sound into waveforms, digital editing software has made mixing something you can do in part visually. But while you want to be looking at the waveforms as you cut and adjust levels, your ears should be the ultimate arbiters of sound quality.

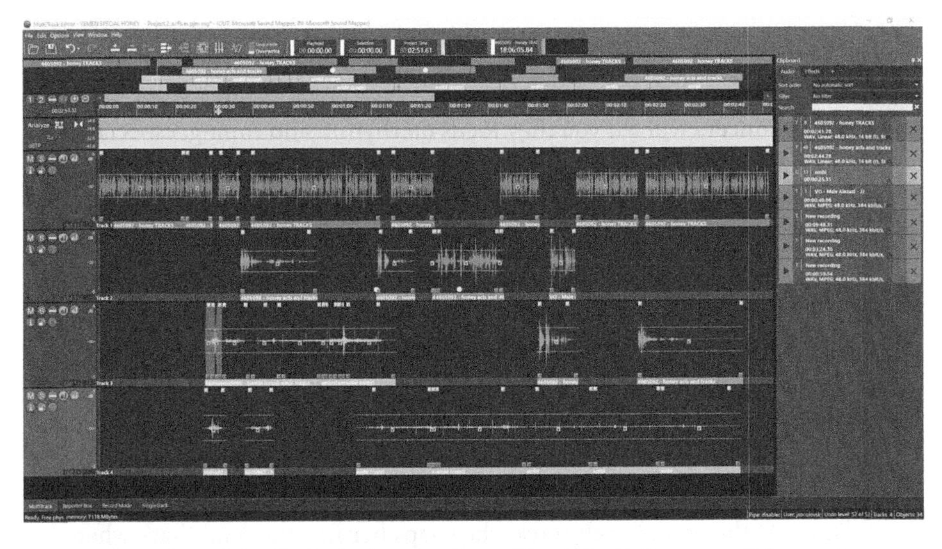

A screenshot of the final mix that producer Claudette Habermann did on a story by Fatma Tanis about honey production in Yemen. Claudette uses color coding while mixing; because this figure is reproduced in black and white, the colors are indicated parenthetically here. In the top panel, all the tracks are Fatma's narration (blue). In the second panel, from left to right, the first two cuts are foreign-language actualities (green); the third is ambi of Fatma tasting the honey and talking about it (pink); and the fourth is the English voiceover (pink) for another foreign actuality (green) that is directly below it in the third panel. Also in the third panel, the cut on the far left is ambi (orange) and the cut on the far right is another foreign-language act (green) with no voiceover. The entire bottom panel is ambi (orange). To find Fatma's story online, see note 1. Adobe product screenshot reprinted with permission from Adobe.

When Claudette makes an edit, she listens not only to the change she made, but scrolls back for up to a minute to check for consistency. And she closes her eyes. "Because sometimes, when you're looking and hearing, you're not sure which one is playing tricks on you."

And speaking of visual editing, many editing programs allow you to assign colors to different pieces of tape. Claudette says color coding helps her keep track of her elements as she mixes them using multiple panels in the program. She makes the reporter's narration blue, English acts and voiceovers pink, foreign language green and ambi usually orange.[1]

CLEANING UP

It's natural for people to stumble or hesitate a little when they talk. They might take half-breaths or make swallowing noises. When these sounds get in the way of a clean actuality, they need to be taken out. And when reporters record ambience for a piece, they will inevitably also pick up sounds that distract from the story. Those also need to be taken out or dampened.

"If we're talking about a new food that bunnies are eating," Argin says, giving a hypothetical scenario, "and there's sirens in the background, but we're not talking about sirens, you should remove those sounds." For a sound that can't be removed cleanly, you can lower the level on them "so it's not as obvious or intrusive."

He compares the work to a photographer adjusting the color of a picture, or cropping a shot. "We're doing that in sound. We're trying to remove distractions."

FADES

While adjusting levels and cleaning up speech is largely a technical task, mixing in ambience and music beds is to a great degree an editorial one. You are essentially adding content to the story and making decisions about how to use the sound. "When does it deserve a starring role, or when is it a bit player?" Barry says. As discussed in chapter 4, ambience can function as a post or a bed (and occasionally both). Something that works well as a post usually won't as a bed. "If it's just traffic sound? I'm not sure why that needs to be heard in the clear. But if it's traffic sound and at some moment somebody says, 'Hey, get out of the road!' that might work," he says.

Another decision is when to fade ambi in and when to fade it out. As mentioned in chapter 3, ambi is often used to set a scene or signal a transition from one location to another. "Listen to what the reporter is saying, look at their script," Barry says. He tells producers to note what the reporter is saying about where they are. "If they are referring to the location where that happened, that audio should stay there," and when they've moved on to a new location—"this is in the jungle, now we're in the city"—then it should be faded down, he says.

The change of location may be conceptual. The reporter is "in the jungle, talking about something, but then they say, 'But back in Washington, they're having a different debate . . .'" without physically going to the U.S. capital. "Conceptually," Barry says, "you've moved to a new location. It's time to fade away" from the jungle ambi.

Fades should be smooth and gradual. "When you are mixing audio, if the sound suddenly jumps in," says Barry, "or even worse, if the sound that's underneath a reporter's voice suddenly disappears very abruptly, it draws your attention to that production moment and away from the content."

GETTING WHAT YOU NEED FROM THE REPORTER

A good reporter's script will come with instructions for fades. Follow them, but let your ears ultimately decide. And if there's something missing, or if you can't make it sound good with the audio the reporter filed, don't be shy about contacting them. It's your responsibility as a producer to get what you need to make the best piece you can. If Claudette is missing an element such as room tone, she will call up the reporter and ask, "Do you have anything, even 10 seconds, you can send us? And nine times out of ten they say, 'Oh, yes, I can send you something!'" In that tenth time out of ten, it's likely they have a good reason for not having it, and at least you tried.

Most reporters will send you files with 30 seconds or more of ambi, of which you may only need to post a few seconds in the clear. Don't just randomly grab a piece of it. Claudette will listen to each file in its entirety and identify the best sound. "Maybe they are at a shopping mall or supermarket and they give you this piece of ambi, and towards the end you hear somebody pushing a shopping cart. Then I would grab that."

Another reason to contact the reporter is if they've filed an actuality in a foreign language and it's not clear where to bring it up. The post should correspond either to the first words of the voiceover or translation, or to the last words of the reporter's previous track. As I noted in chapter 11, tens of millions of Americans speak languages other than English, and we come across as sloppy if what they hear doesn't match the translation.

Here's how I gave instructions to the producer who mixed a story I did about fado singers in Lisbon:[2]

JEROME: Jose Matoso says most of what people hear abroad is not the real thing.

(*Act—Matoso: bring up at "No ensayado, todo spontaneo . . ." then fade*)

JEROME: The real fado is not rehearsed, he says. It's spontaneous. None of this business of, "Let's practice in the afternoon for a performance in the evening."

My spelling in Portuguese may have been incorrect, but it doesn't matter. The point was to help the producer identify the post, and they did it without having to call me up.

MIXING WITH EDITORIAL INTENT

Ideally, reporters should use sound to help tell their stories, not as a kind audio frill or embellishment. The best sound replaces a track or actuality—we hear the funeral dirge for the elder statesman who just died, or the bulldozers removing the rubble of the home destroyed in the hurricane. And because we hear these things, the reporter doesn't have to describe the scenes in the same way, or with the same details. Our ears will fill in the gaps in the narrative; the sound helps us see the funeral and the lost home.

Audio editing, Barry says, is "not just about sounding pretty. It's about mixing and using those elements with purpose." Fading in or posting ambi, shouldn't be done just because it sounds good or because that's what the reporter said to do, but because it makes sense in the context of the meaning of the piece. "If your attention suddenly becomes focused on a production device," Barry says, "you've pulled the listener away from the editorial content. That's a problem."

That principle also applies to how much ambience you mix in. Producing pieces that are rich in sound is desirable, but only when it all helps tell the story, not overwhelm it. "Ambience is like seasoning on the steak," Barry says. "If you get too much A1 or too much salt on it, all of a sudden you're not appreciating the steak, and you're noticing the sauce."

Cutting Interviews

Although many interviews are done live, it's not uncommon for one to be recorded (or "prerecorded"—which is not quite as odd a term as it may seem, since the "live" show is itself recorded). *Morning Edition* is fed to stations at 5 a.m. ET (2 a.m. PT), and it's difficult to find many potential guests other than reporters who are awake and lucid at that hour. Even the shows whose broadcast schedules are more in sync with people's biorhythms record many of their interviews. Those interviews will be edited, sometimes rather dramatically, before they go on the air. So producers need to develop the technical and editorial skills required to cut, say, a 25-minute conversation down to four minutes—sometimes under intense deadline pressure.

At NPR, an editor will usually be on hand when an interview takes place and will make suggestions about how the audio should be cut. But as a producer, you should understand that you are an editor too. Your judgment is an indispensable part of the process of deciding which parts of the interview—which ideas—are essential, and your skills will be called on to package those ideas so that the final product is coherent, gripping, even beautiful.

Cutting interviews well is an art, one that is all but invisible to the audience. If you take a block of rough marble and chisel it down until you've made a polished bust of Plato, people will marvel at your creativity and skill. But if you take a wide-ranging, forty-minute conversation on the federal government's financial priorities and cut it down to five coherent minutes neatly focused on tax cuts, no listener will be the wiser. "You're working to make it sound like you did nothing," says Barry. But if you make even a small mistake, your work will be noticed, and for the wrong reason.

THE PROCESS

After taping, the producer, editor and host usually huddle and discuss what to leave in and what to take out. For obvious cuts, Barry might tell the producer to "just take a whack at it. That taped for six minutes, bring it down to four minutes." Other times, he says, the cuts may not be so clear. "That was a mind-blowing conversation that lasted 30 minutes and we have to bring it down to five," and a lively editorial discussion will ensue. Whatever the case, you'll save yourself a lot of work if you have general agreement on what should be the final mix before you start cutting.

If it's a long interview, many producers start by cutting questions and answers that were dull, redundant, off-topic, or confusing. Whether you remove entire questions and answers, or make cuts within them, always think about the ethics of the cuts. Newspaper quotes have ellipses to show that words have been removed, while a TV news interview can do the same with a dissolve or a jump cut.* But radio and podcast interviews don't have those options. So be very careful that you don't change the meaning of what someone said. Former NPR senior editor Sara Sarasohn says the producer has a duty to respect the speaker's intent. "You have to be able to think, 'What was the host getting at with this question, and how can I make the answer be that?' And you also have to think, 'What did the guest really want to say in this answer, and how can I cut it to make it really reflect that?' If you keep those things in mind, it takes care of a lot of other ethical problems."

As their edits bring the interview piece closer to the desired length, many producers will make a rough cut and share it with the editor before doing the final trims and cleaning it up. *Code Switch* producer Christina Cala takes a different approach, cleaning up the full interview first before making cuts. For her, it's easier to judge all the parts of the conversation when the stumbles and other distractions are removed. And she won't lose something of value by cutting it prematurely. "I've had multiple times, editors come back to me and say, 'I thought that section was going to drag too long because it felt like it did in the taping.'"

That way of working grew out of all the same-day interviews she produced at *All Things Considered*. "Coming from daily news, I needed to make sure my first pass was good because I didn't know if I was going to have a second or third pass," Christina says. "It was always a balance between getting something cut that was to time and getting something that could air right then if it really, really needed to."

As you trim, it's tempting to shave off everything that's not verbal. But sometimes you'll want to leave in a silence to signify a genuine hesitation to answer a host's question. At the same time, little pauses and a certain amount of stumbles can make a conversation sound more real. *All Things Considered* producer Jonaki Mehta always tries to preserve the natural ways both the host and guest speak. "I try to listen for how often they

* It's worth noting here that many video producers conceal edits with "cutaways" of footage that is not from the interview.

naturally stumble. And I'll try to keep at least a third of that most of the time, unless it's distracting."

If you remove too much, it can sound stilted. "Even though our listeners don't have the ear training to know that something's been edited," Barry says, "they can sense something doesn't sound right there. Something sounds artificial about it."

"Breaths are your friends," Barry adds. They are easy places to cut from, since they often come just before the start of a new sentence. In fact, you may find an edit needs a breath. "Sometimes we insert a breath from one place to another to smooth out a cut," says Jonaki.

EYES AND EARS

As with pieces, a lot of editing can be done by sight. "When you've been looking at the audio waveform over time," says Barry, "you can actually recognize with your eyes what certain words are. You can certainly pinpoint what an 'um' or an 'ah' looks like with your eyes, but you can also see breaks and separations."

Visual editing can boost your proficiency, perhaps even more so in cutting interviews than mixing pieces. "The more you get practiced in whatever software you're using, the more you're able to rely on visual aids to speed up working with the audio," says Jonaki. She can sometimes visually identify an entire sentence, such as when a host is asking the same question several times as a follow-up.

And just as with mixing pieces, you should always go back and listen to your edits. Christina says, "I don't just go back one second, but go back to the beginning of the answer when you're making a cut to make sure it all makes sense."

Interviews are usually recorded in split-track stereo, with the host's and guest's voices recorded separately on the left and right channels. That way each participant can be isolated in postproduction. Jonaki says the two tracks can sound very different "in this day and age, where you hear a host in a really clean, sterile-sounding studio" and a guest who is in a remote location with background noise coming down the line. "I do a lot of extra work to make sure that there are fades each time there's a transition from question to answer, to make it sound as seamless as possible. Not to trick people into thinking they're in the same room, but to not make it distracting."

Editing interviews can be nerve-racking, especially if you work on a daily newsmagazine where you may have only an hour or so to do your work. But it can also be extremely satisfying. A good interview never goes on too long; it amuses us, or informs us or touches us; and like any good story, it has a beginning, middle and end. Veteran producer Neva Grant says that in a way, a successful two-way is like a great conversation in an elevator—one that ends just as one person steps out. "If it works well, there's a line, and a laugh, and then the elevator door closes."

MUSIC INTERVIEWS

Show interviews usually require straightforward cutting. There are usually no sounds other than the voices in the conversation. There's one major exception, though: the music interview. There, you will be fading music in and out of the conversation, and you should be "using the rhythm and the sound to your advantage," says Christina. She often posts a bit of music in the middle of a sentence so that the listener can hear a part of the song that the artist is talking about.

Music functions pretty much as ambience, and Barry's question of whether it deserves a starring role applies here too. "Music can get a starring moment," he says, "but even within the moment, don't just bring up in the middle of the music, come up on an interesting lyric or come up on an interesting beat or moment in the song. Be deliberate."

Sometimes the pace of the conversation doesn't match the music. "Someone's talking really, really, really fast. The song is really, really, really slow. It's confusing." By breaking up a few sentences, the music posts serve to slow the conversation down.

The success of the interview will depend mostly on the questions that the host asks. But producers can play a big role in making the conversation memorable and esthetically pleasing. Many producers work with hosts ahead of time to sketch out the structure of the interview. Often the "peg" for the conversation is a new release, but that doesn't mean the interview has to be tied closely to the album. Rather than go through a laundry list of songs, you want to research the musician's biography and previous work, and come up with questions that might lead the guest to say something about a certain song.

Then the producer will queue up the songs to be ready to play them for the guest, or the host, to react to in the moment. Sometimes one of them might even sing along. "If you're playing the music during the conversation," Christina says, "often what's happening is, they're listening to [it] and it's like you're in a movie, like they're reacting to the scoring of the movie of their life."

But be intentional when you mix. Just because it's a music interview, doesn't mean the music should be heard throughout. Fade it in and bring it up when it complements the discussion. Fade it out when it doesn't. And as always with music or any sound that may be subject to copyright, make sure you have the rights to use that music in your piece or interview, or that it is covered by "fair use," a legal doctrine that allows copyrighted work to be used in the interest of freedom of expression. Familiarize yourself with the principles of fair use, and when in doubt, consult your organization's legal counsel or, if there isn't one, an outside lawyer.

Using Music in Podcasts

The digital revolution has not only made audio editing easier, it has also enabled complex mixes in which music is added to enhance the storytelling. "Music just helps so much when you're trying to do something longer," says podcast producer Brent Baughman. "So much of doing long narrative stuff is understanding music—understanding how not to overuse music and understanding how to use it to create scene changes."

To be clear, we're no longer talking about interviews with musicians or stories about them, in which music referred to in the conversation is added. Now we're talking about music that has no direct connection to the story or people in it but is used to create a soundscape that helps the listener follow the narrative.

Ramtin Arablouei was a musician before he became a journalist. He was a drummer and composer for the band Drop Electric, which got its break when Bob Boilen, creator and host of NPR's *All Songs Considered* and *Tiny Desk Concerts*, went to see one of the band's performances and mentioned it on his show. A couple years later Ramtin joined NPR as a producer for *How I Built This*, and then went on to become co-host and producer of *Throughline*, together with Rund Abdelfatah. He says there are four main ways to use music in a podcast.

TO COMPLEMENT OR HIGHLIGHT THE RHYTHM OF A STORY. "If you have an upbeat story and you want to make the audience feel like you're flying through a hectic moment in the story, you might use an arpeggiated piano or an arpeggiated string sound that would kind of create a sense of speed and urgency. In cases where you really want someone to slow down and pay attention to the words someone is saying, what you may do is create little bits of space between phrases, even just a split second, and use music to fill those spaces. You might use a piano playing a particular chord, and then have a five-second pause and another chord. That technique is used when you have a sad moment or a concentrated one where you want to slow things down and get a certain kind of mood created there."

TO PUNCTUATE OR MAKE A TRANSITION. "Generally in our work, you're either explaining concepts or ideas or you're telling stories. You often move from idea to idea or from story to story, or from idea to story or vice versa. Music can be a tool so you don't have to say, 'We've ended this idea and we're heading to another one.' For example, let's say you're telling a story and something sad is happening, and you don't have music because you don't want to overdo it. But once you're out of that and you're heading into a new kind of space, you can use music to signal to the listener: 'I'm now heading into a different thing, a different area. I'm out of the sad story.'"

TO TRANSPORT THE LISTENER TO A TIME OR PLACE. "Music can allow you to put someone in a particular place. However, you don't want to be too literal, too on-the-nose. And obviously, you use this sparingly. A few years ago, I scored a series about artificial intelligence and government surveillance. A lot of the stories were about algorithms and software and programming. And so we wanted the music to feel like you were in a computer, to give you that feeling that things are happening in a very complex, digital way. So all the music I wrote was synth heavy. And I played analog synthesizers as a way to achieve a soundscape that puts you in a particular place."

TO PUT THE LISTENER IN A PARTICULAR MOOD. "This is the most controversial of the four. But you can use music in such a way that creates a certain emotional space that's not changing the listener's mood, but it's creating a kind of undertone for the story to sit on top of. But if you're starting to feel that it's being manipulative, err on the side of caution, switch out to another piece of music or just don't have music. Your piece will survive. The only thing your piece can't survive without is a story."

NPR has one set of standards that applies to all our journalism. "We may well take differing approaches to storytelling in news and podcasts, but in no case do we manipulate the audience," says Tony Cavin, NPR's standards and practices editor. So while podcasts may benefit from musical scoring to help the listener follow the story, it should never be done in a way that's aimed at influencing the listener's point of view.

Ramtin believes music should play a supportive role in storytelling, and that it should always be there for a purpose. "A mistake a lot of young producers make is, they want their thing to pop. They want it to sound like *Throughline* or like [the WNYC podcast] *Radiolab*. And so they overcompensate often by including too much music." But if you listen to those podcasts, he adds, "you'll notice there are large swaths without music. The music is used strategically and with restraint. And that's when you know you're getting good, is when you're able to have the confidence to hold back."

Ramtin Arablouei on Composing Podcast Music

The great challenge of being a composer for film or for radio or any kind of storytelling is that you have to see your work as a part of a whole. If you listen to a Beatles song, it's really hard to pick out one reason why the song is good. They wrote their songs in such a way that the drums fit with the bass, and the guitar and piano, and it all worked together to create a complete thing. If I respect that truth, then it's much easier to swallow the idea that what I'm doing is not the main character of this thing.

One way to look at it is that you are trying to do something that's boring. But another way to think of it is you're trying to make something that's incomplete. Michael May, who used to work at NPR, said you want to make something that doesn't quite finish itself as a piece of music. And when you're listening to it, and you're like, "I can hear someone singing over this," or "I can hear a melody over this," you're going in the right direction. As a composer, what's really hard to do is to hold back from that last step. It's a skill that you develop over time.

Sound Design

At NPR, we generally don't score news stories, though in some cases, such as when a podcast episode is adapted for a show, you might hear a piece with music. But scoring has been embraced by a number of NPR's podcasts. *Life Kit*, *Throughline* and *Planet Money* as well as other productions have allowed and even encouraged producers to experiment with soundscapes that include music and other types of audio not gathered as part of the reporting.

That has led to a raft of ethical questions. As I alluded to above, it's one thing to mix in music by an artist being interviewed about their work. It's another thing to score the story with music that a producer has created or pulled from a library. And then it's a further step to go from that to using sound effects. Whereas it's usually pretty obvious that the kind of subtle, incomplete musical compositions that Ramtin described above are a soundtrack and not something happening in the background, that's not always the case with audio that is manufactured or taken from a sound effects library. Certainly, in the early days of radio theater—think *The Lone Ranger* and *Gunsmoke* before they were TV shows—and even now with the whole gamut of Foley effects, the reproduced sounds that go into the making of Hollywood movies—the point is to fool the audience into thinking fake sounds are real.* Can sound effects be used journalistically in a way that's ethical and transparent? This is a question you should always talk to your editor or newsroom manager about, but what follows is a discussion of some of the ways it can be used and the ethical parameters to keep in mind.

HELPING TO TELL THE STORY

Fundamentally, sound design is about arranging sounds in a way that helps take listeners to a place or create a mood so that they can *feel* the story. We use sound design regularly in audio news stories, though we may not call it that. We do it often to compress time. If you put the moose's

* *The Big Broadcast*, on NPR Member station WAMU, airs *The Lone Ranger*, *Gunsmoke* and other programs from the golden age of radio every Sunday night. "I often find myself listening to them for fun," producer Adelina Lancianese says, "but also as a reminder of what sounds good and what sounds painfully fake in the kind of work we still do today."

howling right after the hiker's footsteps go quiet, even though in reality there was a gap of 15 seconds, that's designing sound. So is the butt cut discussed in chapter 9, in which actualities are played back-to-back. The two people didn't really say those things consecutively, but it's an effective way of contrasting points of view, and listeners understand why they're hearing it that way.

In podcasting, sound design gets even more creative with edits such as these:

- Overlapping voices
- Panning a clip left and right
- Slowing or speeding up a clip
- Making elements louder or softer than usual
- Changing the tone to make it sound as though it's coming through a phone or speaker

Changes like these liven up the storytelling and are often a better way to make a point than saying it literally. For example, Adelina Lancianese produced a story for *Invisibilia* about a man who spent years emailing a friend who had ghosted him. She made a sound collage of him reading from the emails. It was an effective storytelling device because it conveyed the extent of his unrequited efforts to reconnect with that friend, without the story having to lay that out with additional copy. "And to me," she says, "that was totally acceptable because it's clear to the listener what it is."

Indeed, it often serves the listener to have these devices, so long as no one is being deceived. Adelina says she follows two core principles when designing soundscapes: transparency and editorial purpose. The audience must always be clear about what's real and what's not. And the sound should be designed to serve the story, not added just to be "cool for cool's sake."

In another story she produced, this one for *Rough Translation*, Adelina used panning as a way to suggest motion. The episode was about the Lindy Hop, a dance that originated in Harlem and also became popular in Sweden (you might remember the discussion about this episode in chapter 9). Adelina had tape of two dancers recounting what was going through their heads the first time they did the Lindy Hop together. "And I tried to make it sound like movement. He's on your right side and she's on your left side, and you hear them recounting, I was feeling this way. I was feeling that way. I was feeling this way."

So far so good. The listener knows that both characters aren't dancing together as they speak, even though their voices are mixed with the music, but recalling a dance. But Adelina says she went a step too far when she experimented with applause from a sound library in an early draft. She and her colleagues at *Rough Translation* felt that putting the applause there might lead listeners to believe it was really taken from the scene. So they removed it.

In general, Adelina sees a first pass of sound design as putting together an outfit: "Put on the chunky necklace and the cool hat and the glittery belt." But then she follows Coco Chanel's adage: "Before you leave the house, look in the mirror and take one thing off."

> Hearing the scene stuffed to the brim with all the possibilities allows me and my colleagues to figure out which elements we care about most, what's tasteful and what's not, and which aspects are ethical, and which are not. So that in subsequent passes through that sound design, I can keep shedding layers and elements until the final product is its most lean, its most elegant, its most truthful version of itself.

OK, SO WHAT ABOUT SOUND EFFECTS?

In the case Adelina described, the lines between real and fake were blurred when a sound effect was pulled in. But was it problematic because the effect wasn't authentic to the scene, or because listeners might not know that it was added?

Let's consider two other cases, one where an effect was used, and another where it was rejected. In 2020, a few days after the murder of George Floyd, Christina Cala produced a piece for the *TED Radio Hour* featuring writer and poet Clint Smith.[3] As he read from his essay "My Hopes, Dreams, Fears for My Future Black Son," he recalled riding a bike through his neighborhood as a child. In addition to the score, Christina pulled in some street sounds to create the feeling of being there and "put us in this space in his brain." She says it was obvious the sounds weren't real because Smith was reading from an essay.

In another episode, where Christina was the editor—and editors play a key role in judgment calls on sound design—she thought it was not appropriate to add a sound effect.[4] The story, for *Code Switch*, was about people with albinism, and one of them shared a recollection of one day racing down the driveway to his mailbox to collect a parcel that contained information on the condition. Rather than add footsteps, she just "sat with the moment," she says. "I didn't feel we needed the extra sounds."

"No one in public radio argues that it's ethical to deceive the listener," reads the *NPR Ethics Handbook*, quoting the previous edition of this book.[5]

> What people are constantly trying to define is when deception occurs. After all, the production process necessarily involves a certain amount of manipulation of audio, whether it's simply picking the actualities out of a raw interview or fading the sound of a farmer's combine under a reporter's voice track. Our art depends on a certain amount of artifice. So how much is too much? Does every ambience bed suggest that the reporter is really on site, and not in the studio? Should a host always make clear to the audience when an interview has been recorded? If a live interview is rebroadcast on a "rollover" of a program, should it be preceded by an announcement that it was previously recorded? Should the entire show start with such an announcement?

Questions like these, the previous edition of this book went on to say, "can feel like the debate over how many angels can dance on the head of a pin. But whether you are a producer, reporter, editor, or host, it's worthwhile at least to discuss these issues, and to try to come to some agreement with your colleagues about which production techniques might be off-limits."

A lot has changed in radio and podcasting since former NPR trainer and *All Things Considered* executive producer Jonathan Kern wrote those words. But the principles, and the consideration we give them, remain just as important.

CHAPTER 13

Multiplatform Reporting

If you wanted to be a reporter back when I started out, in the late 1980s, you were essentially limited to one platform: TV, radio or print. The newspaper reporter wrote for the reader, the radio reporter for the listener and the TV reporter for the viewer. A broadcast journalist might write a script, but only their producers and editors would see it. Of course, some crossed over—newspaper reporters could give radio or TV interviews, a nightly news anchor might write a column or a book. But these were not the norm.

News consumers, as they are now sometimes called, might have encountered a mix of media during the course of their day, but their experiences were for the most part discrete. The car or kitchen radio could not show images, and the TV had no text; the newspaper on the breakfast table may have had pictures, but they were static and—for most of newspaper history—black and white.

Now, to quote a Hollywood hit, it's everything everywhere all at once. If you want news, you can open an app and become a reader, click on a video in that app and instantly become a viewer, then switch to a podcasting app (or perhaps not even that) and, *poof*, you're a listener.

Journalists are under pressure to be platform agnostic. Problem is, the way the stories are told is as different as it was a generation ago. You can't just take a radio or TV script and post it as an online news story, or just take a text-based story and play it as a podcast. Radio and TV scripts don't translate perfectly either. Readers, listeners and viewers have different needs and experience stories differently, and it's up to us to tell those stories in a way that works best on each platform.

"Journalism is still journalism," Pat Wood, senior producer handling digital stories for *All Things Considered*, reminded me when I spoke to him for this chapter. That's true: The fundamentals are the same—robust story ideas, thoughtful interview questions, shoe-leather reporting—whether you're wielding a notebook, microphone or camera. And yet, as you will hear from Pat and other NPR journalists I talked to for this chapter, the building blocks and the writing are different. The more you develop a sophisticated appreciation of the differences and are smart and efficient about "feeding" the platforms differently, the better a journalist you'll be, and the easier your job will become.

When NPR started putting stories online years ago, we would often just "webify" them. That usually meant little more than converting actualities to quotes by adding "so-and-so said," and adding punctuation to the reporter tracks. " 'Oh, I just put periods at the end of my sentences and put it online, and then it's a digital story.' Nope!" says Rachel Martin, adding that the text was unreadable—and if it was readable, that meant "there's a problem with your radio script."

The good news at NPR is that we are learning how to work across platforms rather well. And while this book is about how to tell stories with sound, as its title proclaims, it would be woefully incomplete without a discussion of how to present those stories in other formats.

First, the Math

I studied engineering in college and still think about words and stories in mathematical terms. So when NPR and its member stations made a push to put more reporting online, and I was asked, together with NPR digital journalism trainer Holly Morris, to do a series of sessions on multiplatform reporting, I began with a little calculation.

I gave everyone the lede of a *Washington Post* story and asked them to time themselves reading it. Most were in the range of 200-300 words a minute, which is about the average silent reading speed of adults in English.[1] Then I showed them *Up First*'s intro for the same story and timed how quickly hosts A Martínez and Rachel Martin read it. They were at 170 words per minute.

That makes sense, because a reader processes information more quickly than a listener. So I told people to imagine reading the *Up First* script silently at 170 words per minute. It felt way too slow. And even at 200-300 words per minute, it still felt as though a lot of information was missing. Why?

FACT DENSITY

To answer that question, I took the *Up First* text and underlined each group of words that expressed a discrete fact. This is what it looked like:

> After decades of inaction—Congress could soon take action on gun safety legislation.
>
> On Sunday, a bipartisan group of senators announced that they had reached a deal on a bill that has the support of 10 Republicans—which of course is the critical number needed to get any major bill through the 50-50 Democratic-controlled Senate.
>
> So, this is actually not a done deal yet, though President Biden praised the early agreement as quote "important steps in the right direction."

Counting up the underlined words, I get a total of 17 facts in this text. Taking the total of 81 words here and dividing by 17 gives an average of 4.8 words per fact. I'm going to call this number the fact density. And what it says is that with about every five words in this audio piece you get another fact.

I did the same for the *Washington Post* story:

> A bipartisan group of senators announced Sunday that it had reached a tentative agreement on legislation that would pair modest new gun restrictions with significant new mental health and school security investments—a deal that could put Congress on a path to enacting the most significant national response in decades to acts of mass gun violence.
>
> Twenty senators—10 Democrats and 10 Republicans—signed a statement announcing the framework deal. The move indicated that the agreement could have enough GOP support to defeat a filibuster, the Senate supermajority rule that has impeded previous gun legislation.

Here I counted 31 separate facts. Since there are 95 words in this text, the fact density comes out to 3.1. So, about every three words you get a new fact. This text is denser in facts.

Of course, I underlined the facts as I understood them; someone else might get slightly different numbers. But I'd be willing to bet they would also show the newspaper text to be denser than the audio script.

That tells me two things:

- Audio uses more words to convey fewer facts.
- A digital piece uses more facts to tell a story, and the wording is more compact.

These back-of-the-envelope calculations are not to be taken as scientific proof. But they do explain why, when we do two versions of the same story, we need to flesh out the digital version with more details. I heard this from many NPR reporters I talked to. "In order to tell the story on the radio, you have to move pretty slowly through the concepts," says Rebecca Hersher. Online, though, "if somebody is trying to understand the thing, they just read the sentence again."

MORE DETAIL AND MORE SOURCES

There are two key distinctions between audio and digital journalism, which also apply to video. If you put too many facts and details in an audio or video story, the listener has to make an effort to keep up with it all. That's why a lot of the work of writing a broadcast piece is shaving away details that may be interesting and support the story's thesis but clutter the narrative. In digital by contrast, having more facts and details shows that you've done your reporting, and the reader can take it in at their own pace.

The same applies to sources. Having too many voices in a radio or video story makes it hard to keep track of them all, and the listener is likely to become confused. By contrast, in digital, the more people you quote, the more solid the story becomes. And the reader can always go back if they don't remember who someone is. You can see how this works in the table below.

	AUDIO AND VIDEO	DIGITAL
More facts, details and sources	Makes information harder to keep track of; risks confusing the listener	Makes story complete and more authoritative
Fewer facts, details and sources	Enhances clarity; makes narrative easier to follow	Makes story feel underreported; risks leaving the reader unsatisfied

This isn't to suggest that audio is less factual; it's just as factual, but uses fewer facts. Another way of thinking about this is that audio is a more focused way of telling a story. But if you're also going to do a digital version, which at NPR is almost always the case, you will need to go into it with a plan to do additional reporting. And chances are that will benefit the audio story as well.

How NPR Journalists Work in Different Platforms

Although audio and digital are NPR's biggest platforms, video is becoming increasingly important (NPR's Visual Newscast gets more than a million monthly downloads). Sometimes reporters and producers also have to file visual elements, and that comes with its own set of challenges in how the information is gathered. You may not be able to take photos or record videos while gathering audio, so maybe you'll stay longer to do it. And telling a story with images may require a storyboard that lists the sequence of shots. Sometimes, NPR will send a photographer to focus on the visual storytelling.

Here's how some NPR journalists say they juggle the demands of working on all three platforms.

REBECCA HERSHER, REPORTER, CLIMATE DESK

Gaming Out the Story

It's very rare that I do a radio story that doesn't have a digital version, but sometimes I do digital stories that don't have an audio version. That's because the shows have limited space, while the internet is endless. But there is an expectation that we as reporters will make the most of any tape we have, that we'll feed as many mouths as possible.

Rather than trying to game it out, by saying I'm definitely going to feed this many mouths with this project, I just tell my editor, "I'm working on this story." She knows it could produce anywhere from one to five versions. Once I'm done reporting, I come to her and say, "I'm seeing a 500-word digital story. I'm seeing two radio stories. I'm going to pitch it to *Short Wave*. I'm going to do a newscast spot. I don't see Instagram for this one." Or I come and I say, "The tape really isn't that good, so I think it's 2,000 words for digital and a two-way." And I find that she really trusts me.

Sourcing Digital Stories

In a stories with similar length on the same topic, you cover more ground in digital. If you just covered the ground that was in the radio version, it would be really short. And your editor would look at it and say, "Why is it so short? Flesh it out." And that requires more sources.

I just assume I'm going to be doing digital for everything I do. That way I'm never caught flat-footed, with not enough reporting. So, when I'm making that list of sources, unless I know it's a very simple news piece that won't have more than 500 words on the website, I'm making a long enough list of sources for a digital treatment. Even if that's two more experts than I need for the radio version. And it makes my radio better because then I have three experts to choose from instead of one.

Which One Takes More Work?

It's a lot more work to write a 1,200- to 2,000-word digital story than it is to do a five-minute audio piece. And that is where 90% of my work falls. I try to get as high-quality audio as possible for everything I do. If the digital deadline is tight because it's pegged to the news, which happens often—the story may be wanted for the website the next morning—I try to get high-quality audio with the understanding that this might also go on the radio or podcast. But in terms of the questions or the actual reporting, there's not really a big difference for a weekly feature.

With bigger stories, there is a difference. You can construct a digital scene without ever going anywhere a lot easier than you can construct an audio scene without going somewhere. Also, I do a fair amount of stuff that touches on events that have already happened, where I get archival tape for the radio piece. But for digital, archival tape is useless. You're not going to use the sound of water from last year on the internet. And it's pretty time-consuming. Part of reporting an audio story is finding that stuff, figuring out what it is and where it is, and getting the permissions sorted out.

Doing Instagram Stories

We do Instagram versions of our stories for many, if not most, of our features. It's an important way for us to get our work in front of a different group of people. Ryan Kellman is our photo editor on the climate team, and either we'll be traveling together or he'll be working with me on archival imagery. The Instagram story is usually really different from the digital. It's anywhere from five to 25 slides; the text is really tight, one or two sentences per slide, and some slides don't get any text. It's interesting because it has audio sensibilities in the storytelling. It's always linear. The number of words is lower. You can cover fewer concepts, but it's fundamentally visual. If there's beautiful imagery or video available, that's what pulls people into it. And then they stay for the story. You can get 10,000 extra eyes on your story. And they're younger, more racially diverse and more geographically diverse.

Making a Multiplatform Series

We did a series on public parks that had radically different treatments in different media. It had a photo-heavy digital story. It had two radio stories around four minutes each. It also had a 15-minute podcast episode that was largely unnarrated, and instead of being about public parks around the country, it was about 24 hours in one park.[2] There were two Instagram stories, one that had images from around the country and one that focused on Phoenix, Arizona.

The 24-hours-in-a-park episode was really cool, and everyone loved it. It was a pitch that immediately got a yes. When we got to the finish line, there was a feeling that everything should be on every format, but we had to say, "Because we love this, the answer is no." If we add photos at this point, it will take away from what we're trying to do in the audio. We had three people (a producer as well), which was barely enough to make what we were trying to make. So there were some hard conversations about sticking to the vision of feeding each medium, but not taking ideas that were meant for one medium and adapting them for another.

Giving Credit

And the last thing that was challenging, but really fun, was deciding who gets credit for what, and in what order. We talked explicitly about it, and it worked out well. Ryan was the primary author on Instagram and on digital. I was the primary author on the radio. On the *Short Wave* episode, we brought in a producer, Margaret Cirino, who had never been in the field, and she not only recorded with us, she co-reported it and her voice was in the episode. So we had three bylines with no primary byline. It was really fun to take a project that had many forms and say everyone's working on everything, but different people are the leaders on different things.

SHANNON BOND, CORRESPONDENT, DEMOCRACY AND DISINFORMATION TEAM

Saving Outtakes for Digital

After a decade as a print reporter, my instinct is writing for print. Sometimes, my first crack at a radio piece actually works really well in a digital story, and it's still worthwhile to hold on to. So I'll have a Google Doc open and add outtakes to it. And those outtakes can be really helpful later.

Using the Transcript

One of the things I had to learn early on was to listen for different things in an interview for your radio piece than you would for your print piece. If it's going to be an actuality in a radio story, the way the person says it matters, the emotion in their voice, the expression. But that might not translate onto the page, so I often find myself going back to my interview transcript. If that was the best way they said it for the radio, maybe they also said it a little bit differently that will hit a bit better on paper. And maybe I can write into it a little bit differently and use a slightly different quote. Or maybe I'll use a more extended version of what they said, to let things breathe or pack a slightly more complicated idea in text.

SARAH MCCAMMON, NATIONAL POLITICAL CORRESPONDENT, WASHINGTON DESK, AND CO-HOST, *NPR POLITICS PODCAST*

Prioritizing Platforms

For me, our primary platforms are the web, the radio and our podcasts. What I'm usually thinking, both in terms of how to tell the story and what the story is like, is whether it's a quick hit. Is this just something that needs to be spotted for our newscast—a 40-second update that can just live and die, be heard by millions of people and then be done? Or is this something that I need to file as a newscast spot and then take that copy and flesh it out a little bit and make a short web post? When it comes to longer pieces, features for the newsmagazines, I've experimented with occasionally writing the digital first and or just in tandem with the radio piece.

The other day I literally had both open at the same time, and I was going back-and-forth, and I found that to be really efficient. Occasionally I've written a digital story first and then cut it down for radio. And it seems like the radio write goes faster than it would otherwise. I don't think there's any one right way to do it, but if I feel something is important enough to be on a newsmagazine, I always think about whether there's a longer treatment that I could put on the web. Even a two-way can be written as a piece. And I like that, because it gives the story multiple lives, so your reporting goes further.

How Digital Can Help Your Source Relationships

If there's something I can't fit into a radio version, often I try to get it into the web piece. Sometimes you interview five people, and you only have room for three voices. I'm not saying that we should ever quote somebody just to quote them; I never put someone in if I feel it's redundant or immaterial. But you can't use every valuable piece of information that you have in every format. And sometimes you do have more good information you wish you could fit in. Having spent that person's time and yours, it's nice to be able to convey that reporting to the reader since you can't get it on the air. It makes a better digital story. And it helps the source relationship to be able to say, "I'm sorry, I have three and a half minutes, but you're quoted in the piece online." I do that fairly often. And I love the flexibility of the web for that reason.

FRANK LANGFITT, CORRESPONDENT, INVESTIGATIONS DESK

Doing Short-Form Video

I just think it's a natural part of what we should be doing, if we've got these tools and we can reach a different kind of audience. The tools are very, very creative and a great way to tell stories. When I'm on assignment, I'm thinking, "What's a really good visual?" Just like if you're in an audio scene, you're thinking, "What's the sound that really brings the scene to life to your ears?" I'm thinking visually, "What brings the scene to life in your eyes?"

What became very clear in the first few months of the war in Ukraine is that there were two countries. There was a country that wasn't near the front, that was occasionally hit with a barrage of missiles but otherwise was mostly normal. How do you convey this to people? Because you'd be going to Ukraine and people would say, "Oh, it's so dangerous." Well, it depends on where you are.

So I decided to shoot a video of a big boulevard right outside the hotel where we were staying. There are rusting tank traps along the median; it's rush hour, and there are tons of people driving by. And except for the tank traps, it looks pretty normal. So I did a standup, and I shot video of the grocery store, which was jammed. It actually had more products than most grocery stores in London where I used to go. And then the second scene is in a bombed-out village right near the front, at the moment that the counteroffensive in the south began. We had gone to this village, and I'm standing in front of a bombed-out town hall, explaining that this was a village of 4,000 people and now it's down to about 700 or 800 people. I put those pieces together, and I think that that captured the two Ukraines better than anything I could have written or put on audio.

Scripting Video

Some Instagram videos are really fun to do. London has great Christmas markets—I've been going to them for years—so I thought I'd do a tour of the Christmas lights. I knew the areas I wanted to go to. I started off on Regent Street, with these gigantic lit-up angels, and then over to Piccadilly, Covent Garden, and Hyde Park, where there's a giant winter wonderland.[3]

You have to time the video to about a minute or so because of the viewer's attention span (which I don't agree with, but apparently that's what they want). And so I just storyboarded the whole thing. I knew roughly what I was going to say. And if I couldn't remember everything, I'd read a little bit and then put my laptop down and do the standup. I think I had five scenes, and I shot all the B-roll and sent it to the editors to stitch it all up.

Drawing the Line

For that story we were going to do an audio piece, but I ran out of time because I was so busy with other things. I just did it as an Instagram piece, which I think got a million views. An audio piece would have been fun, too, because you could have done this sort of audio tour. But I see our jobs now as a double-edged sword. If you're constantly trying to do all these different things, you might not do any of them terribly well.

But I'd also say, if you're going to spend the money covering Ukraine or working in Europe or China, maximize it. There's great stuff, incredibly visual stuff, you can do. I sometimes have to restrain myself. When I was in Santorini it was hard not to do a story on the history of the cave dwellings there, because it's so visual. But I was on vacation—I've got to stop!

NICOLE WERBECK, DEPUTY DIRECTOR, NPR VISUALS

Scripting a Basic Idea

One of the reasons they're so successful is that most of these are scripted. Some people have the false impression that you can just stand there with the video and talk off-the-cuff. That doesn't work.

In most of these videos, you have one basic idea. You break it down and make three points about it in a clear and concise way. The team's gotten really good at writing these scripts and boiling them down, editing and reediting.

Making What Audiences Want

It's just one of those things where you have to do it and then see the results and where the value is. In the beginning, we had some people who said, "I don't want to do that [a short-form video]. I want to do a documentary." And then, after doing one 60-second video, and seeing that it gets two or three million views on one platform, they're convinced. And you're getting immediate audience feedback in the comments.

We all want to do the long documentaries, but right now audiences are really looking for short videos. It's where people are in their lives right now. We are still trying to explore longer-form things. But for right now, this is where audiences are leaning into. I'm sure we'll evolve with it. I think we do a good job of evolving. It keeps us on our toes.

PAT WOOD, SENIOR PRODUCER HANDLING DIGITAL STORIES FOR *ALL THINGS CONSIDERED*

Planning for Different Platforms

What you really want to do is think about how this story is going to be told on each platform before you begin, rather than gathering everything you need for radio or for digital, and then going, "Now I need to somehow make this work for another platform." You really want to consider everything: What's thc framing of the story? What are the elements you need? When do you expect this to air or publish? Is it a day-two story, a day-three story? Will the story have moved on? And how many voices? You might need more voices to quote in your article than for the broadcast element.

The way I generally think about the story is as a set of building blocks. Think about the building blocks, and the writing will come later. So if it's a really visual story, do you need to get photos? And that's not going to work for a broadcast, where you're going to need more description, so who's going to provide that? Or is there really good sound, but sound doesn't translate to an article, so how are you going to fill that gap? So you need to sort of think through all of these planning things before you actually start, rather than rush off in one direction and later realize you don't have what you need. And there may be times where a story just really doesn't work for both. And you make that decision early, and that's fine.

Organizing a Multiplatform Project for the Show

All Things Considered sent a team to Senegal who then traveled through Morocco and into Spain.[4] They were interested in reporting on three distinct things and how they intersect: climate change, migration and far-right politics. This was going to be enterprise reporting for the radio, a podcast and a fully realized digital build for the website as well.

It was going to be broken up into 10 or 12 different radio stories, each little bit along the way, and spread out over about two weeks on the radio. At the start of each one, you could be picking up on the story again that we've been bringing to you, because we have a semiconsistent audience on radio for *All Things Considered*. For digital, we wanted to have one long feature that brought all of these threads together. So the narrative arc was always going to be different than it was for the 12 or so stories spread over two weeks. And even though we were doing multiple digital pieces, we couldn't assume that anyone will have seen any of the others. Each one had to stand on its own. And that's a function of the way the digital audience is these days. Traditionally, if you had a newspaper series, you could publish a story each day and have a fair degree of certainty that people kept coming back to that newspaper. That system is being dismantled with digital. People are coming to it from search or from Google or Facebook or the NPR website or maybe any of the other thousand different avenues. You can't ever anticipate that people have seen the other reporting in the series. Each piece has to stand on its own.

And so what were the elements we needed? Well, for the radio, you need a lot of really good sound texture from places, people, the ocean noise, all of those things. For the digital, what we needed was really strong visuals and maps zooming in on different places. We had a drone fly overhead and give a really good sense of perspective of the places we were talking about. And there's overlap: The people you speak to for the radio can also be in the digital. Because journalism is still journalism. It's still the basics of asking people to tell their story, with open-ended questions so you're not getting yes-no answers.

Turning Host Interviews Into Digital Stories

We might have an interview between a host and a politician, and that works really well for broadcast. You have this natural back-and-forth between the two people. The host can push someone if they don't answer a question; they can follow up. Those interviews alone don't translate well to a digital page. You lose the pushback from the host, and unless you bring a lot more to it—a lot more context, other voices—you end up with an article that is just a single person espousing their views.

Where the interview might work in digital is with a celebrity or other type of guest, where there's not a contentious point of view being put forward. It's just someone talking about a particular piece of art they've been working on, whether that's a book or a song or a movie. Or it might work with an expert, where you're not needing any extra context to understand the story.

We set those interviews up with a few paragraphs of copy, and then there's usually a line that says, "This interview has been lightly edited for length and clarity." There's also a big subhead that says, "Interview Highlights," which makes it clear that what you're reading is this single person's interpretation or point of view on something.

The interview highlights don't need to match exactly the edit that was broadcast. Generally speaking, I'm less edit-heavy for the interview highlights than I expect the radio editors are, because digital doesn't have the same time constraints. If you need to hit that four-minute mark, the knife needs to be wielded a bit more. But with digital, you can run a little bit longer. Having said that, I definitely wouldn't just dump it all on the page, because that makes it a really terrible reading experience.

But in digital I would be clearer about what has been edited than perhaps on the radio, because it reads like a transcript. So if something is not entirely needed, I'd take it out and put an ellipsis. The general bar is to be very careful that any edits I make don't change the meaning or the understanding of what the person is saying. That's job number one.

Switching Platforms

As a reporter, when I needed to adapt radio stories for the web, and later as an editor, when I helped reporters do it, I noticed a pattern that worked for many stories. So did other NPR staff members who had been print reporters before coming to radio. Frank spent 18 years at newspapers—he was filing stories on floppy disks when he started at the *Philadelphia Inquirer*—and when I interviewed him, he had been a radio reporter for just as many years. "If I write a radio piece first, I lop off the intro, and I usually go for a scenic, anecdotal lede," he says. "And then I rework the intro as the nut graf."

Shannon Bond spent 11 years as a reporter and editor at the *Financial Times*. She also thought about how to go back-and-forth between the platforms when she joined NPR in 2019. "Often what works very well as your host intro is your nut graf: This happened, and here's why you want to keep listening," she says. "So your radio news story can have more of an anecdotal lede," something you might use for a feature that will draw people in.

GOING FROM RADIO TO DIGITAL, AND VICE VERSA

Using insights such as these, I've drawn up a road map to help journalists go from radio to digital:

1. Cut the intro and save it for step 4.
2. Try turning your opening scene into a lede.
3. If that doesn't work, find a more compelling scene or anecdote and use that as your lede. You should go back to your notes and interviews because the scene or anecdote you need may not exist in your radio story.
4. Plug the intro back in a few paragraphs down, working it into a nut graf that makes clear why the story is important.

5. Check all your quotes. Do they make sense in print? If not, go back to your interviews and find better ones.
6. Fill out your story with more context, background, detail and voices.

This method won't work for every kind of story, particularly newsy spots where you might want a nut-graf-like lede for both radio and web. And it's not a substitute for the extra reporting needed to add the voices and detail that the digital version requires. But it should help you restructure the story in a way that satisfies the needs of the digital reader.

If you've written your digital story first and want to turn it into an audio script, try these steps:*

1. Use the nut graf as the basis for your host intro. Most likely you will want to rewrite it to make it catchy.
2. See if your digital lede works as an opening scene. If not, try another anecdote or vignette from your story. You may need to go back to your interviews and find something that humanizes the main character or sets the scene.
3. Listen to your actualities. Do they work for radio? If not, go back to your interviews and find ones that do.
4. Trim voices, detail, background and context not needed to sustain a linear audio narrative.

DRAMATIC LEDES

Listeners don't get hooked the same way readers do. You might start an audio story by humanizing the main character, but that may not work in the digital version. When I was on the national desk, I edited a piece by Emma Peaslee, then a Kroc Fellow. She pitched a story about a group of Roman Catholic peace activists, ranging in age from late 50s to early 80s, who had broken into a military base and were being sent to jail. Her audio piece began with a lovely scene of one of them, Patrick O'Neill, with his teenage daughter on a playground:[5]

O'NEILL: You want to go higher?
DAUGHTER: Yeah!
O'NEILL: All right!

* And consult the sections on intros, scenes and actualities in chapter 9.

EMMA: A white-haired man wearing a black beret pushes a girl on a swing at a playground in Garner, North Carolina. The swing is for children with special needs.
O'NEILL: All right! Hold on!

As listeners, we are introduced to this man as a loving father before we hear about his participation in a religiously inspired act of vandalism. We considered starting the digital story that way, but it felt laborious. So we began it this way:[6]

> Dressed in black, the seven intruders cut through a fence and stole along the perimeter of the naval base, trying to avoid detection from the guard towers, as a loudspeaker overhead blared: "Deadly force is authorized!"
>
> Patrick O'Neill, who had a GoPro strapped to his head, tried to reassure himself by remembering a scene in the Bible where Jesus escapes unscathed from a wrathful mob that wants to throw him off a cliff.

It didn't matter that the scene was in the past. Nor that we didn't have great sound from the night of the break-in. The drama of that night was sufficient to draw readers in. And the great thing about digital is that you can bring a scene to life with words.

Common Pitfalls

As news consumers, we don't really think that much about how good storytelling varies across media. But as journalists, we should. Because they are almost like different universes—what sets up a scene or character well in one won't in another. Here are some of the problems that arise when the effort to work across platforms is not entirely successful.

SUPERFLUOUS WRITING

Journalists prioritize the strengths of each platform and compensate for its shortcomings. This is natural and desirable. In radio, writing visually makes up for a lack of images; in video, the images tell the story best, and the writing plays a supportive role. But if you fail to adjust the storytelling when you switch platforms, it can come across as plodding and even pedantic. If you use words to describe what the viewer of a video can plainly see, they will become bored and may feel you're insulting their intelligence. And while actualities that contain descriptions are very effective, you may not want them in a digital story with photos. "There's no point," says Pat, "in having a quote where they're describing what you can see in that photo."

NOT USING ALL POSSIBLE SENSES

Audio journalists are sometimes so keen on getting good sound that they don't notice details that would help tell a digital story. Climate editor Neela Banerjee remembers editing a story about wildfires that was sonically powerful. "You can hear the crunch of their footsteps. You can hear the birds." But the digital draft was short on the details that would bring the story alive on paper. So in addition to gathering sound, write down what you see in your notebook or on your phone, or say it into your recorder. And if you can't do that immediately, then when you're listening to your tape later, try to remember the scene and "the sensory things that it evokes," Neela says. "Did the air smell like burned rubber? When that kid found a coffee cup, did it say, 'World's Best Mom'? Every sense has to be alive."

SWEEPING STATEMENTS

It's normal to start conversations with blanket statements. "I don't like hospitals" or "Summer here in Seattle is so short" are enough to get a discussion going, and that's why they work at the beginning of audio pieces. Here are some examples from NPR stories: "Americans love gas stoves."[7] "November in Mozambique is hot and sunny with a chance of epic rainstorms."[8] "Oil prices are largely a function of supply and demand."*

But in digital, it takes more to grab people. Let's go back to the example of the gas stoves, and consider the way several newspaper stories began pieces on the topic.

> The *Washington Post*: A new front has opened in the battle over climate change: The kitchen.[9]

> The *New York Times*: In a nation that is already deeply split along partisan lines over the pandemic response, racial equity and abortion, add this: gas stoves and furnaces.[10]

---

* This example comes from a newscast spot that is not available online.

> *Slate*: As a physician and epidemiologist with the US Centers for Disease Control and Prevention, T. Stephen Jones spent his career fighting major threats to public health in the US and globally, from smallpox to HIV to viral hepatitis. But it wasn't until Jones was well into retirement that he learned about a widespread yet widely overlooked health risk in his own home in Florence, Massachusetts, and in most US households: pollution emitted by natural gas appliances.[11]

The approach these outlets took would not work for most people in conversation. Imagine telling a friend: "I love to go to the theater, to restaurants, to the park, but there's one place you won't find me if I can help it: a hospital." Sounds pretentious, doesn't it? But in digital, you need to start with a surprise, paradox, compelling anecdote, or something that hints at the tension in the story and piques the reader's curiosity, and often that requires elevating your prose and using nonconversational phrasing.

Read, Listen and Watch

I hope this chapter has convinced you that news stories need to be told differently across platforms. But don't just take it from me. See how others do it. Take a news story that got wide coverage. Read a story about it on a news site. Then listen to a public radio or podcast version of it. Then watch what a TV news broadcaster did with it. Pay attention to the nuances involved in tailoring the story to each medium. And learn to recognize what works and what doesn't. As with so much in this craft, the more you do it, and the more you pay attention to how others do it, the easier it gets.

CHAPTER 14

Hosting

To be a host is to live out a jumble of paradoxes. You are a leader, but you don't manage staff. You sit in a bubble—actually, a soundproof room—and you talk to people from the four corners of the earth. You seem to wield near omniscience on a dizzying array of topics. Yet most of the time you are not out in the world reporting, aside from the occasional host trip or lunch with a source. And you are an intimate friend to millions of people, though the stories you tell are not about you.

How can any human embody such contradictions? Our hosts do it, most of the time with aplomb. To understand how, consider the role they play for the audience.

Friend and Guide

Melissa Block began hosting *All Things Considered* in 2003 and did it for 12 years. As a host, she says, you serve as a guide to what's happening in the world—leading people through the latest news developments. At the same time, you're a part of the listener's life. "You're welcoming people into your space. You're with them in their kitchen or in the bathroom when they're taking a shower, or you're sitting with them when they're feeding their kids dinner. You're in their car with them on their commute."

You're doing more than relating news and analysis. "Obviously, we want to be informative. We want to be journalistically sound. We want to make sure that we have a deeply reported story," says Rodney Carmichael, co-host of NPR's *Louder Than a Riot* podcast. "But as the host, as the person who is responsible for carrying the story across and for getting the listener to ride with us on this journey, you've got to be an engaging storyteller."

When hosts are engaging, listeners feel like they are their friends. Ayesha Rascoe had been hosting *Weekend Edition Sunday* for about a year when I spoke to her about it. Just a few days earlier she had received handmade potholders in the mail. A listener had sent them "with a card that said, 'Thank you so much for this show!'" Another listener sent Ayesha an envelope with $200 in cash to give to a mother she had interviewed about her difficulties affording school supplies for her kids. It just demonstrates the intimacy of NPR hosts' relationship to the audience. "I don't think people send potholders to Wolf Blitzer," she says with a chuckle, referring to the CNN TV anchor. Ayesha sees herself as a friend and tour guide. "You're holding their hand. You're leading them all the way through the show, and it's almost like you're assuring them. You're saying, 'It'll be all right.'"

REPRESENTING NPR

NPR hosts have long been known for being someone listeners can relate to, rather than look up to. Which also means that listeners really care how a host sounds, in good ways and bad. In 1972, when *All Things Considered* had been on the air for just a year, it needed a new host. Bill Siemering, NPR's first director of programming, came under pressure to hire an established network TV reporter, and in particular a man. Instead, he chose Susan Stamberg, who not long before had been cutting tape as a production assistant. Siemering told Fresh Air's Terry Gross on NPR's 50th anniversary that he chose Susan because he liked how she sounded. "She has this wonderful voice that is expressive, had rich tone color, and it's the sound I really wanted for NPR. It's the sound that I still think represents NPR the best, this insatiable curiosity."

Nevertheless, some people didn't like it. "He batted down criticism when it started coming from station managers opposed to a woman doing this," Susan says, recalling the early years. Listeners also wrote in, complaining that she sounded "too New York-y." Frank Mankiewicz, NPR's president from 1977 to 1983, told her: "That's translation for 'being Jewish.'"

Fast-forward half a century, and the story is not all that different. Ayesha says that no one at NPR ever told her to not sound like "a Black woman from the South." But there are listeners "who just cannot stand my voice, and they make that known over and over again on Twitter. I get people who say I'm too loud, to which I go, 'Well, you know, you can turn the volume down on the radio!'" she says with a hearty laugh. "And some people will say that I may be too exuberant. I'm just too upbeat, like it's just too much for them. Some of that," she says, "absolutely is sexism and racism." But she prefers to focus on the many others who love the way she sounds. "A lot of people say that it's joyful and that I bring some joy to their life."

SHOWING VULNERABILITY

It's impressive how hosts can hold their own in virtually any conversation. And yet, they are often at their best when they don't pretend to know everything. "I'm not trying to sound like the smartest person in the room," Ayesha says. "I don't mind asking basic questions, or a question that may sound silly" to experts in the field, because the responses tend to be more enlightening. Even though a host has to keep up with the news, they still have to imagine "the person at home who's yelling at the radio. What would they want to know?" she says.

Rachel Martin, who hosted *Weekend Edition Sunday* and then *Morning Edition* for more than a decade, says the host is a proxy for the audience, and that means "being able to admit when you don't know something."

> Because chances are there's someone out there who also doesn't understand the jargon that that person just used, or the obscure reference that you didn't get. And a lot of times people who are new to hosting get really frazzled by that, because they think they're supposed to know everything and they don't want to be vulnerable. The most effective hosts can channel what someone who's sitting at home, maybe engaging with this topic for the first time, would ask in that moment. And not being afraid to raise that follow-up question in a way that a normal person would.

That can also mean reacting emotionally when you hear something sad. Early in her hosting career, Mary Louise Kelly interviewed a reporter in the Middle East who was telling a wrenching story about the death of a refugee child. As a mother, Mary Louise was affected, but she suppressed her emotions. "And I just moved on and asked the next question because I was just trying to get through the facts. And I remember getting emails from people saying, 'How could you not react? That was heartbreaking, and you just moved right on.' And I thought, 'You're right! You needed to hear me just sit with that for a second, because that is a human reaction.'"

It's reactions like these, which may have nothing to do with obtaining additional information, that turn a host interview into a conversation, as though it were "unfolding at your dining room table or in a living room,"

Mary Louise says. Instead of continuing on to your next question, sometimes it's better to react "as a human" to what someone just said, because that will put them at ease and may lead them to become even more forthcoming in telling their story. "Sometimes," Mary Louise says, "the best questions are not even a question."

The Host Interview

Many NPR hosts come into the job after having been a reporter, which makes sense because the host is a kind of reporter, doing interviews to get a story. But a host interview differs in structure and purpose from what a reporter does. Let's first define both.

REPORTER INTERVIEW. A recorded conversation with an interviewee, often conducted on location with ambient sound, aimed at getting statements, or actualities, that will form bits of a story that will most likely include other points of view.

HOST INTERVIEW. A conversation with a guest, usually in the studio or done remotely over a high-audio-quality connection. The conversation will be presented as a self-contained story, including both the host's questions and the guest's answers. It may be prerecorded and edited down, or it may be aired live in full.

Most of what I and my NPR colleagues said about interviewing in chapter 6 applies to host interviews too. Both the host and the reporter want their interviewees to feel at ease and open up about their ideas, feelings or experiences. Both make an effort to challenge false claims and inconsistencies and ask follow-up questions when things aren't clear or to get someone to go deeper. But the key difference is that while the reporter's questions mostly get cut, the audience hears the host asking questions.

This requires the flexing of a different kind of brain muscle. A reporter can come up with questions on the fly, if necessary. And they can formulate those questions without having to worry about coming across as ill informed. They can ask the same question in different ways. In this sense, it's a little more like an interview for a print news story, in that when you come across a quote, you don't read the question the reporter asked to get it. That's why Sarah McCammon said earlier that she can be a bit "messy and rambly" when she asks questions as a reporter, because it can make the other person feel more relaxed. But as a host, she says, "you have to go in with a plan. Now, you have to be willing to change your plan based on what they say, and listen. But you have to at least have an agenda of, 'This is why I'm talking to this person. This is what I want to find out from them.'"

Ayesha agrees. "Have a very real, laid-out, particular game plan," she says. Because you have so much more going on as a host. True, a reporter needs situational awareness, and has to keep an eye on the recording levels. But as a host, you have to talk to your guest while watching the segment clock tick down, and while communicating with the folks in the control room: with the director, who is signaling when to start and stop talking, and with the editor and producer, who are fact-checking and helping to make sure you've asked all the pertinent questions. Ayesha learned the hard way. In one of her first host interviews, the conversation ended before she had a chance to ask a question she really wanted to get to, "because I was still learning how to communicate with the control room."

The Conversational Arc

A game plan like the one Ayesha recommends means thinking ahead of time about the conversational arc, giving your interviews a beginning, a middle and an ending. That may seem obvious, but let me specify what each part signifies:

BEGINNING. A natural entry point or on-ramp to the conversation.
MIDDLE. Drilling down on the issues, mining an interesting aspect of the story, or challenging a false claim or inconsistency.
ENDING. Having the guest look forward or offer a takeaway.

"I'm constantly thinking of that arc," says Mary Louise. Knowing that she only has, say, four and a half minutes to talk to someone, she needs to make sure that before the time is up "we make sure to get however many competing views there may be in there, and that I'm challenging somebody."

WORK TOWARD THE CENTRAL QUESTION

Unlike in a reporter interview, you can't prepare a list of twenty questions and see where the conversation takes you. You will only have time for three or four in a live conversation, maybe a few more if it's prerecorded. And while that small talk or softball question that I recommended starting with in chapter 6 might work if you're taping an interview, hosts don't do them. It may warm up the guest, but it doesn't serve the audience. They want you to get to the point.

And there are other reasons to get to your most important question as soon as possible, says Mary Louise. "The phone line might die, or the senator might get called away to vote. You just don't know if you're going to get the full time."

So identify what is your most important question by asking yourself, "If I had one shot, if I were only going to get one question, what would that question be?" she says. That question is not necessarily the same as the purpose of the interview. You may want to talk to the U.S. transportation secretary about a major train derailment, but the most important question might be whether they believe railroad companies' pledges to improve safety in the wake of the disaster. Rachel Martin says the one-shot question may be the pushback question. "It's the question where everyone's ears will perk up. It's going to put the person off their talking points." Mary Louise suggests writing that question first, "and then the scripting should be all about getting to that place."

Getting to that central question as early as possible makes for an efficient conversation. When *Morning Edition* host Leila Fadel interviewed a Southern Baptist pastor whose church was expelled by the denomination's leadership in a crackdown on women's ordination, Steve Inskeep thought the guest would need a little warming up.[1] "I was deeply worried about that interview," he says. "How are you going to get that in four and a half minutes?" As the show was about to begin he even looked for something to cut in the next segment in case the interview ran long. But Leila started with her key question: "What went through your mind when you got the news that your church was expelled because you, a woman, [are] the pastor?"

"And the pastor is like, boom, she's off," says Steve. "Because the conversation began at the high point, Leila had time for a couple of follow-up questions, and they finished right on time."

DON'T OVERLOAD YOUR QUESTIONS

Because an interview has to tell a complete story, there's a tendency to want to put a lot of background in the questions. "You end up having a lot of you explaining and very little of the guest talking," says Leila. She said one of the reasons the interview with the pastor went so well was that even though the status of women in Southern Baptist theology has a tortuous history, she let her interviewee give the needed context. "I find that the simpler the question and the more direct, the easier the conversation. 'What happened here?' 'Why?' 'What are you going to do?'"

While a reporter tries to find out everything they can about their story, the host needs to see the big picture. During her years as a White House correspondent, Ayesha's work was featured frequently on the newsmagazines, but she didn't really pay much attention to the rest of the show. "As a reporter, all you care about is your piece."

As a host, she was suddenly thinking about wider coverage. She had to record the billboards—the minute-long introduction aired before the newscast that highlights some of the hour's most interesting stories. She had to read intros to reported pieces on an assortment of topics and make promos for other shows. She delivered some of the stories herself, but those had mostly been reported and written by producers and editors. And most crucially, she had to stay up on everything that was happening in the world in order to be able to interview guests on a breathtaking variety of stories. "It's a totally different type of job," she says. "It gives you a wider sense of what's going on."

BUILD ONGOING CONVERSATIONS

How can a host be knowledgeable on so many fronts, when they barely have time to do their show? I put that question to Mary Louise on a pretty frenetic news day. Former president Trump was about to be arraigned on criminal charges before a court in New York City, and Mary Louise was preparing to interview a former colleague of the district attorney who served the indictment. "In an ideal world, before you go into an interview, you've spent hours reading every article and thinking about your questions and talking them through," she says. "But the reality is, I don't have that kind of time."

Instead, she aims to build up knowledge over time. "Every interview is preparing you for the next," she says. Before a conversation, she thinks back on the previous interviews she's done on the topic. "And every one you're remembering, 'Oh, yeah, they made this really good point. I wonder if what this guy told me yesterday on this story is worth putting to this next guest?' It becomes part of an ongoing conversation, and each question and story informs the next."

Another way of building knowledge is to get it from your team.

Collaborating With Producers and Editors

As I walked into Mary Louise's office, she was texting with her editor and producer about the questions they were going to ask the former prosecutor about the Trump case. Hosts typically work with at least one producer and one editor, and "the three of you together are helping figure out what is it we're really trying to get out of this interview?" she says.

Producers and editors work through many of the key decisions regarding an interview. Is this the right person to talk to? Does it need to be live? What if the news shifts in a different direction? And often they will think of things that didn't cross the host's mind. "As a host, sometimes you may feel like you know the questions," says Ayesha, "but other people have some really great questions too. So if you're only relying on yourself, you're not going to get as good a product."

After the interview, the host may go off to their next interview or story, while the producer and editor take the conversation from there, cutting it and mixing it into its final form. "They're the ones who are really putting everything together," says Ayesha. "I can't really do what I do without them."

Collaboration is also at the heart of the making of NPR's podcasts. At *Louder Than a Riot*, the writing was done as a team. "A producer was definitely doing the early rounds," says Rodney Carmichael, who co-hosted the podcast together with Sidney Madden, "and as hosts, we would eventually come in and we would co-write a little bit or co-edit." That was a departure from his previous work for NPR Music, and at newspaper gigs. There it was mostly a solo effort, where he worked to develop a distinctive voice in his writing. The transition to hosting a podcast meant "learning to cede some of that ground for the betterment of the story." But adaptation was a two-way street: with each episode, the producer would increasingly mimic the way he speaks, favoring contractions, dropping the T or G at the end of certain words, even suggesting ad-libs, which he really appreciated. "You develop that trust between you and a producer and editor, and you all start to kind of like hum and sing on one chord." At *Pop Culture Happy Hour*, the whole team obtains screeners for upcoming TV shows and movies and watches them to see if they might be worth covering on the podcast. The team also looks into the background of the creative talent, in case there's anything noteworthy or controversial. "And that's where our producers really come in," says Aisha Harris, one of the co-hosts. She says the producers spend a lot of time researching and collecting articles, interviews and anything related to the production and put it in a planning document for the hosts.

While her producer helps her with her hosting, Ayesha often thinks about production needs. During a remote interview, Ayesha pays attention to the guest's mic placement because she knows the producer doesn't want to interrupt the flow of the conversation. "They may message me, but if I see it myself, I'll say it myself. 'Hey, can you say that again? Because you were kind of moving and we're not going to really get that.' Or you hear a sound in the background. 'I think you have to stop talking right now.'" If you think it's going to be a problem, address it right away, says Ayesha. "Nobody wants bad sound."

NPR hosts may have a lot of influence in the newsroom, but they are not managers. "I don't sign anybody's timesheet," says Steve. "But you are called upon to be a leader." They often set the direction of coverage, because producers and editors will pitch stories they know the hosts will be interested in. Which makes sense given that, as Mary Louise noted in chapter 2, it's the host who has to sell the story to the audience.

Still, the host should take producers' and editors' ideas seriously. And when it comes to pieces filed by NPR reporters, the host's goal is to get the audience interested in what the reporter has to say. "If I'm doing my job correctly, I'm part of a team," says Steve. "I think a good host is a team captain who will try to lead by example, and also think about the broader team and other people and how to help them do their best."

Steve Inskeep on a Typical Day Hosting *Morning Edition*

I'm not on the show every day, because there's four of us that rotate. But on a show day I get up just after 3 a.m. I need to be in to work by 4, whether it is at NPR or sometimes still at home. And the show goes on right at 5.

I have a very busy time from 4 to 5. And in that time, I'm desperately struggling to find an extra two minutes, three minutes, five minutes to reflect on the things we're about to do, because the show begins at 5 with three straight live interviews that become *Up First*, which is the absolute front page of the network. It is worth spending a few extra minutes, when you can fight for them, to make the writing a little bit better. At the same time, you're also doing some of your own writing. You're writing these billboards, you're getting your head around the show, you're doing preproduction. There's a lot going on.

And I'm just trying to enjoy being there. I like my work. And then there might be three or four or five or six live interviews, and a lot of live intros, until 7:19 a.m., to be exact. And then the rest of the morning is different every day. It could all be on tape the rest of the morning, or it could all be live the rest of the morning. There have even been a couple of days when we've gone past noon Eastern time for a couple of extra hours of live coverage on extraordinary events. And then there's a lot of days where it's something in between. There's a live interview or a couple of live interviews in those hours. And the rest of the time you have to try to think about things coming up, or try to book an interview that you want, or plan a trip or edit an interview you've already done.

I don't work the very worst hours in the building because there are people who work like 11 p.m. to 7 a.m., but they're on a rotation and maybe don't do the job all that many years. So over a long period of time, I think that *Morning Edition* hosts have definitely worked the worst hours at NPR, which is fine. I'm not complaining, but you need to take care of yourself. So I try to discipline myself. If I'm successful, I'll get to bed at eight, I'll get up at three. I'll get seven hours of sleep.

Warm Up Your Guest

As I mentioned before, you don't have the luxury of easing into a conversation during a live interview. But what you can do is have a chat moments before the interview starts. "I'll say, 'Hi. How's your day?'" says Mary Louise, and then she'll set out the parameters of their conversation, stressing the lack of wiggle room. "I've got five minutes and I have a hard out. So, if you hear me interrupting and jumping in, I'm not trying to be rude. It's that I got to keep this train moving." She says it's mainly a way of stressing that "they can't give a first answer that's five minutes long."

When Steve joined *Morning Edition* in 2004, some hosts didn't talk to guests before the interview out of principle, he says, because they thought "the 'Good morning' wouldn't be genuine." But Steve talks to the guest before every interview as a way to seek their cooperation. "I'm setting parameters and rules for them and inviting them to collaborate with me in making good radio and getting their viewpoint out. It is a complex role, of course, with a newsmaker or a policymaker or any number of people who are not an NPR reporter, because I also need to be independent of them." Even in an adversarial interview, which I'll talk more about later, you want the guest to work with you at least on telling the whole story in the allotted time.

A warm-up chat is also advisable for a reporter two-way. It's an opportunity for hosts to make sure the reporters know the answers to the questions the host wants to ask, because reporters' perspectives are often limited by where they are. "They might be able to tell you what's been happening in the streets right in front of them," says Steve. "But they might not know what's happening in the capital of the country they're reporting on because they're in some other city." Remember, the point of a reporter two-way is not to make anybody seem ignorant or uninformed.

As a reporter, I always appreciated when hosts checked in before a live interview, even when it was just small talk or a way to get to know each other a little better. Especially when I was starting out, it helped calm my nerves. In 2007, I was covering a meeting of the Intergovernmental Panel on Climate Change in Valencia after the U.N. organization won the Nobel Peace Prize. The IPCC had just issued a report, and I was about to discuss its findings in a two-way with Andrea Seabrook, then host of *Weekend All Things Considered*.[2] Before the interview, she and I chatted about life in Spain, and that conversation somehow made it easier for me to talk about the IPCC's dire predictions about climate catastrophes.

Go Off-Script

Now you're ready for the conversation. You've worked with your team to find the right guest, and you're sitting down with a script in front of you.

"You have that first question," says Steve. "And you have a strategy for how you would like the interview to unfold. But as soon as they say something interesting, you grab that and go with that." Because the last thing you want to do is just rattle off the list of questions on your script. That would not be a conversation, but rather an interrogation.

I'd like to go back to the idea of reacting like a human to what another human says. Earlier in this chapter, Mary Louise talked about taking a moment to respond with some emotion. Now let's talk about formulating a question in response to a previous answer.

You may want to recap something they just said as a segue into a question you were already going to ask. "So you're saying the president had authorization to order military action. What does that mean for Congress . . ."

Or you can ask a question that was not on your script but comes to you in the moment. When I spoke to Ayesha, she had just done an interview with Tim Nelson of the Australian band Cub Sport.[3] At one point, he talked about a song he wrote during the COVID pandemic, when he missed being close to fans. Notice how Ayesha doesn't respond with a question, but by expressing a thought.

NELSON: . . . I felt pretty lost. I think I felt kind of needy, and like I really just needed some love.

AYESHA: I mean, because it is that thing about being an artist and connecting with people on the stage, which is a different connection than a one-on-one connection. But it's also something deep. It seems like artists really need and crave to have that connection with this larger audience, with the people that listen to your music.

NELSON: Yeah, big time. And I feel like there are also these moments of one-on-one connection in the show as well.

(*Sound of Solange singing and fans cheering at a concert.*)

NELSON: In 2018, I saw Solange play at the Opera House in Sydney. She got down onto the edge of the stage, and I was in the front row. She sang a whole verse, just looking into my eyes. And it changed my life.

That anecdote, mixed with sound of the concert, was a highlight of the interview. That question was not on her script, and Ayesha says she would have never thought to ask about it. "It just came from us talking." But it did come naturally from all the preparation she and her colleagues had done for the interview, and the mental process of formulating the questions that weren't asked. As a result, Nelson opened up about himself in a way he wouldn't have, Ayesha says, if she had just stuck to questions in a list—"and this is my question, and this is my next question, and this is my next question."

Melissa Block, the longtime *All Things Considered* host, says one of the most gratifying things to hear is when someone says, "I've never I've never told anybody this before, or I've never had somebody asked me that before." And you get that by letting the conversation flow and "not just asking them about their latest project."

As part of a series *All Things Considered* did on winter music, Melissa interviewed dance choreographer Bill T. Jones.[4] While talking about a work by Franz Schubert, Jones said the music took him back to a day in fourth grade, when he looked out the classroom window and saw his father walking alone, his back to the biting wind.

> JONES: He had to get to this very insignificant job in a factory miles and miles away—a Black man with no car, trying to hitchhike and no one picking him up, and he has to walk that 10 miles to get to the factory.
>
> And I'm sitting in this warm classroom getting educated, not paying attention to the teacher, and suddenly feeling torn between two worlds. And this music, when I hear it, I feel for my father. And there's something about art that can be, yes, depressing, but helps us bear the pain through just sheer beauty and intensity.
>
> MELISSA: I'm imagining that your impulse would have been to run out of that classroom and get your dad inside, warm him up somehow?
>
> JONES: No. It was more complicated than that because that was my job, to be in school. One of the reasons I was in school was so that I did not have to be out there with him. And that was the painful thing about this sort of class-climbing that we all, in this country, are subjected to. We're supposed to do better than our parents.

Melissa's going along with his story let him unspool a story that was "layer upon layer of heartbreak and beauty and depth." The interview was scored with the hauntingly melancholic music of the Schubert song cycle *Winterreise* (*Winter Journey*). So sometimes it's just about setting aside the script for a moment and letting the guest go a little deeper into their story.

Producing a Host Interview

Search online for "Prisencolinensinainciusol" (don't try to say it) and you'll find a song that went viral in the 2010s, with lyrics that, like the title, are sheer gibberish. When it lit up the internet, forty years after its original release, host Guy Raz and the team on *Weekend All Things Considered* set out to find the Italian pop star behind the hit, Adriano Celentano. Brent Baughman, Guy's producer, says they wanted to "unpack that sense of curiosity and wonder" that Celentano evoked with the song, which mimicked what he thought English sounds like to Italians.

There was a problem, though. "We didn't even know if the guy in that video was still alive," Brent recalls. They finally managed to track him down. "But then he didn't speak English, so I'm thinking, *Is this going to work?*"

There was no guarantee it would. Producers and bookers usually pre-interview guests, unless the guest is too busy—a senator or a famous musician, for example—or you know they will sound good on air. In this case, a pre-interview would be too expensive since Brent would need an interpreter. So, without having talked to the musician, Brent drafted the script and scheduled the call, which brought its own set of difficulties. Celentano would be on the phone from Italy, and Guy and the interpreter, a former State Department translator, would be in the studio. Brent and the editor would be in the control room.

When the connection was made, Guy asked Brent's first question, about how the song came about, but then riffed most of the rest, even trying to sing parts of it.

"Is this what American English sounds like to you?" Guy asked.

"Yes, exactly like that!" answered Celentano, who also sang parts of it.

It was a delightful conversation, but Brent says it was stressful for him during the entire hour that it lasted. "That's usually the hour as producer when you feel the most helpless. Because you know how you want it to sound in your head. But at that point, you have the least control that you will ever have in the process."

The most a producer can do during the interview, he says, is whisper in the host's ear or shoot them a message: "Hey, let's ask this." But that risks endangering the chemistry that you want the host to have with the guest. "You hope that it unfolds well. And hopefully you've done all your work ahead of time to make sure that it does."

So during a conversation, Brent focuses on following the conversation. "Some producers will take detailed notes, and editors will keep a log too. I've never done that because I feel like it interferes with my listening. I prefer to just listen closely and actively. And then afterwards think about what has stuck with me and zero in on those moments."

Afterward, he took the hour of tape, and began cutting it down toward the segment length of seven minutes, mixing it with excerpts from the song. "That's when the fun really starts. Because you just have to decide what delights you the most."[a]

a. The segment begins with Guy's conversation with Cory Doctorow, whose blog post helped the video go viral. You can listen to the story here: Guy Raz, "It's Gibberish, But Italian Pop Song Still Means Something," *Weekend All Things Considered*, NPR, Nov. 12, 2012, https://www.npr.org/transcripts/164206468. Afterward, you'll understand what Brent and many other show producers mean when they say, "If you do it right, no one hears your production."

Empathy Interviews

When a person has been through a tragedy or loss, your aim as a host is just to let them tell their story. When Sarah McCammon interviewed Patricia Oliver, whose son, Joaquin, was killed in the Marjory Stoneman Douglas High School shooting in Parkland, Florida, the aim was "not to watchdog or fact-check her, it's really just to get a snapshot of this really human moment." Here are some of the questions from that interview:

- "What do you want people to know about Joaquin?"
- "The next phase of this trial is the penalty phase. What do you hope will happen?"
- "I know the Parkland community is still moving through this grief, and there are, sadly, far too many families like yours all over the country who are dealing with the aftermath of gun violence. Is there anything you would like to say to them?"

Sarah says she prepared for the interview more than she would have for a reporter interview, with scripted questions based on a pre-interview that her producer had done. But as she spoke to Oliver, Sarah says she was thinking about "allowing a lot of space for the person to say what they want to say, and approaching it with empathy and compassion."

"We want the audience to understand what it's like to have suffered that kind of loss," says Rachel. "And so it's really for us to get out of the way." Showing empathy as a host is something "you can't practice and you can't calculate. And when you do, it goes horribly, horribly wrong." Still her advice is to keep your questions short and basic, and respond the way a normal person would. "Now, if you're a big crier like me, you do sort of have to keep it in check because you just can't go around crying all the time," she says, with a chuckle. "Just try to be there as authentically as you can. Just imagine how you would want to be treated if you were in a similar situation."

Adversarial Interviews

The conversations where you may *want* to get in the way is with someone whose goal is to make themselves or their organizations look good or to promote a political agenda. The host must balance letting that person make their point with setting the record straight on anything they might say that is dubious or untrue.

"If we invite somebody onto the show to talk to them," says Mary Louise, "it's because they have a point of view we feel is important to get across to our listeners." It's important to get that person's perspective, says Steve, but he wants to hear it in context. "If you have something to say that is false or arguable or just kind of your very strong perspective, and there are other perspectives out there, I want in some way or another, whether it's live or recorded, to be sure to get that full perspective across."

A controversial viewpoint doesn't disqualify someone as a conversation partner. But your objective is accountability, and you try to get it by challenging them respectfully. "If I ask a question and they don't answer it in a way that's satisfying or feels complete or feels honest, then you push them on it," says Mary Louise. That doesn't mean playing "gotcha." "But to try to understand what they really think, and why, and what evidence they can marshal to support their argument, how they reconcile what they're saying now with what they said last year." You might want to know how what they're saying jibes with what other people in their party are saying, if you see space between the two.

This pushback is a key reason people listen to NPR. "We live in an era where you can look up what anybody has to say on Twitter and their social media," says Mary Louise. "There has to be a value for both sides in their coming in and having a conversation. So I'm there to push them a little beyond their talking points."

Adversarial interviews are tricky because the host can't be rude, but at the same time they shouldn't let themselves be railroaded by the guest. "It's better to keep your cool and be calm," says Leila Fadel, "because if you get riled up, then the interview becomes about that." During the 2023 debt ceiling negotiations, she interviewed a Republican congresswoman from South Carolina who accused Leila of ignoring the facts, saying, "This is the problem with the media."[5] But rather than get pulled into a shouting match, Leila replied, "I don't appreciate that," noted that her questions were based on facts, and moved on to the next question.

Learning to deal with mendacious or troublesome interviewees, such as those who try to run out the clock without being challenged, is one of the most difficult parts of hosting. "I'm still learning how to do it," Leila told me a year and a half after becoming a *Morning Edition* host. One of her learning moments was another conversation on the debt ceiling, also with a Republican lawmaker from South Carolina.[6] Well into the five-minute segment, he called the Biden administration "compromised and corrupt."

> We're three minutes into the conversation, and I don't want to bust the segment, and if I ask him, "Hey, wait a second, you need to explain that," that's going to take everything. So what I did was—and I regret this—I said, "Well, the debt was compiled under many administrations, both Republican and Democrat." What I should have done, now that I look back on it and I've had time to think about it, is said, "That's a big accusation that you presented with no evidence in this moment. But we're not here to talk about that. We're here to talk about the debt ceiling and why you don't want this deal."

As a reporter, you don't need to worry about setting the record straight in the moment, because your questions aren't on the air. Which is why pretty much every reporter who becomes a host has to find the right balance—being strident when necessary, and pushing back when appropriate, but most of all staying in control of the conversation. "With hosts, that is our constant battle: to be honest and respectful, but also hold people accountable," says Leila.

Live or Prerecorded?

When the previous edition of this book came out in 2008, most NPR show interviews were recorded prior to broadcast. Within the next few years, the tables turned and most interviews were being done live. In the case of a major news event, a show may do wall-to-wall live coverage. The increase in live segments has been partly out of necessity: Amid the sped-up news cycles of online media, a live conversation cannot be overtaken by developments.

A major advantage of a live interview is that it can convey what's happening just by dint of being live. The atmosphere captured by the sounds in the background sometimes tells the story better than anything a reporter or guest might say. *Morning Edition* host David Greene was talking live with Africa correspondent Eyder Peralta from Nairobi about the 2017 presidential election in Kenya when a protest Eyder was covering turned violent. Eyder had prepared a script, but, he says, "the script goes to hell when the rocks start flying and the glass starts breaking." Not following a script didn't matter, because you could hear what was happening live. "The thing that listeners were left with is, the opposition leader is causing a mess in this country," Eyder says. "What kind of an election can this possibly be if you have people breaking into polling places, and you have police firing at them? Ultimately, you are left with the most important part in a supervivid way."

In a situation like that, Eyder says, the host needs to "show the way." Which is what David did. Even though there was a long delay on the line, David asked questions that were not on the script—"What is that sound behind you?"—and was telling Eyder to be careful. "If you listen to David, he is letting this play out," Eyder says. "He knows there's a massive delay. He knows that there's something that we're not getting to," but he lets Eyder focus on his surroundings and brings it back at the end with a quick summation of the context in his own words. Even when it's not strictly necessary, a segment done in real time makes a show feel more urgent. "I think the more live the show is, the better," says Steve, whose tenure as host has spanned the change from taped to live. You can hear the difference in energy in a live interview, he says. "They force you to focus." But there are some types of interviews that are better prerecorded. One is with a public figure whose whole bread and butter is to misinform. "If you know somebody is going to come on the air and they're going to run out the clock spewing a bunch of bullshit and actively trying to spread disinformation on your airways," says Leila Fadel, "you don't let them use the airways for that. You pre-tape those." And even then, preparation is key so that you can respond to whatever they say.

A great example is Steve's 2022 interview with Donald Trump, in which the former president repeated his lie that the 2020 election was stolen, and then abruptly hung up after Steve challenged Trump's claims of fraud. Prerecording the interview allowed *Morning Edition* to precede the interview with tape from other Republican officials contradicting Trump.

You may also want to avoid going live with interviews that are sensitive or emotional. Asma Khalid wanted to prerecord her interview with Bozoma Saint John, a marketing executive and author who wrote about losing a child, in order to explore the topic with her guest without any pressure.[7] "There was this degree of intimacy as she was describing it that I have just rarely heard women publicly discuss," Asma says. "And I really wanted her to talk about that and why she decided to make that decision."

Another reason to prerecord is if the subject is wonky or complex, like Asma's conversation with Jack Lew, who was U.S. treasury secretary during the Obama administration.[8] The broadcast version of the interview begins with Lew describing what it was like to be in the Oval Office 12 years earlier, in 2011, when the country came close to defaulting on its debt:

> LEW: There were nights when we were negotiating literally through the night, you know, at multiple levels. I was with the vice president then, and Senator McConnell, on calls. We were in and out of the Oval Office. And there were nights when you would check the Asian markets in the middle of the night to make sure the world didn't think that was the day we were going over the line. And this was not at a random time. It was right as we were emerging from the great financial crisis.

When the interview was taped, that clip actually came at the very end, as they were wrapping up the interview. Asma decided to restate her first question, since he seemed a little stiff at the beginning. "I think he felt more comfortable after talking to me for 20 minutes. And because it was a pre-taped conversation, we could shift it, and we could begin the interview with it. And that was really a better conversation."

There's a danger with pre-taped interviews of asking a million questions just because you can. Remember that a host interview is not a fishing expedition. Try to capture the freshness of a live interview by talking as though you have only one go at it, and focusing only on the most important questions. Your production team will thank you for not overloading them with tape that needs to be cut and cut and cut.

In Studio or Remote?

Another major change of the last couple decades is the increased frequency of guests and reporters connecting from their home or office and talking to the host via a video link. It used to be the norm for our daily news shows that the guest would either come into the studio and sit across from the host, or go to a local studio that NPR's engineers would connect to with an ISDN line. Or, a third alternative was that a producer would go on location to do a tape sync, holding the microphone and recording as the interviewee spoke over a landline or cellphone to a host, and then sending the high-quality audio file to the studio.

Now it's much simpler to connect remotely, not only because the technology has advanced, but because virtually everyone has it in their pocket. All it takes is a good broadband connection and a little fumbling with an app to do a live hit. While that has made it possible to reach many kinds of people we might not have before, it's harder for the host to establish the kind of personal connection that can sometimes enrich the conversation.

It even affected the quality of interviews with NPR reporters. As the COVID pandemic was subsiding, Rhitu Chatterjee went into the studio for the first time in more than two years to tape a *Short Wave* episode.[9] She sat across from the host, Emily Kwong, to talk about her latest reporting on school shootings and mental health. And toward the end, Emily looked at Rhitu and asked a question that was not in the script.

> EMILY: Rhitu, having done this reporting, how do you look at school shootings now?
>
> RHITU: I—you know, in many ways, I look at them as society's failure to protect youth, the perpetrator as well as the victims.

"It was an amazing question," Rhitu recalls. It came from an unspoken connection the host and reporter made in person, and it drew out a deep personal reflection from Rhitu. "And that's harder to do when we're sitting in our closets or desks and home studios and doing it remotely."

The Tyranny of the Clock

You may be getting the idea that there's little about hosting that is firm or predictable. But one thing is constant: the clock. On a show, at least. Each segment has a set length, so you have to make sure you get the entire conversation in during that time. "You're looking at the clock at all times," says Ayesha. "You're listening to what the person is saying, but then you're also looking at the clock and you're trying to figure out, do you have enough time to ask one more question?"

And then there are the posts, set moments when you have to close out a segment. "When you're doing live radio," Ayesha says, "at this exact second, you have to say: 'You're listening to NPR News,'" She didn't have to think about things like that when, as a Washington desk political reporter, she occasionally hosted the *NPR Politics Podcast*.

BREVITY IN PODCASTING

And yet. There may be no clock in podcasting, but it doesn't hurt to use a stopwatch. Aisha Harris says she and the other rotating hosts of *Pop Culture Happy Hour* always keep one running. "We have to be mindful of making sure that we're not going too long, because then that's more work for our producers." A typical taping lasts 30-40 minutes. Every additional minute means more editing and more tape from music, movies and TV shows that need to be mixed into the conversation, which gets cut down to 25 minutes or shorter per episode. And the producers and editor may be working on several stories at a time.

Hosts of roundtable and other conversation podcasts intervene in much the same way as radio hosts do when guests go on too long: "Not cut them off, but try to jump in when you can, when they take a breath, and steer the conversation," Aisha says.

On podcasts, you may not have a control room with a director giving you signals. Many podcasts are recorded with some or all of the staff remote. Still, Ayesha says that when she hosted the *NPR Politics Podcast*, the editors would be sending her messages while she was "really trying to make sure that we hit the points we needed to hit and that we [did] them on time."

INFORMALITY

At NPR, we've found that our podcast listeners tend to be younger and more diverse than our radio audience. That has opened the door to more informality and experimentation.

On the *NPR Politics Podcast*, there is a basic script with questions and an arc for the conversation. "But none of the answers are scripted," says Asma Khalid, who has hosted both the podcast and news shows. "The most we'll have scripted in the answer is who's going to talk into or out of a cut of tape." And the conversation is "a lot more free-flowing." The panelists might stop to discuss who takes which question. "Or we'll sometimes drop the last two questions because we feel the podcast was lively enough as it was."

That informality has seeped over to the news shows. On *Morning Edition*, hosts typically engage in banter while the engineer opens their mics to test levels. "I had long thought that that banter ought to be on the air somewhere," says Steve, "and I'm delighted that now it sometimes is."

The *Up First* podcast, produced by *Morning Edition* staff, comprises three reporter two-ways that are taped live at 5 a.m. every day, in the opening segment of the show's first feed. A producer then adds a prerecorded intro and outro. Sometimes, they tack on some of the banter Steve was talking about as a cold open. "We're just sitting there bullshitting, and they will take 11 seconds that are really funny and just stick it on, and you don't find out about it until later in the day when someone writes you about how funny that was."

The Host as Reporter

In 2019, after the Trump administration withdrew from a deal restricting Iran's nuclear program and reimposed economic sanctions, *Morning Edition*'s executive producer at the time, Kenya Young, got in touch with Steve. "I told him, we can cover Iran sanctions. We've been doing it for two weeks," she said, stressing that so many of the show's top-of-the-hour segments had been about Iran. "What I need you to do," she added, "is go into Iran and tell me, show me, report for me, how these sanctions, how this news that we've been reporting on for the past two weeks, is playing out in people's lives."

Sometimes a host trip is called for on major stories. For breaking news, it often means the host will go out and do their own reporting while also hosting the show on location. "When we turn the mic on in the morning, we should be where the story is," Kenya told me in 2022, just before she left NPR to become senior vice president of WNYC Studios. Even on stories where NPR has reporters on the scene, Kenya, referencing a former host and another major story, one of the deadliest mass shootings in the U.S., felt "*Morning Edition* should open where this story is and be able to say, 'I'm David Greene, and I'm in El Paso in front of the Walmart.'"

Because the host has a show to think about, they go with a field producer and sometimes an editor, both of whom pick up a lot of slack. "The host is inevitably pulled in 700 different directions at once," says Lauren Migaki, who spent seven years as a producer at *Morning Edition* and has traveled alongside NPR hosts and reporters to tell stories in Crimea, Israel and Brazil. "You have to be in the background, doing the things they don't have time for, and get as many obstacles out of the way as possible," she says. As discussed in chapter 3, the producer might be doing some of the reporting and interviews, or ordering takeout and getting a shirt laundered.

Why send a host when reporters are already there? Kenya answers that question by recalling another former host, Noel King, whom she sent to Minneapolis after the murder of George Floyd. Kenya told her at the time: "I don't need you to go do the protests. I don't need you in the streets talking to marchers. We've got reporters there for that. I need you to go into the communities, into people's homes, sit down and just talk, and just open the mic."

One question that arises for the local reporter is: when a host is assigned a trip in their region, will the reporter be eclipsed on an important story on their beat? "I've heard those concerns, but I've found it's a great opportunity to be a part of some lively storytelling," Middle East correspondent Daniel Estrin says. Hosts on trips are allotted longer-length segments, which allows for more atmospheric, take-you-there stories than a typical correspondent package.

Daniel finds that traveling hosts tend to do "journey stories," recording their drive and the various people they meet along the way. He recalls that in 2018, Steve took listeners on his drive through Gaza Strip checkpoints to a hospital in the border zone, where patients injured by Israeli soldiers were being treated.[10] Steve brought the fresh eyes of a visitor, while including Daniel's local expertise and reporting for context. "It's a chance to shadow a great storyteller, and observe new ways of approaching your beat," Daniel says.

As I mentioned before, many NPR hosts were previously reporters, and so the role comes naturally. And they relish it. But when you're reporting as a host, you no longer have the opportunity for long casual conversations with people on the street, says Don Gonyea, who has served as fill-in host on many NPR newsmagazines. "It's one thing to just be somebody with your handheld mic and a notebook in your pocket. But as soon as you've got your producer with the fishpole who's got their headphones on and they're next to you as you walk up, it's a whole different ballgame."

As a host, Steve has traveled to Iran six times and twice to China, and he's done scores of voter stories around the country during U.S. elections. He avoids standing with the TV trucks and "hovering over the breaking news," and shares Kenya's preference for getting into people's homes and hearing their stories. "You want to hear someone unwind a narrative or give a perspective we could not get from the studio," he says.

Like a good reporter, the host should look for things that give a sense of place: "finding sounds, stories and descriptions," Steve says, "that transport the listener to somewhere other than where they are." And perhaps more than a reporter does, the host needs to seek out local help. That could mean, when going overseas, working with an NPR correspondent or hiring a fixer. "In the U.S., we very often collaborate or at least go to dinner with someone from the local station," Steve says. In 2016, he and a producer arranged for a Spanish-speaking KCUR reporter to go with them to visit immigrant communities outside Kansas City. And in 2022, Stephen Fowler of Georgia Public Broadcasting "served as a sounding board and gave lots of advice" before a trip Steve took to interview voters in that state.

To Steve, the ultimate purpose of a host trip is to "add depth and insight to our coverage of a big story, or else to elevate a story that we feel has not received sufficient attention." I would add that it's also to play that role that we talked about at the beginning, where the host is the trusted friend, bringing the world—or, in the case of a podcast host, a certain part of it—to the listener's ears, whether it's from the studio or from the place where news is happening.

CHAPTER 15

Shows and Podcasts

At the start of his workday, Nathan Thompson takes out a pen and paper and sketches out two columns. On the left, he lists every hour between 5 p.m. and 9 p.m. On the right, he writes 5:30 p.m. through 9:30 p.m. He puts several hashmarks under each heading. At this point the scribbles, which only he has to decipher because they will be transferred to a computer later, look a little like a quickly drawn football gridiron, or a food menu without the dishes. Then he starts filling it in with the surnames of reporters—Khalid, Kelemen, Peralta, Rott, Frayer.

Nathan is a senior producer in NPR's Newscast unit, which produces NPR's hourly news broadcasts. He's responsible for programming the hours on his scratch sheet, which he draws up to plan the lineup of reporter spots. The five-minute broadcast at the top of the hour has room for up to four spots, and the three-and-half-minute broadcast at the bottom can have up to three. "It's a constantly evolving problem-solving exercise," he says. That's because there are variables galore: News happens all the time, reporters have other obligations that could delay filing a spot, all the spots and anchor copy need to be edited, and each newscast has to be timed to a tee to end before an NPR show or a member station's programming resumes. "Imagine a jigsaw puzzle that you're attempting to solve, and you have the perfect piece and you place the piece on the puzzle, only to find that the other pieces have disappeared or shifted in some way," he says.

Previous chapters described how an audio news story is created from the moment it's pitched until it goes to air, and all the editorial decisions involved in each stage of that process. But creating the vehicle that carries the stories to air, whether a newscast, a show or a podcast, also involves a great deal of editorial thought and judgment that begins long before showtime.

Show producers work with editors on the national desk, international desk, business desk and other teams to bring in stories that are a good fit

Notes by Newscast senior producer Nathan Thompson, who plans out the reporter spots that will be featured in upcoming broadcasts on a scratch sheet he draws up at the start of his workday. After the plan is firmed up, he transfers his scribbles into a computer program that displays each broadcast rundown on a newscast board, as shown in the figure on page 311.

for their particular production. On podcasts, the reporting and hosting are usually done by their own people, though it's not uncommon to tap expertise and even a voice from one of the desks. And both shows and podcasts hold their own pitch meetings, brainstorming the creation or acquisition of a broad mix of stories while trying to maximize resources and meet deadlines.

Newsmagazine and podcast deadlines may not be quite as relentless as those facing the Newscast team—which broadcasts every hour, and every half hour during drive times—but all of them have a similar puzzle to solve: how to present a selection of stories that will fascinate listeners and leave them feeling smart, informed, up to date and coming back for more.

Vision and Curiosity

Putting the production together starts with a vision for what it should accomplish. At *Morning Edition*, executive producer Erika Aguilar says she wants her show to help people feel "motivated, smart, strong and ready to tackle their day." Imagining herself as the listener, she tells her staff, as well as reporters and editors around that newsroom, the things their stories should accomplish:

- Catch me up. Tell me what I need to know, so I'm in the know.
- Give me astute, insightful analysis. Make me think about something I had not considered yet.
- Remind me that it's a privilege to be alive today. Despite our problems, there is much to fight for. Show me someone doing something about it. Show me hope, courage and conviction.
- Leave me with a smile. Make me shimmy. Make me laugh. Be my friend.

When deciding whether to bring in a story, she wants her staff to think about it from the listener's point of view and ask, Does the story change my access to power? That can mean explainers and updates on political and electoral developments, or about money or health. "Does it affect how you pay your bills? If you don't have good health, you really don't have that power." She wants stories that appeal to a wide range of audiences, including those that NPR has targeted as part of a strategy to increase Black and Latino listenership. But as a Latina herself, she says that doesn't just mean reporting on those communities, because they are "just as interested in all of the similar things that we might consider white listeners might be listening to."

Morning Edition programmers are "deeply curious about the world," she says. "And that doesn't just mean, 'I am so curious about those sea turtles' or 'Isn't it interesting that microbes in your body do *x*, *y* and *z*?'" When we spoke, *Rolling Stone* founder Jann Wenner had just been removed from the board of the Rock & Roll Hall of Fame after denigrating Black and female artists. "Blowing up his career in a ten-minute interview" triggers all sorts of questions, Erica says. "I want to know, is he going to double down? What does that say about the industry moving forward?" She was also interested in the decision to remove Wenner from the Rock & Roll Hall of Fame. "I want to go talk to the Rock & Roll Board. I want to know why they did that. Let's be deeply curious about the world. Even when we may disagree with it."

Curiosity is also what drives programmers on other shows. "*All Things Considered*, true to its name, is about exploring the incredible diversity in the world," says its executive producer, Sami Yenigun. He wants to inspire listeners by telling them "character-driven storytelling that shows people avenues, maps towards ways that they can live their own lives."

Because of the timing of his show, he assumes listeners already know something about the major stories of the day, and may have read an article or two about them. "And so, what's an angle that we can bring that moves the story forward in some kind of a way—or just, not even forward, but just an angle of the story that you hadn't even considered?" he says.

Putting the Story First

In contrast to shows like *Morning Edition* and *All Things Considered*, many NPR podcasts—which include some that are also broadcast as weekend radio shows—can take a step back and spend more time refining the storytelling.

Planet Money, which was born during the financial crisis of 2008, has as its subtitle, "The Economy Explained." When I asked its executive producer, Alex Goldmark, to describe his vision of what the podcast should provide listeners, he said: "You enjoy listening to the end because it's just a good story or it's a fascinating listen. It doesn't really matter what we're about, or how capital-I-important our topic is, if it isn't actually enjoyable to listen to." Afterward, he added: "It does have to explain something about how economic forces shape your life. About how some concept or system works that connects to you as a listener, to a decision you face, or an experience you've had, or maybe to your confidence, explaining something you've been ashamed to admit you didn't know, like how interest payments work on your credit card."

In podcasting, listeners always start at the beginning, so you don't have to interrupt periodically as in radio to recall the topic or reintroduce the guest. That gives producers more narrative leeway to be creative with their storytelling. "We can build suspense, we can hold back information," says Alex. "We can intentionally not tell you something in the intro because we want it to be a reveal or a surprise later."

For instance, *Planet Money* host Alexi Horowitz-Ghazi reported an episode about Hollywood studios competing to make movies based on real events. The event in question was the meme-stock phenomenon, which hit the headlines when Reddit users fueled a run on shares of GameStop in 2021. At one point, there were nine competing GameStop projects in Hollywood: It was a winner-take-all contest. "Because plenty of people might be interested in buying a ticket to relive the GameStop saga on the big screen," Alexi says in his radio version of the episode. "But how many are going to go for version number 2 or 3?"

The movie that won the race was *Dumb Money*. "In the podcast episode, we don't tell you that," says Alex. "We tell you we're following the making of a movie, of one of the nine, and we don't let you know if it won the race or not until late in the episode."

But *Planet Money* recast the 28-minute podcast episode to also run as an eight-minute segment on *All Things Considered*. To do that, they didn't just cut it down; rather, they told the story in an entirely different way, in which *Dumb Money* is identified from the get-go. "We give it away because we feel on the radio, on that given day when *Dumb Money* was just being released, people would lean in more if we said, 'Hey, here's the behind-the-exciting-scenes race to make the movie you've already heard about.'"

Many listeners will already come to a podcast with an interest in the subject matter. But that's not a license for flat storytelling. A bored listener could very well leave your podcast faster than they would have turned a radio dial, says Alex. "The skip button is probably actually closer on a podcast to somebody's finger than a radio listener who has it on all day and they just grumble at the station but will stick with it."

Segments

VARIETY IN A NEWS SHOW

On news shows, programmers are always looking at the whole production, trying to ensure a variety of segments that come together for a satisfying and enriching listening experience. For the Newscast team, that happens 37 times a day on weekdays, and 24 times on weekend days.

The newscast is like a minishow. It leads with the most important stories and then moves to lighter fare—a movie review, a sports result, an obituary. But not always. When I spoke to Nathan Thompson, the senior producer, a newscast had recently led a broadcast with the death of singer Tina Turner.

Every broadcast has its own board, which lists the items in the order they will be aired. The top-of-the-hour broadcast always starts at one minute past the hour, and runs until six minutes past. One complication for anchors is that many stations air only the first three minutes. "So at 5:04, or 5:03:59, really," Nathan says, using the 5 p.m. ET broadcast as an example, "Jack Speer or Lakshmi Singh or whoever is doing that newscast needs to be aware of the clock and not speaking at that time, because many local stations will cut away at that post to do local news broadcasts."

In those first three minutes, producers will commonly put up to three spots and anchor-read copy. On the back end, the remaining two minutes, there will be room for another spot, more anchor copy, and a funder. Producers have different types of spots and other elements they can choose from to vary the texture of each broadcast. Here are the main ones:

VOICER. A 40- to 45-second reported spot focused on a single news development and composed entirely of the reporter's voice. The voicer also comes with an anchor intro, which the reporter is expected to draft, usually made up of two lines—a grabber sentence, which, Nathan says, explains "why should somebody listen to this," and a second sentence that identifies the reporter and gives a little context. For example, here's an intro I drafted for a spot I filed during a stint on the culture desk: "Juries at a couple of international film festivals delivered their verdicts this weekend on the best movies of the year. NPR's Jerome Socolovsky reports the awards highlight promising young directors."

WRAP. Same as a voicer, except that it includes an actuality or sound in the body of the spot.

Newscast (2023-06-05) 1700

05:01 PM - EST

01:00	Newscast I	02:59	Live	(+00:16)
	KELEMEN #1 BLINKEN SAUDI			00:58
	SAUDI OIL			00:29
	SMITH #1 IOWA BUILDING COLLAPS...			00:55
	GANNETT-WALKOUT			00:37
	STOCKS EXPLAINER			00:16
04:00	Newscast II	01:39	Live	(+00:10)
	UKRAINE PUSH			00:26
	PERALTA #1 MEXICO ELECTIONS			00:56
	EARLY HUMANS			00:27
	NC014		Live	00:19

A newscast board for the 5 p.m. ET broadcast on June 5, 2023, which Nathan Thompson planned in the upper-left corner of his scratch sheet in the figure on page 306. The spots are slugged with the surname of the reporter and the topic. The remaining items indicate segments of anchor copy that weren't on Nathan's scratch sheet. (Numbers next to reporters' names are meant to prevent confusion in case those reporters filed more than one spot that day.)

Q. A cut of tape, about 20 seconds long, taken from a reporter or producer asking themselves a question and sending the answer to the Newscast team. These are especially useful for the broadcast at the bottom of the hour because it's shorter and can't fit as many elements.

READER. Anchor copy of varying length about news stories and market updates that aren't covered otherwise in the broadcast.

For producers, the trick is to take all the incoming spots, weave them with anchor copy and program them in a way that maximizes their shelf life but allows each broadcast to sound fresh. "It usually is not a question of not having material to schedule the newscast," says Nathan. "It's how are you going to solve the puzzle, and how will I maybe use that spot again, two and a half or three hours from now, maybe for the bottom-of-the-hour anchor?"

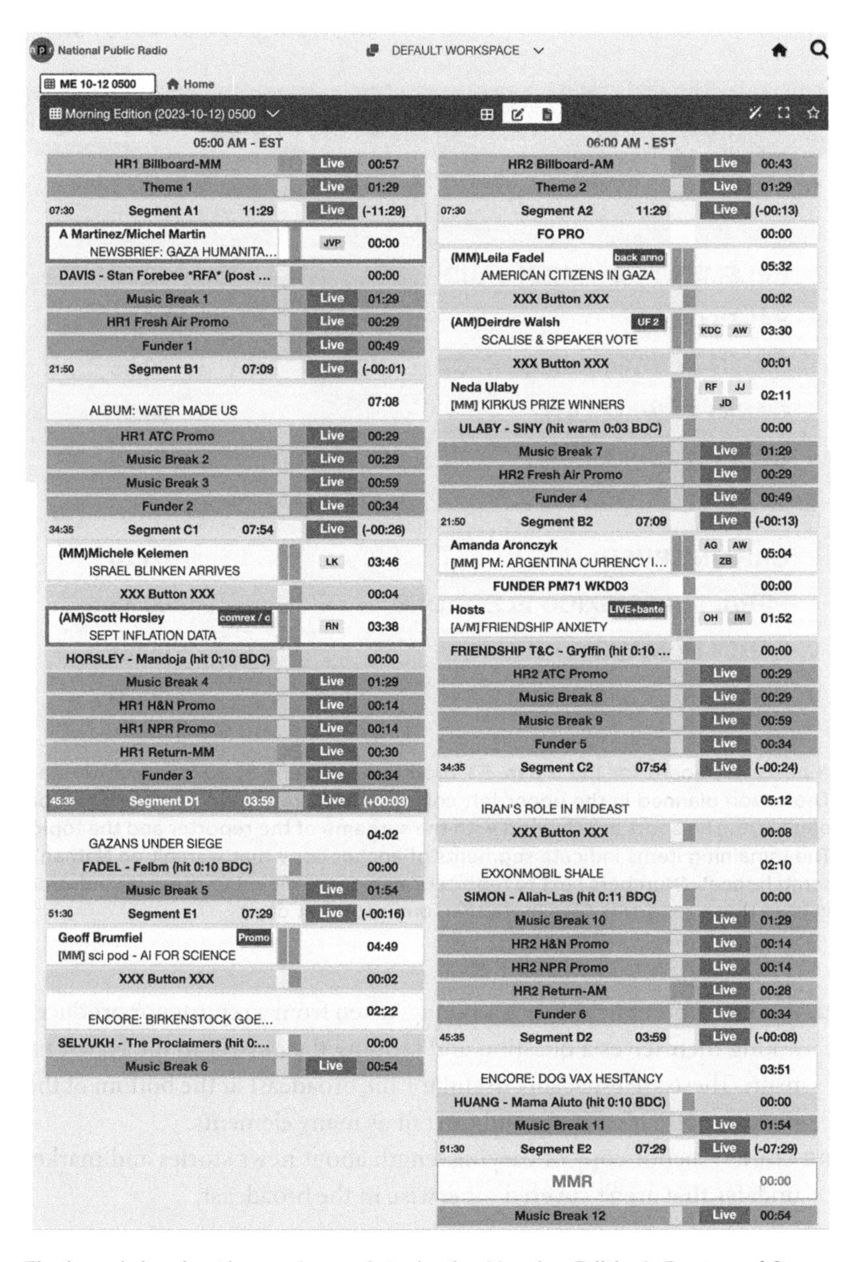

The board showing the rundown of stories for *Morning Edition*'s 5 a.m. and 6 a.m. ET feeds on Oct. 12, 2023. Each hour has five segments, labeled A through E; each segment contains one or more stories and may be followed by music breaks, promos, funders and so on. A billboard and theme music appear at the top of each feed. Each item has its own naming conventions according to the type of story (interview, reporter piece and so on) but will always have the slug in capital letters and the length of the story; it may also show the host's name or initials, the reporter's or guest's name as well as the initials of producers and editors and other flags such as the connection type and the status of the audio and script.

Newsmagazine programmers have an assortment of longer segment types to work with. On the daily newsmagazines, each hour contains five segments, ranging in length from four to 11½ minutes. On the weekends, the segments are longer, up to 16 minutes long. The shortest segment usually fits just one story, while the longer ones typically have two or more, though it's not unusual to have a single long-form report or a shortened podcast episode in one of the longer segments. Here are the main types of stories that newsmagazine producers can choose from:

TWO-WAY. A host interview with a guest or a reporter, either from NPR or another news organization. In rare cases, there may be more than one guest, such as for a music band interview. Or there can be two or more reporters giving multiple perspectives on a story, usually referred to as a *roundtable.* Two-ways and roundtables usually run anywhere from three and a half to eight minutes long.

REPORTER PIECE. A produced story, with a mix of ambient sound, interviews, and standups and narrated by a reporter, and typically three and a half to five minutes long.

HOST PIECE. A produced piece narrated by a host. Often runs an entire segment.

SUPERSPOT. A short news report that is meatier than a newscast spot, and usually runs one and a half to two and a half minutes.

TAPE AND COPY. A host-narrated segment, typically up to three minutes in length, reported and mixed by a producer who also does the interviews and drafts a script that a host may tweak or rewrite to put it in their own voice.

Each of these categories is its own art form. For instance, tape and copies, originally created to fill holes in the board, have come to offer the listener a breather and inject some lightness or wonder, allowing the show to cover quirky or off-the-beaten-path topics. *Morning Edition* producer Barry Gordemer recalls one he made about a tuba museum in Durham, North Carolina, and another about a YouTube channel that featured Steven Spielberg-themed cat videos.[1] "They can take all frickin' day to do," says Barry, even though they are short. But he calls them a "great playground for audio storytelling techniques" that often stay with the listener. "At the end of the day, what are people talking about? Was it the six-part series about the Middle East or the piece about the tuba museum with all those funny people and great tuba sounds?"

VARIETY IN A PODCAST

Podcasts don't work with segments the same way the shows do. But producers do think in terms of genres. *Planet Money* does a lot of explainers, and Alexi has different approaches he'll take. One is historical: "Here's the last time this thing happened in a big way that sheds light on what's happening now. It doesn't answer where things are going, but we're going to tell you this parallel story that's going to explain the dynamics of the moment." Other times, he'll do an origin story, going back to "the moment an idea was invented: the open office or the bond or double-entry bookkeeping, cryptocurrency or the internet cookie.[2] Whatever it is, there's something inherently exciting about talking to a person who was in the room when the thing was invented."

And yet another type is the anatomy story, which is how Alexi classifies the *Dumb Money* episode. Unlike the origin story, it doesn't go back to the first time a studio made a movie based on real events. Instead, it focuses on one "particularly newsy and high octane" example as a way to explain the phenomenon. The approach worked, he says, because he was able to talk to the movie's producer and co-writers. "In the editorial meeting, there was a lot of excitement around doing it as a 'horse race' style story, keeping people in suspense about which project would win. If I hadn't got the inside players, I might've had to do a more expert-driven explainer."

Alex says each of the seven or eight co-hosts has their own repertoire of genres. And that, combined with their varied reporting interests and strengths, allows the podcast to tell all kinds of stories. "We are a naturally varied show because we have an ensemble cast of hosts who are charged with following their curiosity," Alex says. "The team is somewhat intentionally assembled to have a diversity of expertise, backgrounds and interests."

Alex Goldmark on "Doing the Thing" You're Reporting On

This goes back to the founding of *Planet Money*, and I believe in it deeply: We should be constantly surprising our audience with form and craft in the service of explaining the most complicated, confusing, but still important and relevant ideas from economics and finance and business and money. And maybe the most *Planet Money*–brand attribute is the projects where we do the thing ourselves in order to explain the system.

When we did a story to explain how does oil get from the ground to your gas tank, we bought oil. We literally incorporated ourselves as oil buyers in the state of Kansas with the secretary of state in Kansas. I had to get the NPR comptroller to fill out forms that were very confusing as to why we had to actually do this. And we got asked over and over again, do we need to do this, or can we just follow an existing oil buyer? And I was like, I need to do the buying. We learn details and understand the experience we'd never have otherwise about how profit is made in the oil industry, and what affects the fractions of a penny per barrel sometimes or how oppressive weather will eliminate your whole enterprise maybe for that week. Our truck got stuck in the mud and then the oil price changed and then we lost money. And that made the story better.

We wanted to do the economics of superheroes, so we found a public-domain superhero, and we reclaimed him and turned him into a product that we could monetize in the same way that Disney monetizes Marvel. We started a record label. We wanted to get a million streams and see how much we got paid by Spotify. Once, *The Indicator* bought a junk bond and just followed it to see what happened to it. And we didn't just cover how AI clones voices, but we found somebody to clone one of our own voices, former editor of *Planet Money* and longtime correspondent Robert Smith. And then we actually played that as a podcast for the audience and then responsibly destroyed it so that it couldn't be used again to clone Robert's voice. So we want to do the thing. Robert himself said, "If there's a thing to taste, taste it. If there's a thing that makes sound when you knock on it, knock on it."

And sometimes when we do this, we have a little investment fund in a jar with a label on it that we pull out when we need to for those kinds of things. We're usually buying very small amounts, like a few hundred dollars. We don't want a listener of a public radio station to think, "I can't believe you bought a toxic asset," even if it's not that much money. Because we buy one share, or the smallest quantity we possibly can, of something. Oil was different. We had to use company money. We had some assurances that we weren't going to lose it all. Right. In the end, I think we lost less than $1,000, which is less money than we would have spent on the travel and hotels for our team to get there.

Producing Shows

Although most of the time listeners hear just the voice of the reporter or host telling a story, it's always a team effort. *All Things Considered* has a staff of 50 people who tend to work in teams of three: host, producer and editor. In some roles, there are specific skill sets, such as book editors and features editors. And while most producers are general assignment, working on stories assigned to them in the morning that will air later that day, there are also producers who focus on assembling the show. In chapter 3 you learned about the different kinds of roles that producers play. Here's how the producers of *All Things Considered* put together the show each day, as outlined by producer Jonaki Mehta.

TEAMWORK ON A NEWSMAGAZINE

The line producer begins the day by running the pitch meeting at 9:30 a.m. ET, after assessing what various desks have to offer based on the news of the day. Show producers, editors and managers pitch stories during the meeting, and then the line producer considers those pitches, along with managers and whichever members of the team choose to stick around and weigh in. The line producer is on the lookout for two-ways with people in the news, lawmakers, scientists with new discoveries and people affected by things happening in the news. They're thinking about the most important news of the day, but also considering the show's composition in terms of length of each element and the nature of the content. If it's a very Washington-heavy news day, for example, they may want to balance the show with international news, culture, arts, or other nonpolitical stories. The line producer is monitoring the news and making changes to the show board based on the news cycle throughout the day. They are also the final set of ears on every mix that goes to air. They continue to steer the ship, working with hosts, other producers, editors and desks, until the first feed of the show airs from 4 to 6 p.m. ET.

The update producer is essentially the line producer for the later two feeds of the day, which air from 6 to 8 p.m. ET and 8 to 10 p.m. ET (to serve stations in the middle and west of the U.S.). They begin their day at 1:30 p.m., by which time the first feed is getting into shape. They have a conversation with the line producer about which stories are expected to continue developing throughout the evening and may need an update in the second or third feeds of the show. The update producer, in consultation with managers and desk editors, plans the update coverage with the hosts who take over after the first feed, the midday producer, and the late producer on the show that day. They assess and rearrange the show board accordingly, keeping in mind that breaking news can always reshape the show after the first feed. If live coverage is required during the second or third feed, the update producer helps with scripting two-ways, arranges audio connections and coordinates the work of hosts, editors and reporters. They also direct any live coverage that happens during the second or third feed.

The show producer's day varies greatly, but it begins with the pitch meeting. Once they're assigned a story, they are usually either coordinating same-day host interviews, mixing stories that come to the show from desks, or writing and producing tape and copies, which are similar to reported pieces but usually shorter in length.

A same-day host interview can be either internal, meaning with an NPR reporter, or external, with a guest who may be a subject-matter expert or person affected by a particular news story. If the producer is coordinating an internal interview, the reporter and desk editor normally draft a script, and the show editor edits that script to ensure that it's the desired length and that it matches the tone and style of the show hosts. The show producer is responsible for coordinating connections between the host and the reporter and for pulling any tape the reporter might need played during the interview. For an external interview, the producer vets potential guests, pre-interviews them, writes the script and records the interview. Then, after the taping, they confer with the host and editor to decide how to shorten the interview to the desired length, then proceeds to cut and mix the interview for air. If the external interview is live, the producer primarily coordinates the technical and audio aspects of the interview.

TAPE AND COPIES AND ENTERPRISE STORIES

Show producers may also be assigned to mix reporter pieces and put together tape and copies; they may even be responsible for the bigger enterprise stories that *All Things Considered* airs. In the case of a reported piece, they communicate directly with the desk editor or another producer to cut and mix the piece. For a tape and copy, they find people to interview, write the scripts and mix the piece after editing and taping with hosts. Sometimes the host's tracks in a tape and copy are done live, and the producer will line up the cuts, Jonaki says, "so that the director and engineer can fire them during the show."

For enterprise stories, she adds, the producers do "long-term planning with hosts and editors, identify and pre-interview guests, arrange logistics for field reporting and producing, handle all the technical aspects of field producing, and ultimately help to script and mix the final piece for air." That will involve traveling with a host and carrying out duties such as setting up and recording interviews, attending press conferences and liaising with shows. It could also involve logistics planning, as discussed earlier.

PROMOS

Producers on shows and podcasts also write promos. A promo is a short script highlighting one of the most compelling stories on the next show or the next episode, which will run on another NPR show or podcast. Here's a promo for a *Morning Edition* story about airline economics that aired on the previous day's *All Things Considered*, hosted by Michel Martin:[3]

MICHEL: You get to the airport . . . and no Delta, no American, no United. That's what you can find at some regional airports across the country.

UNIDENTIFIED SPEAKER: I mean, they're literally abandoning rural America.

MICHEL: Why more small cities are being cut off. On the next *Morning Edition* from NPR News.

A show promo has to be 20-30 seconds long and must "get across complex ideas in a very short window," says Sami Yenigun, executive producer of *All Things Considered*. "And you have to have very tight writing that weaves pretty naturally in and out of that piece of tape that you've selected."

A senior producer planning the next day's show who "knows the content well enough to highlight the piece that we're most excited about" will select which story should get the promo. Then that person confers with the update producer, "whose hours are better aligned to write the copy," Sami says.

Booking Guests

When producers at the newsmagazines book guests for host interviews, it can look like football players breaking out of a huddle. After the pitch meeting is done and the pitches are accepted, the producers get to work on snagging the best people for their hosts to interview.

It wasn't easy back when Melissa Block, who would later host *All Things Considered* for more than 12 years, was hired as a desk assistant on the show in 1985, booking guests for hosts Susan Stamberg and Noah Adams. The work was done with worn-out Rolodexes and voluminous phone books that were often out of date. "It was always a big deal if a producer went out on a reporting trip. 'Get the get the phone book from the town that you're in and bring it back!'" Melissa recalls hearing.

And if you were trying to reach someone who was away from their phone, you just had to wait until they came back, or find someone else. Now we can search online for people and find their contact info. And virtually everybody carries a cellphone. What hasn't changed is the art of finding the right person: someone who is informed, vibrant, compelling; someone you can learn something from, be entertained or inspired by—or possibly be provoked or angered by.

Before you make any calls as a booker, you must understand the conversation you're booking for. Write a one- or two-sentence summary and share it with the editor, host and anyone else working on the interview. Then find a few articles that will help you come up with smart questions, and mine them for names. They may not be good guests, but they will likely be able to suggest other people who could be.

"You're basically a reporter," says Melissa. "You're figuring out the best people to talk to and whether they're not just versed in what they can say, and smart, but also, can they talk on the radio" or on a podcast?

GOOD TALKERS

When you get a prospective guest on the phone, you'll want to do a pre-interview. That doesn't mean testing them on all the questions the host will ask. This is just a chance to get a sense of who someone is, how they talk, what they know. And you should have a robust BS meter. They may be a consummate conversationalist, but if they don't know the topic well, thank them and move on to someone else. The only times you can get away without doing a pre-interview is when it's a known news maker such as a high-profile politician, business executive or celebrity; you pretty much already know what you're getting, because they've spoken publicly before, and they probably don't have time for a pre-interview anyway.

One of the cardinal principles of booking has traditionally been to find a "good talker." Are they, says Rebecca Ramirez, producer for the science podcast *Short Wave*, "the person who is going to get the listener to continue listening?" That is still the aim, but unfortunately, she adds, the term "good talker," especially when applied to experts, has often justified going back again and again to the same sources, favoring a narrow slice of not-very-diverse experts. That's why, even though you are usually looking for a single guest, it's important to talk to an assortment of people—including those who haven't been on air before. And that's where the journalism is done: All those conversations inform the questions that will make the interview shine.

Bookers research their guests and sometimes also prepare a focused summary of the background of a story, any previous public statements by the guest, and proposed questions for the host to ask. For next-day interviews, the booker will prepare a packet, sometimes with books, for the host to take home and read that night.

As Melissa says, a booker must be a reporter, fact-checking and calibrating the original story idea as you go. You may discover that the premise of the pitch was a little off, or find more interesting pieces of the story, or dig up an unheard-from stakeholder. A good booker is resourceful and tries to find people in places they may not themselves be comfortable: in certain Facebook groups or subreddits, by searching online for businesses and calling them. A good booker also knows how to read the room. "Be personable and human and gentle in some cases, or funny and a little disarming," says Alexi, who started his NPR career as an intern at *All Things Considered*.

RELUCTANT GUESTS

And be persistent. Once, Alexi wanted to do a story on joke theft, and he tried to get an interview with an internet personality who was the target of much online ire because he reposted other people's jokes and made money off them.[4] It took "six weeks of emailing and long conversations trying to convince them that if there's any platform to kind of tell your side of the story, this is it."

With reluctant guests, it's best to make your case in two ways. First, by conveying why you want them on your show or podcast. Here are some possible reasons:

- This is a very important story for others to hear.
- This is a very interesting story.
- No one will have the perspective that they have.
- They can trust NPR to provide them a fair and balanced platform.

And second, by conveying why they should tell their story to NPR's audiences:

- Because it's the right thing to do.
- Because it will help others in a similar situation.
- Because their story will make a difference to others.
- Because interview requests will come from all over the place; once they've told their story on NPR, they will have already reached a huge audience.

During the Trump years, many Republicans refused to talk to NPR, following their leader's vilification of news media. But NPR still tried to book the party's lawmakers, in the interest of fairness and presenting all sides of political debates. Krishnadev Calamur, the head of the Washington desk, says he convinced some of them by highlighting our ethics and our format. Unlike some conservative talk shows on cable TV, interrupting all the time is not our style. "What I would urge you to do is listen to the conversations your party leaders have had on our air," he would tell them. "Quite frankly, even when you go on Fox, you're not getting that much time to talk."

During the pre-interview, make sure to set realistic expectations for your guest. You can never be certain if the segment will be three or five minutes long, or when it will air. You can say you'll try to mention their book if time allows, but don't make promises.

If they will be connecting to the show or doing the podcast interview remotely, ask them to record themselves during the pre-interview and send you the file. You don't want to do the interview, says Rebecca, "and then find out that the guest didn't realize how to record themselves."

Regardless of how the actual interview goes, send a thank-you note afterward. Guests take the time to speak with us and share their knowledge for the benefit and enrichment of our listeners. So show them gratitude. And tell them if they have any feedback for you or your team, you'd love to hear it. Then save the guest's contact info along with any notes that will be helpful for future bookings.

Editing Shows

As I said in chapter 10, the editor's role in putting together a story cannot be overlooked. The same applies in a show or podcast.

In podcasting, the editor's role is similar to a desk editor. They often work with the host or reporter from the moment they pitch the story, throughout the reporting phase and until it's ready for air, though sometimes a story has to be handed off to another editor for reasons of availability. The main difference in podcast editing at NPR is an additional phase in the process known as the group edit. I'll talk about how that works a little later.

All editors have the same overarching goals: ensuring accuracy and the highest standards of storytelling and journalistic ethics. But show editors, instead of working with reporters, mostly edit hosts and producers. The process and the dynamic are different, but the aims are the same.

There are times when the editorial jobs resemble each other. That can happen when hosts travel as reporters and file produced stories with sound and actualities. There are also show-created pieces, reported by producers and narrated by the hosts, which also require the same kind of editing as a reported story. Those include tape and copies that are up to two and a half minutes long or the host-reported pieces I just mentioned that can fill a whole segment and run 10 minutes or more. When reported pieces come from desks, editors are responsible for rewriting the draft intros, if necessary, to match the latest news developments or to fit in with the flow and tenor of that day's show. The line editor, a role I'll talk about later in this section, gives the whole piece a final listen.

Unique to the show editor's responsibilities is the management of the host's on-air conversation with guests. Editors make sure the host is prepared for each interview with a script and necessary background on the guest and the issues they may bring up. Editors or producers typically draft the intro and questions for the host, but the editor is responsible for the journalistic integrity of the outcome. The editor listens for accuracy, fairness and completeness, and suggests cuts to make the story fit the length allotted to it on the show. "It's not strictly a management role in the definition of labor management," says *Morning Edition* senior editor Jacob Conrad. "But you're managing stories, and you're managing the flow of the stories to air."

Show interviews can be tricky. They must tell a whole story, but unlike in a reported piece, there is usually just one interviewee. Any needed context and balance should be either written into the intro, or made clear in the host's questions and spur-of-the-moment responses. Sometimes you may need to reach out to a third party for reaction to be added to the script.

If a guest just spouts talking points, the editor can help the host steer the exchange toward a more productive conversation. When *All Things Considered* host Ailsa Chang interviewed Texas Democratic Rep. Greg Casar about the debt ceiling standoff in 2023, he criticized a deal President Biden had just struck with Republican leaders. But Casar dodged the question of whether he'd vote for it. "I think Ailsa had to ask a total of four times," says Ashley Brown, the editor on the segment. Ashley was sending messages to Ailsa to make sure she kept on him. It's not that a host will forget, but much of their mental energy is focused on other things such as questions they need to get to before the clock runs out.

Although the host should and often does prepare for an interview, sometimes the host has to go in cold, possibly because of a late-breaking story or because another all-consuming story is taking up their read-in time. Or, in the case of *Morning Edition*, the host hasn't been awake very long. The editor's help in this case is indispensable. "No host should be in a position of having to do the basic editorial work at three in the morning," says senior editor Jacob Conrad. *Morning Edition* goes to air at 5 a.m. ET, and all live interviews are done during the first feed and the first segment of the second feed, which ends at 7:19 a.m. Some of those interviews may be about news that developed overnight, while the host was asleep. On *All Things Considered*, staff have the benefit of reaching guests during normal business hours, so most of those interviews are recorded in the morning or early afternoon before the show goes to air at 4 p.m. ET. And on the weekend shows, many of the non-time-sensitive interviews are taped late in the workweek.

Editors work with producers to prepare a script with a draft intro and suggested questions, along with recommended readings and other guidance such as a bullet-point summary of the main themes and possible points to push back on. "A lot of it comes down to doing your homework and seeing what this person has said previously," Ashley says, referring to interviews of lawmakers and other public figures. She'll especially flag talking points that someone has made in the past that have been proven to be wrong or not representative. She will also try to anticipate facts or personalities the guest may assume are public knowledge. "They might rattle off a name, and it's the host's job to help clarify it. 'This is who you're talking about when you mentioned this person.'"

The point of doing all this research for what may be a four-minute interview is not only to prepare the host. It also allows the editor to listen in, give guidance in and fact-check in real time. If it's live, any error or misstatement has to be corrected immediately. If it's prerecorded, it will need to be dealt with before the guest leaves the studio or the call ends.

Just as the desk editor helps a reporter see the big picture, a show editor may need to do the same with a producer cutting an interview. The conversation should tell a story from beginning to end, and it shouldn't have too many twists and turns. But often there is so much good material, and producers sometimes know the subject matter so well (as they should!), that it's hard for them to determine what will interest listeners the most.

Neva Grant once produced a two-way edited by the late Neal Conan before he became the longtime host of NPR's *Talk of the Nation*. They butted heads over a section in the two-way that Neal wanted Neva, who was a young producer at the time, to remove. "I was freaking out and saying, 'You can't take that out! That's cutting the heart out of the story.' And he just looked at me and said, 'Just remember, the audience doesn't hear what isn't there.'" Neva saw his point. "He's right! *We* know it isn't there. The audience doesn't know." When you need to shorten a long interview, there are clues to help you figure out what to cut, she says.

> Often in a long interview, you hear a question where you're like, "Oh, I didn't think of that." And unless the answer is just amazing, that's usually the Q&A you should take out. The question that you didn't think of might be a great question, but it really may not be something you need if

> you're trying to take a ten-minute interview down to four minutes. That's a luxury question. It's like, if you only have an hour with a friend, someone you haven't seen in a long time, you're not going to tell them how your bus ride was getting there. You and that friend are automatically editing and curating because you have such a limited amount of time together. And so you have to just think of it in those terms. I've got four minutes with the listener.

THE LINE EDITOR

While most show editors work on individual broadcast segments, the line editor edits the show as it's being broadcast. And they work as a partner to the line producer. "The line producer is more focused on audio, whereas the line editor is more focused on copy," says Ciera Crawford, who was a regular line producer for *All Things Considered*. "And in a perfect show, they're working in tandem: They are the last eyes and ears of what's going on the show."

The line editor reviews and updates intros as the segments go to air and listens to them live. Although other editors prepared the stories, the line editor, says Jacob, "is looking at the show in a more holistic way, thinking more in terms of hours than in segments."

Despite all the work done to prepare scripts, the hosts sometimes depart from them to put them in their own voice, or because they catch a mistake or have their own expertise to add. Entire segments may be dropped, added or moved around in the middle of programming. That's because every NPR show, says Jacob, is ultimately an exercise in improvisation.

> And the role of the line editor is to be the sort of conductor of that sort of organized improvisation. That's us at our best. We never want the hosts to rely too heavily on a written intro or written questions. They should be responding to the story in real time, responding to the person they're interviewing. And so to a certain degree, my job is to give them the latitude to do that.

The line editor may decide that a conversation needs a little more time, and that may force the show to bump the next piece, or "dump out" of it, which means to end it early. The decisions have to be made quickly, and they are never easy. "What's the value of this interview versus that piece? That piece might be something that brings joy or enlightenment. And to drop it might affect the overall texture of the hour," Jacob says. It's a collaborative decision, made with the host and the line producer. If the host asks for it, the editor can overrule them, but the decision is undergirded by a sense of mutual respect. "The host has to trust that I am coming at this not in a malicious way, but with the goal of making that host sound the way they intend to sound, and beyond that, making the show sound as we all intend it to," he says.

GROUP EDITING

Planet Money actually grew out of a program in 2008 called *A Giant Pool of Money* that was a special collaboration between NPR and *This American Life*. And one practice that *Planet Money* inherited from *This American Life*, and which has since spread to many other podcasts, is the group edit.

The group edit at *Planet Money* has evolved over the years. Long ago it involved the whole team, who met in person; now it's about five people who gather virtually for a listening session after the reporter has planned and worked through the story with their designated editor. The team can include the host(s), the producer(s) and others, including an intern if there is one, and maybe someone from *Planet Money*'s sister podcast, *The Indicator*. They listen to the story and take turns giving feedback, always working on the assumption that the best idea can come from anyone. "Even if they're the intern and they say, 'I'm confused at the top,'" says Alex, the executive producer, "that's a valid piece of information that we want to address."

The group edit actually works on two levels. Besides the listening session, which is ideally about the bigger picture, there is a shared online document where people can make specific line-editing comments. While the document is good for detailed reactions, the discussion in the listening session can be more general and impressionistic. "It is a place to find a better structure or sharpen an idea or give feedback," Alex says, "on whether or not the tone or the pacing is working or a joke is landing or a metaphor you're using holds."

In the newsroom, the business desk also does group edits, but the group comprises only editors. It can be a scheduling challenge to get them together, but they do find weaknesses in the story that the original editor might have missed. As former tech correspondent Shannon Bond says, that often happens because her editor is already familiar with the story.

> But the other editors, maybe they've only seen the pitch, or they've heard me talk about it once or twice in group meetings; they really are the freshest possible sets of ears. And so if they say, "I didn't really get this," or "The structure here doesn't work for me because I need to know this before we set up that," I try to really take that to heart, because I want it to be as clear as possible.

There are some reporters who prefer to work with a single editor. And Shannon acknowledges that group edits can get "pretty nitpicky," with editors who are not on her beat sometimes asking for too much background. Sometimes she feels like saying, "Everybody knows this. We don't need explain this." The edit may go on for several rounds, but in the end, the story is almost always stronger than it would have been with a single editor, she says.

Directing

One of the most exhilarating jobs on NPR's shows is that of director. They are a kind of traffic cop-*cum*-orchestra conductor who makes sure everyone in the studio is in their lane, making sonically beautiful programming.

"It's a high-wire act, and it can be really fun. But it's not for the faint of heart," says Melissa Block, who was also a director before she became a longtime host. That was back in the analog era, when "doors were being held open as people ran in the studio with reels of tape in their hands."

The job now is just as fast paced. "Having been on the other side of the glass as a host, to have a calm and really confident director is gold," says Melissa. Here is how one of those calm and confident directors, Lindsay Totty, describes the job he's held since 2010.

LINDSAY TOTTY, DIRECTOR, *MORNING EDITION*

The Director's Responsibilities

As the director, you have two main responsibilities. One, you're in charge of making sure that everything runs smoothly in the studio. And two, you're responsible for choosing the music for the show.

The cardinal rule of directing is that nothing happens until you say it happens. Everything that the engineer does to put sound on the air, almost all of it requires an instruction from the director. That can be something like "Open microphone," or "Close microphone" for the host or for the guests. Or there's music that should be playing underneath the end of a story and you tell them to fade it out.

If you're doing a live interview, you can see the script. The engineer has other things to worry about—they have that whole console in front of them—so that's why they need you, as the director, to give them those verbal commands. When a reporter or a host is going to call for a clip, you say, "Ready. Hit the clip." If the clip has the sound of a politician giving a speech at a big rally, there might be some applause at the end, and you need to tell the engineer to fade it down and open the reporter's mic so that they can continue their answer and describe how that clip illustrates their story.

There are also commands you give the host. When the host is working remotely, you may press a button to talk down the line to them and say, "Go!" when it's time for them to start reading an intro or a backannounce. If the host is in the studio, you give them hand cues. I'll usually hold a hand straight up to show them that their mic is open, and then point at them to start talking.

And if there's a live guest on the show, you're talking to them, making sure that they are comfortable and prepared to be interviewed. You'll ask them to confirm certain details, how to pronounce their name, what their job title is. Sometimes, the host wants a quick chat with them beforehand. So you want to get the guest situated early. Also, you want to make sure that the engineer or technical director has time to talk to the guests as well, to make sure they're getting a good level on their voice and, if it's a remote connection, making sure we've done everything we can to make it sound as good as possible on the air.

Choosing Music

On most shows at NPR, the director comes in maybe four or five hours before the show and starts choosing the music. You listen to the stories on the rundown for a sense of the tone of each story. Some stories have a sad tone, some are more upbeat. We have stories that have a lot of different viewpoints and different concerns. Maybe you're going to choose music that has a more neutral tone. And if a story has a very specific cultural context, maybe it's set in a certain region of the world, and you may search for music that is from that culture or from that country, but not necessarily.

A couple of years ago there was a story by our media reporter David Folkenflik about a newspaper in Youngstown, Ohio, that was closing down. There happens to be a song by Bruce Springsteen called "Youngstown" that has a downbeat tone, lamenting economic decline. And the deadroll—the music that plays right after a piece as the host is closing the segment—is usually instrumental, but when a story calls for it, you can use a song with lyrics. So I chose it for the deadroll.

Every director has to walk a tightrope between choosing music that reflects the tone of the story but doesn't tell the listener how to feel. You need to make sure the music you choose isn't going to make light of a serious story or come off as editorializing. We had a story on a member of the Japanese royal family who decided to abdicate or step down from the line of succession so they could marry someone who was not of royal or noble lineage. And I found a song by a Japanese composer who works in soundtracks for video games. The tone was both celebratory and wistful at the same time, because this was a person doing something for love, but also saying goodbye to a major part of their life. If it were a story about political turmoil in Japanese politics, I wouldn't necessarily use that music. I might use jazz that's busy or has a tone of conflict or controversy.

Directing "Live Special Coverage" for Big Breaking News or Major Events

There are a lot of live interviews on the show on a normal day, but live special coverage can be a very different beast. I've done special coverage of NASA's space shuttle launches and for various presidents giving major speeches from the Rose Garden or in the White House briefing room. You're waiting for the president to give a speech; in the meantime, there's a host or two in the studio talking to reporters about what's expected to happen and what the implications could be. And as director, I'm listening for the sound of the president or whoever it is coming out. And that usually is signaled by the sound of photographers' cameras clicking. And at that point, you'll quickly say to the host, "He's coming out," and tell the engineer to start feeding in that audio source so that the host can wrap up his conversation with the live guests and then transition to the live speech or press conference.

You may be having multiple guests coming into the show at any time, some that are decided just two minutes before. "Hey, this reporter has some context to add about how this is going to affect Congress," or "This other reporter will talk about the scientific implications. They're connected now." So you will quickly check that the reporter is connected and let the host know, "You're good to go."

The Director's Personality

What's important is a strong sense of production judgment and prioritization. You have to be able to keep a level head. You need to be able to communicate effectively and clearly and efficiently. Give the engineers and the line producer the information they need without being too terse or verbose or vague, and respond to changes when they happen. Maybe a guest that was supposed to be in one segment just canceled, or they are having trouble connecting, or their line sounds bad, and we have to switch to another story at the last minute, which may mean that the music that you've chosen won't work for that segment anymore. Or maybe a guest who was supposed to connect five minutes ago gets connected at the last minute. And you have just a few seconds to tell the host something. You may choose to say, "Hey, this is how you pronounce the guy's name." Or you may say, "There is a slight delay on the guest's line." You may not have time to give all the information that you would like to give. You have to prioritize.

It can definitely feel like you're a conductor, making every part of your orchestra come together. The music, the piece that's playing, the clips, the host, you're directing everything, giving the instructions to make magic happen. It is a leadership role in a lot of ways.

Video

What is a show? In broadcasting, it's the presentation of stories in a deliberate order. And arguably, NPR's video production also fits that description. "Much like a show, we have pitch meetings, and all entities—people from the desk, the shows, the reporters, our team—get together and go through and decide, what are the stories going to be?" says NPR Visuals deputy director Nicole Werbeck.

Many of their stories have already aired in audio form, and producers or reporters are asked to do a short video version as well. That doesn't mean they tell the exact same story. "Maybe it's an extension of the storytelling, maybe it's just a smaller piece of what was in a podcast," Nicole says. *Life Kit* has a "Test Tips" video series that offers hacks for daily living and another, "Dear *Life Kit*," that answers letters from listeners. *Planet Money*'s Jack Corbett does often cover the same ideas as the podcast, but with a voice and tone that is very much his own.

The NPR Visuals team creates and produces regular video series on Instagram, TikTok, YouTube and NPR.org from *Planet Money*, *Consider This*, *Life Kit*, *Throughline* and the *TED Radio Hour*, as well as from reporters across the country and overseas. It also produces a visual newscast with Korva Coleman or one of the other regular anchors, which can be watched on smart speakers equipped with a screen.

The visual newscast goes out multiple times every day, while the other video series may be daily or weekly. One thing NPR has learned from previous efforts to produce video is that cadence and consistency pay off. "The more regular you are and the more they understand that you're going to be there," Nicole says, the more reliable your audience will be. In the first two years that NPR released short videos regularly on Instagram and TikTok, she says, they had been viewed 1.75 billion (yes, billion) times.

The NPR App: Algorithm-Assisted Programming

So far, in the shows and podcasts I've highlighted, decisions about what stories to air are fully in the hands of producers. But with the increasing power and complexity of computer algorithms, the team that manages the NPR app has been experimenting with letting them help curate stories for listeners on the app, smart speakers or other digital devices.

"The idea behind the NPR app is that there are things that computers do really well and there are things that humans do really well," says Emily Barocas, deputy director of digital platforms and curation. "And we try to lean into each of those." Computers are superior at determining whether a story is resonating with the audience, while humans can better judge which is the biggest news story of the day or at what point a story becomes outdated. The NPR app mimics the show experience by playing the national newscast followed by the local member station newscast, and then offers a mix of personalized stories. In that way it's also different from a show, in that a listener always starts at the beginning. There's no need to wait until the top of the hour to get the headline stories.

Despite the personalized feed, Emily says the NPR app is programmed to give the listener a complete picture of the news. "We're not trying to shield the audience from difficult news or news that we really think is important." The aim is to prevent the kind of "media bubble" fostered by many social media platforms. "We're very conscious of the fact that we have an editorial responsibility within this algorithm," she says.

A large button in the middle of the app's screen lets people click to the next story when they're no longer interested in the one that's playing; this enables the NPR app to collect data that gives NPR journalists insights into what interests the audience and what doesn't. "We've never had data like this for audio before—moment-by-moment insights into how the audience is engaging with our content," Emily says. "That's one of the most revolutionary parts of this."

As mentioned in chapter 9, most story skips happen within the first 10-15 seconds, which has led the newsroom to prioritize writing strong intros. Another remarkable piece of data is how long people listen. "We see average listening times on the app upward of 40 minutes, which in broadcast or digital is a really amazing engagement time," Emily says. Compared to the half hour or less that listeners on average stay with newsmagazines or podcasts, or the one to two minutes that people spend reading a typical digital story, it is impressive.[5]

Back to the Future?

There's a folding table in the NPR headquarters lobby with freebies for employees. There are stickers, buttons and can coolers with podcast logos and music motifs. And sometimes there are surplus items from the NPR Shop, which disappear quickly. Walking by one day, I saw an item I hadn't seen before. It was rectangular, off-white and smooth. It looked like a smart speaker, but on closer inspection, I saw a round knob with numbers and a telescopic antenna.

It's been years since I've had a radio at home; I've gotten used to streaming NPR on my phone or laptop, or on the wireless speaker in our living room. The device in the lobby was a Tivoli Audio PAL, designed in the early 2000s. It had no Wi-Fi or Bluetooth capabilities. I took it home and plugged it in and was impressed by the pristine FM signal from my local public radio station, WAMU 88.5 FM.

The best thing about it? You turn the switch, and it plays! No unlocking my phone, fumbling for the right app. No waiting for my computer's operating system to boot. No sitting through a funding credit whenever I click Play, and again whenever the livestream sputters.

Don't get me wrong. I'm not predicting a triumphant revival of terrestrial radio. My listening habits have changed since I put that radio in my kitchen, but I still listen a lot by streaming NPR on my phone and downloading a podcast. All technologies have advantages and disadvantages. And while we as public radio journalists need to adapt, we also have an obligation to make editorial decisions, with or without the help of an algorithm. We decide what's news, but we should tell it as a good story.

"If we're really doing our job right, it's about storytelling," Kenya Young, the former *Morning Edition* executive producer, told me before she left to focus on podcasting as senior vice president at WNYC Studios. "Because that's what bends the ear. I don't think that's ever going to die."

ACKNOWLEDGMENTS

Without the help and contributions of many people—too many to name here—this book would not have been written.

I first talked to Kasia Podbielski, then head of NPR Training, about the idea of a new edition of *Sound Reporting*, and she immediately went to bat for it. My other training team colleagues Holly Morris, Argin Hutchins, Cheryl Thompson and Sara Richards offered ideas on digital news reporting, audio production, video, social media and more. Julie Rogers, Will Stone and Sarah Knight of NPR's Research, Archives and Data Strategy team dug into the network's past and helped confirm many historical facts.

The team at the University of Chicago Press, including Mary Laur, Andrea Blatz, Carrie Olivia Adams and Christine Schwab, along with Jane von Mehren of Aevitas Creative Management and Kristen Hartmann at NPR, believed in this book and helped make it a reality.

And rather than lament the end of his book's long shelf life, Jonathan Kern, author of the previous edition, was excited to see me update it. Many of his ideas live on in this text.

It took more than a year to write *Sound Reporting*. I am incredibly grateful to Edith Chapin, NPR's senior vice president and editor-in-chief, and Shirley Henry, managing editor for content talent strategy and development, for letting me carve out that time, especially given the unrelenting demands of news coverage.

Gerry Holmes, NPR's managing editor for enterprise and planning, somehow created time in his packed schedule to edit the book, chapter by chapter. Shirley, along with Steve Drummond, education desk executive producer; Tony Cavin, managing editor for standards and practices; and Keith Woods, chief diversity officer, also reviewed it and suggested many improvements. So did Leslie Keros, eagle-eyed copyeditor and jazz radio-show host at WDCB in Chicago.

All the journalists I interviewed influenced what I wrote far beyond their quoted words. In the face of relentless deadlines, they took the time to answer every single one of my questions. There are few things I have enjoyed in my long journalism career more than those conversations. If I knew at the start what I know now as a result of these interviews, I would have been a much better journalist.

Finally, these acknowledgments would be woefully incomplete if I didn't also mention Petra, my wife and a former Swedish Public Radio correspondent, whom I talked to every day about this book. Many of my own stories cited in this book were done as a parental tag team, where one of us did an interview while the other circled the block, first with Mendel and then Nils in a stroller.

On the topic of parents, my own journalism really starts with mine, Alberto and Maria Socolovsky. They embodied the traits I describe in chapter 5 as characteristic of the best reporters: tenacity, curiosity, skepticism, precision and warmth of personality.

APPENDIX 1

Journalism Safety and Support Groups

ACOS ALLIANCE (ACOSALLIANCE.ORG). Coalition of news organizations, freelance journalist associations and press freedom NGOs that champion safe journalistic practices. Offers guidance on digital security and psychological safety and resources for insurance and legal assistance.

COMMITTEE TO PROTECT JOURNALISTS (CPJ.ORG). Nongovernmental organization promoting press freedom worldwide and defending the right of journalists to report safely. Offers a kit with physical, digital and psychological safety resources and tools.

INTERNATIONAL COMMITTEE OF THE RED CROSS (ICRC.ORG). Humanitarian organization that operates a hotline for journalists, their families and the media organizations they work for to request assistance if they are wounded, detained or missing. Call +41 22 730 34 43, write to press@icrc.org, or contact the nearest ICRC office at icrc.org.

INTERNATIONAL FEDERATION OF JOURNALISTS (IFJ.ORG). Organization supporting journalists and their unions that tracks journalist and media staff deaths, provides training and advice to journalists and organizations, runs campaigns for journalist safety and offers financial assistance for journalists facing violence, persecution or other emergencies.

INTERNATIONAL NEWS SAFETY INSTITUTE (NEWSSAFETY.ORG). Global organization dedicated to journalist safety whose members include NPR, Associated Press, BBC and the *New York Times*. Publishes research reports and advisories.

INTERNATIONAL PRESS INSTITUTE (IPI.MEDIA). Global network of editors, media executives and journalists focused on independent journalism. Offers courses for newsrooms to address online harassment.

INTERNATIONAL WOMEN'S MEDIA FOUNDATION (IWMF.ORG). Group focusing on the needs of women and nonbinary journalists and offering fellowships, grants, safety training and emergency aid.

PEN AMERICA (PEN.ORG). Organization promoting freedom of expression that offers training on online harassment and other threats to journalists, novelists and other types of writers and publishers.

REPORTERS COMMITTEE FOR FREEDOM OF THE PRESS (RCFP.ORG). Nonprofit group providing specialty training and legal support for journalists.

REPORTERS WITHOUT BORDERS (RSF.ORG). Nonprofit organization advocating for press freedom and defense of journalists. Offers training and travel medical insurance to journalists working outside their home country.

RORY PECK TRUST (RORYPECKTRUST.ORG). Independent group providing practical and financial support to freelance journalists and their families worldwide, including safety training for freelancers planning to work in hostile environments.

APPENDIX 2

A Producer's Packing Checklist

This is national desk producer Liz Baker's field-producing packing checklist, which she put together based on her extensive reporting experience. It's designed for a wide variety of situations. Before every trip, she goes through it and asks herself, "Do I need this?"

ALL SITUATIONS

Work Gear

handheld digital recorder (small and lightweight, for sound gathering and on-the-go interviews)
two-channel digital recorder (larger, for seated interviews)
extra SD card for recorder (check compatibility)
AA batteries
microphone (shorty)
microphone (shotgun)
cable: USB to micro-USB
cable: XLR to headphone jack
cable: XLR-male to XLR-female
cable: mini to mini (3.5 mm) headphone
backup thumb drive
earbuds
Sony headphones
NPR mic flag
NPR car magnets
NPR hat
power inverter for car (supplies 110V AC from cigarette lighter)
laptop
laptop charger
external speaker for laptop (USB or battery powered)
wireless mouse
Verizon MiFi hotspot
work cellphone
personal cellphone
solar charger for cellphone
small power bank for cellphone
Comrex audio codec for remote studio connections
Comrex battery
Comrex charger
iRig audio interface for connecting microphone to iPhone for live studio connections
first-aid kit
notebook

pens
Sharpie
press ID badge and lanyard

Personal Items

small bottle of liquid soap
toothbrush and toothpaste
deodorant
hand sanitizer
hydrocortisone cream
sunscreen
comfy bra
clothes to wear to bed (in case you're sharing a room or have no time to dress)
sweater
ultralight wool socks and underwear (avoid cotton garments, even in summer; they dry slowly and retain smells)
hat, if cold weather
Pepto-Bismol
Advil
Tylenol PM
Tums
instant coffee, cold-water soluble
travel-sized electrolyte and caffeine drink mix
granola bars
water bottle
fingernail clippers and tweezers
DivaCup
pads
bivy sack
headlamp
sunglasses
watch
passport and cash wallet

Tasks

download area map to phone
follow officials on X
update contacts in phone
notify Mom and landlord
need vaccinations?
need to change phone plan?
need a visa?
need press credentials?
contact bank about overseas ATM usage
obtain proper electricity adapter
check weather forecast

SPECIFIC SITUATIONS

Power Outage

battery-powered fan, if hot weather (if USB fan, beware of draining your laptop battery)
extra cash (more than usual, in smaller bills)
hand warmers, if cold weather
filled 5-gallon gasoline can (in case gas pumps stop working)
lantern and extra headlamp
body and facial wipes
toilet paper
smoke and carbon monoxide detector
food for meals, in case restaurants are closed

emergency radio
YETI cooler and bottle, if carrying meds that must be kept cool (put meds in plastic baggies and place inside bottle with ice and water; good for several days)
water filter

Fire

power outage gear (see list)
helmet
Nomex pants
Nomex shirt
nonsynthetic sweater or jacket, if cold
over-ankle hiking boots with Vibram soles
leather work gloves
N95 masks
respirator
backup filters for respirator
smoke and carbon monoxide detector
granola bars and other snacks
emergency radio with batteries
BGAN satellite terminal
Vicks ointment or eucalyptus oil (to mask strong smells from smoke, death, disease, etc.)
eye protection (sunglasses or clear shooting-range-style glasses)
hand and face lotion
2 bottles eyedrops
toilet paper
neti pot and saline packets
tasks: (1) download wildfire tracking and traffic apps that show road closures; (2) ask team members about respiratory and cardio health and any risks from smoke inhalation; (3) keep car air vent on recirculate

Hurricane

power outage gear (see list)
rain boots
rain pants
2 rain coats
goggles
towel (bath and camping)
minihairdryer
Vicks ointment or eucalyptus oil
DEET bug-repellent spray
battery-powered fan (if USB fan, beware of draining your laptop battery)
extra nitrile or rubber gloves
plastic wrap
gallon and quart baggies
duct tape
garbage bags
life jacket(s)
emergency radio
smoke and carbon monoxide detector
BGAN satellite terminal
50-, 100-, 500-year flood maps
cellphones with two different providers
whistle
mosquito net

water filter and ultraviolet pen water purifier or bleach
tire patch kit
keychain car escape tool
N95 masks
tasks: (1) download NOAA Clime and Red Cross emergency apps; (2) check rental car for spare tire and proper jack; (3) make sure your hotel has a generator, and ask about size and priority (Will it just go to the lobby, or to all rooms? Do they have a plan for refueling it after the storm if necessary?)

Earthquake

power outage gear (see list)
water
food for meals, in case restaurants are closed
Vicks ointment or eucalyptus oil
emergency radio
smoke and carbon monoxide detector
personal locator beacon, if necessary
BGAN satellite terminal
helmet
N95 masks
safety glasses
whistle
leather gloves
cash, as much as possible
water filter and ultraviolet pen water purifier or bleach
tire patch kit
fire-resistant clothing
hiking boots
inflatable camping mattress

Extreme Cold

hand warmers
nylon pantyhose or long underwear
space blanket
mittens
texting gloves
wool socks
ski goggles
warm hat
earmuffs

Extreme Heat

battery-powered fan (if USB fan, beware of draining your laptop battery)
ice pack
YETI bottle or lunchbox
sun hat
any lightweight, UV-resistant fabric shirt
bandana or washcloth
tasks: (1) remind team that some medications increase vulnerability to heat illness, and to ask their doctor about any concerns; (2) review signs and symptoms of heat illness such as headache, weakness, nausea, dizziness, tunnel vision, confusion, rapid heart rate or inability to walk in a straight line

Mass Shooting

dress
dress shoes
nice outfit
sun hat
mic clip
fishpole (boom mic)
cable: extra-long XLR to XLR

Civil Unrest

2 quarts water
granola bars and other snacks
helmet
press ID
body armor, if necessary
goggles
N95 masks
earplugs
extra ID and cash in concealed wallet
hiking boots
Velcro "Press" patches
small first-aid kit
small daypack for gear
tasks: (1) download police scanner app; (2) disable fingerprint and face ID on cellphone

Disease Outbreak or Pandemic

oral thermometer and thermometer covers
extra battery for thermometer
disposable medical gloves
N95 masks
goggles
rain jacket
hand soap
hand sanitizer
hand lotion
Vicks ointment or eucalyptus oil
toilet paper
water purifier tabs
Clorox wipes
bleach
plastic wrap

GLOSSARY

act or **actuality.** An excerpt from an interview, news conference, speech and so on—in other words, the recorded sound of someone speaking. Also known as a *cut* or ***selex***. The plural is *acts* or *ax*.

acts and tracks. A simple **mixed piece** that does not include **ambience**. A reporter might say, "I'm sending you two pieces—but the first is just acts and tracks."

ambi or **ambience.** Sustained background sound that is captured at a remote location. Ambience gathered at the site of an interview—whether it's a relatively quiet office or a noisy police station—permits an audio engineer to fade in and out of an **actuality** as it starts and finishes. Much more ambience is needed whenever the sound is a key part of the report. Good ambience can be used to advance a story and to create a **scene**—you can *hear* the people on the lobster boat hauling in their catch, or the protesters trying to block the motorcade or the stagehands helping a performer get ready for their act. Also known as *natural sound* or *nat sound*. See also **sound effects**.

analog. An **audio**-encoding format that uses continuous electrical signals such as sine waves to transmit sound. NPR stopped analog satellite transmission of programming to stations in the early 2000s and switched to digital means, but many stations still broadcast analog signals. Contrast **digital**.

audio. Sound that is recorded or broadcast. See also **tape**.

backannounce. The **copy** that follows a report or **two-way**. It's usually used to identify the person who has just spoken or to add a bit more information to the report (for example, "You can see a timeline of the Supreme Court's decisions on school integration at **N-P-R-dot-org**."). Also known as *outro*, by analogy with **intro**.

background sound. Sound that is within earshot of a conversation—such as crowd noise, traffic, music or an air conditioner—and is picked up by the microphone in an interview. It's always advisable to record the background noise on its own for at least 30 seconds, both before and after the interview, if possible.

backtime. A calculation to determine when a specific part of a recording will be heard. It's a way to ensure, for example, that a chorus to a song will begin just when a reporter mentions it or that the theme music for a program will conclude precisely at the end of the hour. The word may not be commonly used outside of broadcasting, but the concept is part of daily life. A cook makes sure that all parts of a meal are ready at the same time by backtiming the preparations, starting the meat at 5:00, the dessert at 5:15 and the vegetables at 6:00 so that dinner is ready at 6:30. In a radio **piece**, the sound of a demonstration might be backtimed so that the chant "Keep hope alive!" begins just after the **SOC**. See also **post**; **deadroll**.

bed. Sound running underneath a track or other audio. Not very dynamic—often music or background noise.

billboard. The first minute of each hour of most public radio programs, during which a host mentions some of the items to be heard in that hour, after the **newscast**. Also known as an *open*. Some programs include **teases** in their billboards. Podcasts also use billboards, often as the final part of the **topper** or **intro** to communicate the episode's central themes and questions, and sometimes with **tape** from the episode in the same way a show billboard does.

blade. To cut something out, often something very short such as a cough or a few unwanted words. The term is a vestige of the **analog** era, when razor blades were used, but it's still common at NPR. An editor might say to a producer, "See if you can blade the stutter out of that last question."

board. At NPR, a screen showing the **rundown** of **segments** planned for an upcoming program, which lists host and producer assignments, interview times and **studios** and the names of reporters and their editors.

bust. To accidentally continue broadcasting a **segment** through a scheduled time **post**. For example, if a segment is supposed to end with two seconds of silence at 19 minutes and 58 seconds after the top of the hour, and the host is still talking or music is still playing at 20 minutes past the hour, the program has busted the segment by two seconds.

butt cut. One **actuality** placed immediately after another, without **copy** or **ambience** in between. It's often used to create a **transition**, reinforce a point or demonstrate a contrast. The term can also be used as a verb: "Let's butt cut those two actualities."

button. A short piece of music that creates a transition between two unrelated stories, or stories with contrasting moods and tone.

cascade or **waterfall.** An audio montage involving three or more distinct pieces of audio that are combined by fading one into the next.

clear. Used in the phrase *in the clear* to describe sound that is in the foreground, with no other sound. Often forms part of mixing instructions for **ambience** or **actualities**. For example: "Maintain **ambi** of gunshots in the clear for four secs."

clip (noun). A **cut** of **tape**.

clip (verb). To edit audio badly so that a sound at the end has been cut off abruptly. See also **ambience**; **upcut**.

coda. An extra thought or two tacked on to the end of a podcast episode to give new or updated information or to serve as a kind of epilogue to the story (often after a more artistic ending). See also **backannounce**.

control room. A room connected to the **studio** where the director, engineer and producer manage the recording of a show or interview and watch the hosts through a soundproof window.

copy. Text in an **intro** or story that is not part of an **actuality**.

crash. To work on deadline, often an imminent one. If you're crashing, you don't have time for socializing or to discuss unrelated work. A producer might say, "I'm crashing on a **mix**, so I can't talk right now" or "I'm crashing on an episode this week, so I can't meet" or "That's a story with a crashy turnaround."

crossfade. In **mixing**, the process of simultaneously decreasing the audio level of one source and increasing the audio level of another so that one sound goes away while another comes in. Sometimes a crossfade takes place under a voice **track** or **actuality**.

cut. See **act**.

cutaway. A part of a radio program designed as a module that can be removed by a station and replaced with local announcements, news or other material.

DACS. An abbreviation of the Direct Access Communications System, pronounced *daks*. DACS is the computer system that connects public radio stations. Stations often use the information on the system's **rundowns** for on-air promotion during the day. The DACS line that summarizes each story should be concise, accurate, grammatical and ready for air. See also **line**.

deadroll. Audio that is started at a precise time but not made audible until it is needed on the air. A deadroll is often begun so that it will conclude at a predetermined time. A good example is the theme at the end of most programs. If the theme is two minutes and 20 seconds long, and the program ends at 58:58 (58 minutes and 58 seconds after the hour), the music will be started as a deadroll at 56:38 (56 minutes and 38 seconds after the hour). The theme can then be faded up whenever the host has finished, and it will always end at just the right moment. See also **backtime**; **post**.

digital. An audio-encoding format that uses binary information to transmit sound. At NPR, this term can often mean text-based reporting published on our website; a digital story is what used to be known as a print story. But it can also mean content distributed digitally, such as audio that's available on podcast **feeds** and the NPR app. Contrast **analog**.

donut. (1) A show **segment** that is a cross between a **two-way** and a **piece**. It begins with the host asking the reporter a few questions, **transitions** to a narrated piece and then concludes with another question or two. (2) A **mixing** technique, mostly used with music, crowd chants or movie **clips** in which sound **posts** in the clear, plays under **trax**, and then posts again in the clear (the sound is the donut and the trax are the donut hole).

elements. The component parts of a news report—the **actualities**, voice **tracks**, and **ambience** that will be **mixed** to make the final **piece**. Colloquially called *acts, tracks and ambi*.

fade (in, out, up, down, under). To adjust the volume of sound from low to high or high to low at a gradual pace.

feed (noun). An entire program that is fed to stations—for example, the second feed of *All Things Considered* begins at 6 pm ET—or an external audio source brought in to be recorded by the **ops desk**, such as a press conference or a presidential address. In podcasting, a feed refers to the distribution queue for a single show's productions, including any episode or trailer that goes to any podcast platform. A producer might ask, "When are we putting that episode in the feed?" or "Should we put a message in the feed that next week's episode will be delayed?"

feed (verb). To send **actualities**, voice **tracks**, and **ambience** to a station or network from a remote location. A producer or engineer then takes these **elements** of the report, arranges them in order and produces the final **mixed piece**.

feed drop. Insertion of an episode or episodes from a different show into a **feed**. A producer might do a feed drop because they want to support that show or because they think listeners will enjoy it. The show might also be included as part of a feed-swap agreement.

field producing. Producing with a host or reporter at a remote site.

fire. Synonym for *play*, as when the director or producer plays a **cut** of **tape**. A reporter might ask an engineer, "Will you fire the **clips** while I **track**?"

funder. An on-air underwriting announcement. At NPR, hosts and other journalists do not read funders.

graf. Shorthand for *paragraph*. See also **nut graf**.

hit. A reporter's live appearance on a show, usually for a **two-way**. A producer might say, "You've got four minutes for your hit in the second hour of the show."

hot. (1) Used to describe a **piece** of audio that is louder than other audio in the same piece or interview. A reporter might tell an engineer, "I've got four **acts** in this piece, but be careful because the second one is hot." (2) Used to mean "at full volume." A producer might want one piece of music to be "hit hot"—played at full volume from the beginning—and another to be hit **warm** or as a **sneak**.

incue or **in cue.** The first few words of an audio file. In a reporter's script, it indicates where an **actuality** should start. On a program script, it tells the hosts and the show editor how a report begins. Often abbreviated *i.c.* or *IN*. For example, "i.c. 'For more than thirty years, NPR has . . .'"

intro. The **copy** that a host will read on the air; it usually includes information for the program director about **tape time** and other peculiarities of the audio. See also **lede**; **topper**.

lede or **lead.** The beginning of a **digital** news story. Audio stories don't have ledes; they usually start with an **intro**.

line. At NPR, a short description of a **piece** that editors provide so that program producers know what's in the works on any given day. Lines are used to create the **DACS** and may also be relied on for writing **billboards**. For that reason, it's essential that the facts and pronunciations in the lines be accurate, up to date and in keeping with NPR and AP style. If a story changes over the course of the day, the lines have to change as well. Podcasters use this word the way actors do, to refer to a line of dialogue or **tracking**.

live mix. The **mixing** of a **piece** at the same time that it is being broadcast. Mixing live obviously doesn't allow a producer to fix **pickups** that occur during the course of the mix.

live read. A reading of a script, usually by a host, as the program is being broadcast, while the **actualities** are played off a computer. A live read is usually called for when there isn't time to **track** in advance, which often happens with **tape and copies.**

mix. To combine two or more audio sources to make one finished **piece.** In the **analog** era, an audio engineer would physically control the starting and stopping of **tape** machines and adjust sound levels according to a producer's directions. Now, the producer digitally arranges the **acts, tracks,** and **ambi** (and, for a podcast, **scoring**) on a computer screen before going into the **studio.** The audio engineer in that case may still be responsible for improving the quality of the audio—putting in a fix—and adjusting the levels as the producer directs the mix.

music bed. Music mixed for an extended period under a **track** or **actuality.**

natural sound or **nat sound.** See **ambi.**

newscast. An NPR news program that is five minutes long, though some stations cut away after three minutes. Newscasts are broadcast every half hour when *Morning Edition* and *All Things Considered* are on the air; at other times, they are broadcast hourly. Listeners hear newscasts as part of the program, though they are produced, written and read by a separate group of people. For that reason, information in **billboards** and **pieces**—death tolls, pronunciations and so on—must correspond exactly to what's being presented in newscasts.

Nipper. (1) NPR slang for the one-line network identification, "This is NPR—National Public Radio." (2) The name of the mixed-breed terrier that was mesmerized by the Victrola in the old RCA logo, subtitled "His Master's Voice."

NPM or **National Public Media.** Jointly owned by NPR, PBS and GBH and working independently from editorial divisions, this organization offers on-air and online sponsorships to companies interested in underwriting public media.

NPR.org. The NPR website. Editors often just say "dot org," as in, "Will you write this up for dot org?"

nut graf. A paragraph in a **digital** story that tells the reader why the story is important. Audio stories don't have nut grafs, though the information contained in a nut graf may be in the **intro** or in the story, either in the reporter's **tracks** or in the **actualities.**

ops desk. The central desk for coordinating live **feeds,** special coverage and the booking of **studios** at NPR. ("Ops" is shorthand for "news operations" and is also used in other organizations.) The ops desk routinely records all audio feeds from the White House, including the daily press briefing. Other regularly recorded feeds include briefings from the State Department and the Pentagon.

outcue or **out cue.** The last few words of a recording. In a reporter's script, it indicates the end of an **actuality**. On a program script, it tells the hosts and the show editor how a report ends. Often abbreviated *o.c.* For example, "o.c. '. . . only time will tell.' "

outro. See **backannounce**.

***p* pop** or **plosive.** Distortion caused by air being emitted from the mouth, usually because the speaker is too close to the microphone and is talking straight into it or because the microphone lacks a pop filter.

passoff. At NPR, any information passed along from one editor or producer in the newsroom to another or to a program host. In podcasting, passoffs usually happen in audio-editing sessions, for example, when a junior producer sends an episode to a more senior producer for a final listen.

phoner. An interview conducted over the telephone.

pickup. A place in a sound file where a reporter or host has made a mistake or for some other reason had to restart—pick up—their **track**. One of the producer's jobs is to edit out any pickups before a **piece** airs. There may also be pickups in a **mixed** piece, if a section was recorded more than once in order to get things just right. For podcasts, a pickup is a type of **retrack** at the very end of an episode's production when a final **line** or two, or more, are needed from the host or reporter. "Let's do pickups tomorrow for that correction from the fact-checker, and the line with the door-slam in the background."

piece. At NPR, a news or feature report broadcast in the body of a program. Pieces can be of almost any length, from about a minute and a half to 20 minutes—or even longer. Compare **spot**; **superspot**.

pitch. A description of a proposed story to an editor or host. Making an effective pitch involves examining a possible story from many different angles—anticipating how it might be reported, what the listener would learn, even how the **intro** for the story might be written.

post (noun). Any of several things, all of them related to the timing of a **piece** or broadcast. **Newscasts** and programs have designated time posts—points at which the **segments** have to end (usually so that member stations know when they can begin adding their own material). A post is also the result of a **backtime**—the part of a piece of music or other sound that should be heard by the listener. The producer of a music piece, for instance, may say, "There's a good post at 32," meaning that

32 seconds into a particular selection, there's a place where the music should be heard in the **clear**, without anyone talking over it.

post (verb). To create a post in a piece. A reporter might say, "Post the sound right when you hear the car horns beginning."

postcard. The radio version of a picture postcard. It's short—usually three minutes or less—and the "picture" is a remarkable sound or series of sounds. For example, a reporter may produce a postcard whose **tracks** consist of little more than this: "I came to central New Hampshire to do a story about state funding of public schools. But what I found instead—everywhere I turned—was the sound of the January thaw." The postcard might then have the sounds of ice breaking up in the river, or snow falling off roofs or whatever the reporter found.

presser. Press conference.

promo. A short promotional announcement for upcoming **pieces** or programs.

pronouncer. An informal way of making it clear how a word or name should be read out loud. For example, the pronouncer for "Ivan Basso" might be "ee-VAHN BAH-soh." NPR's pronunciation key appears in chapter 11.

pull tape. To isolate **actualities** or **sound bites**. There was a time when producers and reporters actually pulled from a reel the selections of **tape** they wanted; the phrase survives even though the process today usually involves a mouse and a computer. Also called *pull* ***selex***.

reader. A news story, most often heard on **newscasts**, that is told entirely through **copy** read aloud, as opposed to a **spot**.

ready for review (RFR). Used to describe a **piece** of **tape**, full **mix** or script that is ready for the editor or senior producer to audition and greenlight, either for air on the radio or for the next stage of production in podcasting.

retrack. To rerecord part of the script to update a story or fix an error or audio issue.

return. At NPR, the 28-second block of **copy**, often on a lighthearted topic, read by program hosts in the second half of each hour. A return is sort of a miniature **billboard**.

roll. To play or record **audio**. A reporter may tell a guest, "We're not rolling yet" to indicate that anything the guest says won't be recorded. A **studio** director may instruct an engineer to play a report or **actuality** by saying, "Roll **tape**." Many radio people still use the term even when nothing is actually rolling—or moving at all—on either the digital recorder that captures sound or the computer that plays it back.

rolloff. A **piece** done as a short radio-friendly excerpt or spinoff of a podcast episode or series. Sometimes called a *tie-in*.

rollover. A delayed version of a broadcast. Public radio news programs such as *Morning Edition* and *Weekend Edition Saturday* are generally recorded as they are broadcast on the East Coast so that they can easily be rolled over for broadcast in other parts of the country. At NPR, rollovers are always kept up to date; if the news changes, **segments** may be reworked, interviews may be redone, and **intros** may be read live in an otherwise recorded program. In addition, any mistakes in the first broadcast are fixed for the rollovers.

room tone. Indoor ambience recorded at the place where an interview is conducted or an event takes place. Usually low dynamic level.

roundtable. An on-air or prerecorded conversation, on a radio show or podcast, between one or more hosts and more than one reporter or guest. Sometimes, to distinguish it from a **two-way**, it's called a three-way, four-way and so on, depending on the number of participants.

rundown. The list of items in a program, reflecting the information on the **board**. The rundown is sent out to member stations on the **DACS**.

scene. The location where some action takes place; it has much the same meaning in radio as in a movie or play. The best radio **pieces** are usually built around scenes; they can make even complex, issues-oriented pieces intelligible. For example, a report on how a government policy is affecting immigrants might include **actualities** of a university professor describing their research on the policy and what it implies. But the report will be effective and memorable only if it includes scenes—a worker trying to apply for a job, or an official trying to explain to a non-English-speaking person how to apply for a government benefit. Creating a scene usually involves both good sound and careful, descriptive writing. See also **ambi**; **sound effects**.

scoring. Adding music to a podcast to enhance the storytelling.

script. The text version of an **audio** story, including **intro**, **tracks** and **actualities**.

segment. A portion of a program of fixed length, which may include one or more stories. Segments are designated alphabetically, and their lengths differ within and among programs. For example, each hour of *Morning Edition* is made up of five segments, A through E; the A segment is 11½ minutes long and the B segment is a little more than seven minutes. *Weekend Edition Saturday*'s hours have four segments,

A through D, and the A segment is the same length as *Morning Edition*'s but the B segment is nearly 14½ minutes long.

segue. To go from one sound to another without interruption. A producer may decide to segue from one **piece** of music to another, or from one report to another without a host introduction. Called *transition* in podcasting. See also **crossfade**.

selex. Synonym for **acts** or **cuts**. Usually plural; for singular, you might say, "one of the selex."

signpost. A sentence or a few words that step out of the story (1) to explicitly mark a transition or an important moment in the narrative ("What you're about to hear . . . ," "Next we're going to hear about . . . ," "This is the sound of . . ."), (2) to indicate that the story is taking a step back in time or in the logical flow ("To understand *x*, you first need to know about *y*") or (3) to review or recap something. The signpost is meant to help the listener follow the story, so it doesn't need to include new information.

slug. A word or phrase used to identify a news story, usually set in capital letters. At NPR, program producers assign slugs to all stories that appear on the **board**. A slug goes into the **DACS** and out on the **rundowns**, so it may also show up on the internet. Producers should choose their slugs carefully: A phrase that's intended to be amusing or seen only by one's colleagues, such as "WACKO GEEZER PILOTS," can end up being part of a public document.

sneak. To fade audio in gradually, usually under a voice **track**. Sound that begins as a sneak should eventually be faded up; experienced producers avoid sneaking sound or music in and then fading it out without ever letting listeners hear it in the clear at full volume. Compare **hot**; **warm**.

SOC. An abbreviation of standard **out cue**, pronounced *sock*. Reporters use it to identify themselves at the end of a piece. A typical SOC for an NPR news reporter would read, "Nina Kim, NPR News, Whitfield, Indiana." A non-NPR reporter would say, "For NPR News, I'm Nina Kim in Whitfield, Indiana."

sound bite. See **act**.

sound design. The arranging and **mixing** of sounds in a way that helps take the listener to a different place or create a mood that enhances the listening experience.

sound effects. Canned or manipulated audio; common examples are applause or car horns, taken from a music library, or the chirping of real-life birds that is distorted for dramatic effect. Sound effects are different from **ambi**, which is real-life sound that is not manipulated.

split track. This term comes from the practice of recording one source of audio (such as a host) on one half of the **tape** and a second source (a guest) on the other. Now the sources are recorded on separate **tracks**, with the host panned to one side and the guest to the other.

spot. At NPR, a short news report (under a minute) for inclusion in a **newscast**. Compare **piece; superspot.**

standup. A **track** recorded on location, often without a script; as a production device, it often works best when the reporter's perspective and observations play a critical role in the story. For example: "I'm standing at the corner of Tenth Street and Jefferson Boulevard—which police describe as the most dangerous spot in this city. But right now the street I'm on is empty, and the stores are closed and boarded up."

storyboard. A sketch or outline that shows the sequence of scenes that will tell the story. More commonly used in video.

studio. A soundproof room where hosts record shows and podcasts and conduct interviews and where a reporter can track a piece. See also **control room.**

superspot. At NPR, a piece that is little more than an overgrown spot, up to two minutes and 45 seconds long. Show producers sometimes ask a reporter to carve a piece down to a superspot length so that it can fit a small hole in the program. Compare **piece; spot.**

sweep. To quickly fade up; a quick fade up.

sync or **tape sync.** To record a high-fidelity version of the guest or reporter's side of a remote interview on a digital recorder or phone app, while conducting the actual conversation over a landline phone or cellphone. Producers used to go on location to do tape syncs with a guest in person, but now they more commonly instruct the remote guest on how to record simultaneously on their own smartphones.

tape. Formerly magnetic tape, at NPR it refers to audio that is recorded outside the studio. Though sound is recorded and edited electronically, many people in public broadcasting continue to talk about "tape editors," "tape-and-copy blocks," "**tape syncs**," "great tape" and so on.

tape and copy. At NPR, a story, often written by a producer for a host to read on the air, that includes both text and audio.

tape time. The length in seconds of a piece of **tape**. Usually marked in a script next to each **actuality**.

tease. A short **actuality** that precedes any other information in the story. A good tease makes the listener eager to hear more. Some shows include teases in their **billboards**. Billboard teases should be catchy—their whole point is to make the listener stay tuned—and they should reflect the basic theme of the report.

tie-in. See **rolloff.**

top. The beginning of a recording or script. A **studio** director may ask a reporter to "take it again from the top"—to reread the story from the beginning—or a director may caution the engineer to watch the levels of a **cut** because "the sound is a little **hot** at the top."

topper. The **intro** section of a podcast episode that can run several minutes long. Its aim is to bring listeners in with some kind of hook or a **scene** that introduces the show; often it includes a **billboard** that sets up listener expectations and story questions.

track (noun). A part of a report that is read by the reporter or host. A typical public radio news **piece** may include tracks, **acts** and **ambi**. Also spelled *trax* in plural.

track (verb). To record narration. An editor might ask a reporter: "Have you tracked yet?"

transition. See **segue; crossfade.**

two-way. An on-air conversation between two people, usually a host and an interviewee or a reporter. Often used to describe conversations heard on newsmagazines. A three-way involves a host and two guests, and so on. See also **roundtable.**

upcut. Audio that is missing the beginning or end of a sound element or word. Also known as *clipped audio*. See also **clip.**

voiceover. At NPR, an English translation that is **mixed** over an **actuality** in another language. Often abbreviated *v.o.*

voicer. A news **spot** involving only the reporter's voice with no **actualities.**

vox. Person-on-the-street **actuality**, usually gathered to present a variety of opinions on an issue. Some reporters refer to vox as *MOS*, an abbreviation of "man on the street."

VU meter. An instrument that measures audio volume. VU meters use volume units as a standard of measurement.

warm. Audio that is played louder than a **deadroll** but not at full volume. A producer **mixing** a music **piece** might call for a "warm hit" of a song, allowing listeners to hear the music as the host starts to talk about it.

windjammer. A microphone cover with long hair. Specifically designed to reduce heavy ambient wind. Also called a *dead cat*.

wrap. A news **spot** featuring an **actuality** placed between the reporter's **tracks** (the actuality is *wrapped* by the tracks).

NOTES

CHAPTER ONE

1. Terry Gross, “Gene Simmons,” interview of Gene Simmons, *Fresh Air*, WHYY, Feb. 4, 2002, https://freshairarchive.org/segments/gene-simmons.
2. Penny Abernathy, *The State of Local News*, Northwestern University, Medill School of Journalism, Nov. 16, 2023, https://localnewsinitiative.northwestern.edu/projects/state-of-local-news/2023/report/#vanishing-newspapers.
3. Dan Froomkin, “Soledad O’Brien’s Critique of a Pence Puff Piece Sets Off Anti-NPR Twitter Frenzy,” Press Watch, Nov. 26, 2019, https://presswatchers.org/2019/11/soledad-obriens-critique-of-a-pence-puff-piece-sets-off-anti-npr-twitter-frenzy/.
4. *NPR Ethics Handbook*, https://www.npr.org/ethics.
5. Alicia C. Shepard, “NPR’s Giffords Mistake: Re-learning the Lesson of Checking Sources,” NPR Public Editor, Jan. 11, 2011, https://www.npr.org/sections/publiceditor/2011/01/11/132812196/nprs-giffords-mistake-re-learning-the-lesson-of-checking-sources.
6. Tim Jones, “Dewey Defeats Truman: The Most Famous Wrong Call in Electoral History,” *Chicago Tribune*, Oct. 31, 2020, https://www.chicagotribune.com/featured/sns-dewey-defeats-truman-1942-20201031-5kkw5lpdavejpf4mx5k2pr7trm-story.html.
7. “Quick Facts,” U.S. Census Bureau, http://www.census.gov/quickfacts/fact/table/US/RHI125222#RHI125222.
8. William Bright, “What’s the Difference Between Speech and Writing?” Linguistic Society of America, https://www.linguisticsociety.org/resource/whats-difference-between-speech-and-writing#.
9. Alexandra R. Webb, Howard T. Heller, Carol B. Benson and Amir Lahav, “Mother’s Voice and Heartbeat Sounds Elicit Auditory Plasticity in the Human Brain Before Full Gestation,” *Proceedings of the National Academy of Sciences* 112, no. 10 (Feb. 23, 2015): 3152-57, https://www.pnas.org/doi/10.1073/pnas.1414924112; Beth Skwarecki, “Babies Learn To Recognize

Words in the Womb," Science, Aug. 26, 2013, https://www.science.org/content/article/babies-learn-recognize-words-womb.

10. Robert Krulwich, "Consider Your Ears," *Adjacent*, no. 5 (April 2019), https://itp.nyu.edu/adjacent/issue-5/consider-your-ears/.
11. Tatum Hunter, "The Nonstop Podcast Listeners Are On to Something," *Washington Post*, Aug. 28, 2023, https://www.washingtonpost.com/technology/2023/08/28/podcasts-listening-constant-background-neurological-effects/.
12. This is the story she had heard: Kevin Kling, "Prayer: Once a Last Resort, Now a Habit," *All Things Considered*, NPR, Jan. 10, 2007, https://www.npr.org/2007/01/10/6788291/prayer-once-a-last-resort-now-a-habit.
13. Terry Gross, "Satirist Al Franken," interview of Al Franken, *Fresh Air*, WHYY, April 30, 2002, https://freshairarchive.org/segments/satirist-al-franken.

CHAPTER TWO

1. Jasmine Garsd, *La Última Copa/The Last Cup*, NPR, https://www.npr.org/podcasts/510367/the-last-cup.
2. Elissa Nadworny and Claire Harbage, "How the War in Ukraine Has Forever Changed the Children in One Kindergarten Class," *Morning Edition*, NPR, April 12, 2023, https://www.npr.org/2023/04/12/1161970852/ukraine-russia-war-kindergarten-students-schools.
3. "Our Mission and Vision," NPR, https://www.npr.org/about-npr/178659563/our-mission-and-vision.
4. Michelle Jokisch Polo, "In Michigan, Undocumented Immigrants Form Learning Pod So They Won't Lose Their Jobs," *All Things Considered*, NPR, Nov. 1, 2020, https://www.npr.org/2020/11/01/928319913/in-michigan-undocumented-immigrants-form-learning-pod-so-they-wont-lose-their-jo.
5. John Burnett, "Mexico—Leaf Musician," *All Things Considered*, NPR, June 14, 2001, https://www.npr.org/templates/story/story.php?storyId=1124393.
6. Jerome Socolovsky, "Spanish Town Hosts Running of the Bulls for Kids," *Morning Edition*, NPR, Aug. 19, 2004, https://www.npr.org/templates/story/story.php?storyId=3858653.
7. Jerome Socolovsky, "Spain's Special, Savory Ham Headed for U.S.," *Weekend Edition Saturday*, NPR, Feb. 9, 2008, https://www.npr.org/transcripts/18550917.
8. Jerome Socolovsky, "Time To Give the Siesta a Rest?" *Marketplace*, Dec. 15, 2006, https://www.marketplace.org/2006/12/15/world/time-give-siesta-rest/.

9. Jerome Socolovsky, "In Spain, It Takes a Village To Babysit," *Morning Edition*, NPR, July 17, 2009, https://www.npr.org/templates/story/story.php?storyId=106722343.
10. Alexi Horowitz-Ghazi, "How the Wave of Synthetic Cannabinoids Got Started," *All Things Considered*, NPR, Jan. 4, 2019, https://www.npr.org/2019/01/04/682350000/how-the-wave-of-synthetic-cannabinoids-got-started.
11. Alexi Horowitz-Ghazi, Emma Peaslee and Jess Jiang, "Consider the Lobstermen," *Planet Money*, NPR, Dec. 3, 2021, https://www.npr.org/2021/12/03/1061289894/consider-the-lobstermen.
12. Daniel Estrin, "New Embassy in Jerusalem Attracts Devout Christians From the U.S.," *Morning Edition*, NPR, Oct. 9, 2018, https://www.npr.org/2018/10/09/655757013/new-embassy-in-jerusalem-attracts-devout-christians-from-the-u-s.
13. Shannon Bond, "That Smiling LinkedIn Profile Face Might Be a Computer-Generated Fake," *Morning Edition*, NPR, March 27, 2022, https://www.npr.org/2022/03/27/1088140809/fake-linkedin-profiles.
14. Gregory Warner and Jasmine Garsd, "Intruders," *Rough Translation*, NPR, June 29, 2018, https://www.npr.org/transcripts/624799760.
15. David Greene, "Vatican Responds To Pa. Grand Jury Report on Abuse of Children by Clergy," interview of Sylvia Poggioli and Marie Collin, *Morning Edition*, NPR, Aug. 17, 2018, https://www.npr.org/2018/08/17/639491434/vatican-responds-to-pa-grand-jury-report-on-abuse-of-children-by-clergy.

CHAPTER FOUR

1. Jerome Socolovsky, "Spanish Town Hosts Running of the Bulls for Kids," *Morning Edition*, NPR, Aug. 19, 2004, https://www.npr.org/templates/story/story.php?storyId=3858653.
2. Ryan Lucas, "Ukrainian Men, Manning a Checkpoint for Six Hours, Talk About the War and Their Lives," *Morning Edition*, NPR, March 15, 2022, https://www.npr.org/2022/03/15/1086605712/ukrainian-men-manning-a-checkpoint-for-six-hours-talk-about-the-war-and-their-li.
3. Jerome Socolovsky, "Many Look to Buddhism for Sanctuary From an Over-Connected World," *Weekend Edition Saturday*, NPR, July 7, 2018, https://www.npr.org/transcripts/625332469.
4. Michaeleen Doucleff and Jane Greenhalgh, "The Next Pandemic Could Be Dripping on Your Head," *Morning Edition*, NPR, Feb. 21, 2017, https://www.npr.org/sections/goatsandsoda/2017/02/21/508060742/the-next-pandemic-could-be-dripping-on-your-head.

5. Steve Inskeep and Renee Montagne, "Getting a Time-Saving Jump on Thanksgiving Dinner," *Morning Edition*, NPR, Nov. 22, 2006, https://www.npr.org/transcripts/6518611.
6. Christina and Ari Shapiro went back and redid the interview, and ended up with just as moving a story. Sam Gringlas, Ari Shapiro and Christina Cala, "A Father, a Husband, an Immigrant: Detained and Facing Deportation," *All Things Considered*, NPR, Jan. 25, 2018, https://www.npr.org/2018/01/25/579761240/a-father-a-husband-an-immigrant-detained-and-facing-deportation.
7. Jerome Socolovsky, "Militants Invoke Spain's Andalusian Heritage," *Weekend Edition Sunday*, NPR, April 3, 2005, https://www.npr.org/templates/story/story.php?storyId=4573301.
8. Eva Tesfaye, "Life in the Roaring 2020s: Young People Prepare To Party, Reclaim Lost Pandemic Year," *Morning Edition*, NPR, April 17, 2021, https://www.npr.org/transcripts/987865318.
9. Lulu Garcia-Navarro and Lauren Migaki, "As Brazil's Largest City Struggles With Drought, Residents Are Leaving," *Weekend Edition Sunday*, NPR, Nov. 22, 2015, https://www.npr.org/sections/parallels/2015/11/22/455751848/as-brazils-largest-city-struggles-with-drought-residents-are-leaving.

CHAPTER FIVE

1. Alexi Horowitz-Ghazi, Emma Peaslee and Jess Jiang, "Consider the Lobstermen," *Planet Money*, NPR, Dec. 3, 2021, https://www.npr.org/2021/12/03/1061289894/consider-the-lobstermen.
2. Here's an article I wrote on common math problems journalists have to deal with: Jerome Socolovsky, "Must-Have Math Skills for the Number-Crunching Newsperson," NPR Training, June 17, 2020, https://training.npr.org/2020/06/17/must-have-math-skills-for-number-crunching-newsperson/.
3. NPR Corrections, https://www.npr.org/corrections/.
4. Cheryl W. Thompson, Cristina Kim, Natalie Moore, Roxana Popescu and Corinne Ruff, "Racial Covenants, a Relic of the Past, Are Still on the Books Across the Country," *Morning Edition*, NPR, Nov. 17, 2021, https://www.npr.org/2021/11/17/1049052531/racial-covenants-housing-discrimination.
5. John Burnett, "How the COVID-19 Pandemic Has Disrupted Life in the Rio Grande Valley," *Weekend Edition Sunday*, NPR, Sept. 6, 2020, https://www.npr.org/2020/09/06/910194920/how-the-covid-19-pandemic-has-disrupted-life-in-the-rio-grande-valley.

6. NPR's rules on conflicts of interest are in the "Independence" section of the *NPR Ethics Handbook*, which can be found at https://www.npr.org/about-npr/688405012/independence.
7. Jerome Socolovsky, "Fans Wait: Manchester United Meets Barcelona," *Morning Edition*, NPR, May 22, 2009, https://www.npr.org/templates/story/story.php?storyId=104426607.
8. Robert Siegel, "The View From Madrid in Wake of Bomb Attacks," interview of Teresa Pelegri, *All Things Considered*, NPR, March 12, 2004, https://www.npr.org/templates/story/story.php?storyId=1763658.
9. See "Transparency" in *NPR Ethics Handbook*, https://www.npr.org/about-npr/688413859/transparency.
10. See "Anonymous Sourcing" in *NPR Ethics Handbook*, https://www.npr.org/about-npr/688413859/transparency#anonymoussources.
11. Ruth Sherlock, "In Turkey, a Mother Tries To Save a Son Trapped in the Rubble," *Morning Edition*, NPR, Feb. 8, 2023, https://www.npr.org/2023/02/08/1155335729/search-and-rescue-operations-in-antakya-turkey-are-incredibly-dangerous.
12. Jerome Socolovsky, "Spain's Cyclists Don't Include Average Spaniards," *Morning Edition*, NPR, July 29, 2009, https://www.npr.org/templates/story/story.php?storyId=111273331.
13. Sam Brasch, "One of the Heaviest Snowstorms on Record Hits Rocky Mountains," *All Things Considered*, NPR, March 15, 2021, https://www.npr.org/2021/03/15/977548725/one-of-the-heaviest-snowstorms-on-record-hits-rocky-mountains.
14. Christie Thompson and Joseph Shapiro, "Doubling Up Prisoners in 'Solitary' Creates Deadly Consequences," *All Things Considered*, NPR, March 24, 2016, https://www.npr.org/transcripts/470824303.
15. Jerome Socolovsky, "Sweden Won't Put Public Funds Into Volvo, Saab," *Morning Edition*, NPR, Sept. 21, 2009, https://www.npr.org/templates/story/story.php?storyId=113018114.

CHAPTER SIX

1. Emily Gersema, "The Quality of Audio Influences Whether You Believe What You Hear," USC Today, April 17, 2018, https://news.usc.edu/141042/why-we-believe-something-audio-sound-quality/.
2. Ari Daniel, "Badly Damaged Ukrainian Hospital Struggles To Provide Emergency Services," *Morning Edition*, NPR, April 7, 2022, https://www.npr.org/2022/04/07/1091382698/badly-damaged-ukrainian-hospital-struggles-to-provide-emergency-services.

3. Andrea Hsu, "As Holidays Near, Nationwide Rail Strike Is Still on the Table. Here's the Latest," *Weekend Edition Sunday*, NPR, Nov. 17, 2022, https://www.npr.org/transcripts/1136459343.
4. Andrea Hsu, "Gas Prices Rise as People Return to the Office—and Their Commute," *Weekend Edition Saturday*, NPR, March, 12, 2022, https://www.npr.org/2022/03/12/1086274125/gas-prices-and-return-to-office.
5. Ari Shapiro, "The Addiction Crisis in New Hampshire Shapes Presidential Primary Votes," *All Things Considered*, NPR, Feb. 10, 2020, https://www.npr.org/2020/02/10/804616745/the-addiction-crisis-in-new-hampshire-shapes-presidential-primary-votes?ft=nprml&f=804616745.
6. Rhitu Chatterjee, "Researchers Track the Pandemic's Toll on Health Workers' Mental Health," *Morning Edition*, NPR, March 28, 2022, https://www.npr.org/2022/03/28/1089121175/researchers-track-the-pandemics-toll-on-health-workers-mental-health.
7. Isabel Wilkerson, "Interviewing Sources," *Nieman Reports* 56, no. 1 (Spring 2002): 16.
8. Eyder Peralta, "Uganda's Museveni Faces Tough Challenge in Presidential Election," *Morning Edition*, NPR, Jan. 12, 2021, https://www.npr.org/2021/01/12/955938674/ugandas-museveni-faces-tough-challenge-in-presidential-election.
9. Eyder Peralta, "Behind the Humanitarian Crisis Caused by the Civil War in Ethiopia," *All Things Considered*, NPR, May 11, 2021, https://www.npr.org/2021/05/11/995942182/humanitarian-crisis-out-of-the-civil-war-in-ethiopia.
10. Steve Inskeep, "Returning to St. Bernard Parish," *Morning Edition*, NPR, Oct. 10, 2005, https://www.npr.org/templates/story/story.php?storyId=4952440.
11. Andrea Hsu, "Daycare Is Costly in the U.S.—So Is Biden's Plan To Fix It," *All Things Considered*, NPR, Aug. 12, 2021, https://www.npr.org/transcripts/1026549127.
12. Susan Stamberg, "Helen Hayes, Honored Actress, Profiled," *Morning Edition*, NPR, May 7, 1990.
13. Rhitu Chatterjee, "Farming Got Hip in Iran Some 12,000 Years Ago, Ancient Seeds Reveal," *Morning Edition*, NPR, July 5, 2013, https://www.npr.org/transcripts/198453031.
14. Elissa Nadworny, "Movies, Worksheets, Computer Time: Inside LA Schools During the Teacher Strike," *Morning Edition*, NPR, Jan. 16, 2019, https://www.npr.org/transcripts/685777410.
15. Adrian Florido, "The Evolving Grief of Uvalde Residents," *All Things Considered*, NPR, June 19, 2022, https://www.npr.org/2022/06/19/1106168912/the-evolving-grief-of-uvalde-residents.

16. Nathan Rott, "To Conserve Vast Areas of Land, Biden Needs Help From Private Landowners," *Morning Edition*, NPR, Sept. 21, 2021, https://www.npr.org/2021/09/21/1039191120/to-conserve-vast-areas-of-land-biden-needs-help-from-private-landowners.

CHAPTER SEVEN

1. Jeffrey Gottfried, Mason Walker and Amy Mitchell, *Americans' Views of the News Media During the COVID-19 Outbreak*, sec. 2, "Americans Are More Negative in Their Broader Views of Journalists Than They Are Toward COVID-19 Coverage," Pew Research Center, May 8, 2020, https://www.pewresearch.org/journalism/2020/05/08/americans-are-more-negative-in-their-broader-views-of-journalists-than-they-are-toward-covid-19-coverage/; Matthew S. Schwartz, "1 in 4 Americans Say Violence Against the Government Is Sometimes OK," *Morning Edition*, NPR, Jan. 31, 2022, https://www.npr.org/2022/01/31/1076873172.
2. Caroline recommends the anti-doxxing guide written by *New York Times* employees: Kristen Kozinski and Neena Kapur, "How To Dox Yourself on the Internet: A Step-By-Step Guide To Finding and Removing Your Personal Information From the Internet," Medium, Feb. 27, 2020, https://open.nytimes.com/how-to-dox-yourself-on-the-internet-d2892b4c5954.
3. Alice Park, "Why Working at Night Boosts the Risk of Early Death," *Time*, Jan. 7, 2015, https://time.com/3657434/night-work-early-death/.

CHAPTER EIGHT

1. Steve Inskeep and Sacha Pfeiffer, "The Biggest Banks in the U.S. Are Stepping In To Save First Republic Bank," interview of David Gura, *Morning Edition*, March 17, 2023, https://www.npr.org/2023/03/17/1164146742/the-biggest-banks-in-the-u-s-are-stepping-in-to-save-first-republic-bank.
2. Geoff Brumfiel, "BRUMFIEL #1 NASA UAP," NPR (newscast), May 31, 2023.
3. Here's a podcast exploring this idea: Rob Rosenthal, "Radio Is a Visual Medium," *Sound School*, Transom, Dec. 12, 2017, https://transom.org/2017/radio-visual-medium/
4. Geoff Brumfiel, "NASA Officials Say Its Asteroid Defense Test Was a Success," *Morning Edition*, NPR, Oct. 12, 2022, https://www.npr.org/2022/10/12/1128312531/nasa-officials-say-its-asteroid-defense-test-was-a-success?ft=nprml&f=1128312531.
5. A Martínez, "Inflation Is Coming Down but the Fed Isn't About To Declare Victory Just Yet," interview of Scott Horsley, *Morning Edition*, NPR, Feb. 1, 2023, https://www.npr.org/transcripts/1153150882.

CHAPTER NINE

1. A Martínez, "The 1st Black Woman To Pilot a Spacecraft Says Seeing Earth Was the Best Part," interview of Sian Proctor, *Morning Edition*, NPR, Sept. 24, 2021, https://www.npr.org/transcripts/1040353478.
2. Geoff Brumfiel, "As Saudi Arabia Builds a Nuclear Reactor, Some Worry About Its Motives," *All Things Considered*, May 6, 2019, https://www.npr.org/transcripts/719590408.
3. Sonari Glinton, "Stock Market's Sudden Correction Might Not Impact Most Americans," *All Things Considered*, NPR, Aug. 27, 2015, https://www.npr.org/2015/08/27/435273159/stock-markets-sudden-correction-might-not-impact-most-americans.
4. Joanna Kakissis, "Life on the Greek Island of Evia Will Never Be the Same After Catastrophic Fires," *All Things Considered*, NPR, Sept. 1, 2021, https://www.npr.org/2021/09/01/1033374422/life-on-the-greek-island-of-evia-will-never-be-the-same-after-catastrophic-fires.
5. Jerome Socolovsky, "Spanish Civil War Volunteers Revisit Battlegrounds," *All Things Considered*, NPR, Oct. 8, 2006, https://www.npr.org/templates/story/story.php?storyId=6221378.
6. Justine Yan and Gregory Warner, "May We Have This Dance?" *Rough Translation*, NPR, Jan. 12, 2022, https://www.npr.org/2021/12/22/1066965712/may-we-have-this-dance.
7. Adora Namigadde, "Pride in the Pews Encourages Black Churches To Welcome LGBTQ People," *All Things Considered*, NPR, May 15, 2023, https://www.npr.org/2023/05/15/1176294868/pride-in-the-pews-encourages-black-churches-to-welcome-lgbtq-people.
8. Robert Smith, "You Asked: How Do You Tell a Story in 3 Acts," NPR Training, Nov. 10, 2017, https://training.npr.org/2017/11/10/you-asked-how-do-you-tell-a-story-in-3-acts/. Much of this section is drawn from that article.
9. Frank Morris, "Javelin Missiles Are in Short Supply and Restocking Them Won't Be Easy," *Morning Edition*, NPR, May 27, 2022, https://www.npr.org/2022/05/27/1101701890/javelin-missiles-are-in-short-supply-and-restocking-them-won-t-be-easy.
10. Meghan Keane, "Kraftland," *Invisibilia*, NPR, Aug. 23, 2019, https://www.npr.org/2019/08/21/753114415/kraftland.
11. Roy Peter Clark, "15 Tips for Handling Quotes," Poynter, Sept. 2, 2015, https://www.poynter.org/reporting-editing/2015/15-tips-for-handling-quotes/.
12. Katie Peikes, "Farmland Prices Are Up Sharply. How Did It Get So Expensive?" *Morning Edition*, NPR, Sept. 7, 2022, https://www.npr.org

/2022/09/07/1121427428/farmland-prices-are-up-sharply-how-did-it-get-so-expensive.

13. Nell Greenfieldboyce, "How Mosquitoes Sniff Out Human Sweat To Find Us," *All Things Considered*, NPR, March 28, 2019, https://www.npr.org/transcripts/706838786.
14. Steve Inskeep and Rachel Martin, "New York City's Public Libraries Abolish Fines on Overdue Materials," *Morning Edition*, NPR, Oct. 7, 2021, https://www.npr.org/2021/10/07/1043938102/the-new-york-public-library-system-abolishes-fines-on-overdue-materials.
15. Jerome Socolovsky, "What To Do When People Don't Practice Social Distancing," *Morning Edition*, NPR, April 28, 2020, https://www.npr.org/sections/coronavirus-live-updates/2020/04/28/846684162/what-to-do-when-people-dont-practice-social-distancing.
16. Mary Louise Kelly and Audie Cornish, "British Company Tests a Jet Suit That Could Change Future of Emergency Care," *All Things Considered*, NPR, Oct. 8, 2020, https://www.npr.org/2020/10/08/921782166/british-company-tests-a-jet-suit-that-could-change-future-of-emergency-care.
17. Stephanie O'Neill, "Recent Supreme Court Rulings Encourage Some To Continue In-Person Worshiping," *All Things Considered*, NPR, Dec. 11, 2020, https://www.npr.org/2020/12/11/945578807/recent-supreme-court-rulings-encourage-some-to-continue-in-person-worshiping.
18. Julia Carpenter, "You Don't Have To Own Things To Be an Adult. So Why Do I Feel Like I Do?" *Wall Street Journal*, Feb. 18, 2023, https://www.wsj.com/articles/adulthood-finances-metrics-a4b00909.
19. Asma Khalid, "Biden Wants To Reshape the Economy by Investing in America, Not Unlike Trump," *Consider This*, NPR, April 13, 2023, https://www.npr.org/transcripts/1169841335.
20. Dan Charles, Nathan Rott and Berly McCoy, "Megadrought Fuels Debate Over Whether a Flooded Canyon Should Reemerge," *Short Wave*, Jan. 26, 2022, https://www.npr.org/2022/01/21/1074795925/megadrought-fuels-debate-over-whether-a-flooded-canyon-should-reemerge.
21. Elizabeth Blair, "After Touring With Beyoncé, Divinity Roxx Brings Positive Vibes to Children's Music," *Morning Edition*, NPR, Nov. 4, 2021, https://www.npr.org/transcripts/1051890641.
22. Steve Inskeep, "On Sept. 11, He Checked Hijackers Onto Flight 77. It's Haunted Him Ever Since," *Morning Edition*, NPR, Sept. 9, 2016, https://www.npr.org/transcripts/493133084.
23. Lee Hale and Brent Baughman, "The Lasting Toll of 9/11," *All Things Considered*, NPR, Sept. 10, 2021, https://www.npr.org/2021/09/10/1036039786/the-lasting-toll-of-9-11.

24. According to the U.S. Census Bureau, 67 million Americans speak more than one language. U.S. Census Bureau, "Selected Social Characteristics in the United States," American Community Survey, ACS 5-Year Estimates Data Profiles, table DP02, 2022, accessed Dec. 28, 2023, https://data.census.gov/table/ACSDP5Y2022.DP02.
25. Adrian Florido, "Latino Community in Orlando Bands Together in Wake of Massacre," *All Things Considered*, NPR, June 14, 2016, https://www.npr.org/2016/06/14/482055743/latino-community-in-orlando-bands-together-in-wake-of-massacre?ft=nprml&f=482055743.
26. Alison MacAdam, "Writing Through Sound: A Toolbox for Getting Into and out of Your Tape," NPR Training, June 23, 2015, https://training.npr.org/2015/06/23/writing-through-sound-a-toolbox-for-getting-in-and-out-of-tape/.
27. Alix Spiegel, "By Making a Game out of Rejection, a Man Conquers Fear," *Morning Edition*, NPR, Jan. 16, 2015, https://www.npr.org/sections/health-shots/2015/01/16/377239011/by-making-a-game-out-of-rejection-a-man-conquers-fear.
28. Justine Yan and Gregory Warner, "May We Have This Dance?" *Rough Translation*, NPR, Jan. 12, 2022, https://www.npr.org/2021/12/22/1066965712/may-we-have-this-dance.
29. Nick Fountain, James Sneed and Jess Jiang, "Charles Ponzi's Scheme," *Planet Money*, NPR, Jan. 20, 2023, https://www.npr.org/2023/01/20/1150332566/charles-ponzi-financial-scam.
30. Shankar Vedantam, Thomas Lu, Tara Boyle and Rhaina Cohen, "Finding Your Voice: How the Way We Sound Shapes Our Identities," *Hidden Brain*, NPR, July 15, 2019, https://www.npr.org/2019/07/15/741827437/finding-your-voice-how-the-way-we-sound-shapes-our-identities.
31. Cheryl Corley, "The Buffalo, N.Y., Community Holds Funerals This Week for Shooting Victims," *Weekend Edition Saturday*, NPR, May 21, 2022, https://www.npr.org/transcripts/1100532492.
32. Cheryl Corley, "Pink Cadillacs Will Line Up for Aretha Franklin's Funeral," *Morning Edition*, NPR, Aug. 29, 2018, https://www.npr.org/transcripts/642871634.
33. Jerome Socolovsky, "Lincoln Brigade," *Weekend Edition Sunday*, NPR, July 6, 2003, https://www.npr.org/templates/story/story.php?storyId=1321483.

CHAPTER TEN

1. Emily Harris, "The Mother Who Wouldn't Let a Teacher Shame Her 3-Year-Old," *Weekend Edition Saturday*, NPR, May 7, 2016, https://www.npr.org/sections/goatsandsoda/2016/05/07/477025837/the-mother-who-wouldnt-let-a-teacher-shame-her-3-year-old.

2. Jerome Socolovsky, "Spain's Special, Savory Ham Headed for U.S.," *Weekend Edition Saturday*, NPR, Feb. 9, 2008, https://www.npr.org/transcripts/18550917.
3. Gregory Warner and Catarina Fernandes Martins, "Stuck@Work: Your Country's Brand Is Escape, But You Can't," *Rough Translation*, NPR, June 15, 2022, https://www.npr.org/transcripts/1105201154.
4. Kelly McBride, "NPR's *Code Switch* Is an Overnight Sensation 7 Years in the Making," Poynter, Dec. 11, 2020, https://www.poynter.org/ethics-trust/2020/nprs-code-switch-is-an-overnight-sensation-7-years-in-the-making/.
5. Rachel Martin, "After 24 Years, Scholar Completes 3,000-Page Translation of the Hebrew Bible," interview of Robert Alter, *Morning Edition*, NPR, Jan. 14, 2019, https://www.npr.org/transcripts/684120470.

CHAPTER ELEVEN

1. Julie Rogers, "A Timeline of NPR's First 50 Years," NPR, April 28, 2021, https://www.npr.org/2021/04/28/987733236/a-timeline-of-nprs-first-50-years.
2. Debbie Elliott, "Wade Goodwyn, Longtime NPR Correspondent, Dies at Age 63," NPR, June 8, 2023, https://www.npr.org/2023/06/08/1167837454/wade-goodwyn-npr-correspondent-dies.
3. Chenjerai Kumanika, "Vocal Color in Public Radio," Transom, Jan. 22, 2015, https://transom.org/2015/chenjerai-kumanyika/.
4. Rachel Martin and David Greene, "Lawmakers Start Quoting Meat Loaf Lyrics Just Because," *Morning Edition*, NPR, Dec. 8, 2017, https://www.npr.org/transcripts/569365614.
5. U.S. Census Bureau, "Selected Social Characteristics in the United States," American Community Survey, ACS 5-Year Estimates Data Profiles, table DP02, 2022, accessed Dec. 28, 2023, https://data.census.gov/table/ACSDP5Y2022.DP02.
6. Ibid.

CHAPTER TWELVE

1. The figure reproduced in this chapter shows the final mix that Claudette did on a story by Fatma Tanis. You can find the story here: Fatma Tanis, "The Honey Industry in Yemen Is Feeling the Impacts of War and Climate Change," *All Things Considered*, NPR, Aug. 21, 2023, https://www.npr.org/2023/08/21/1195134501/the-honey-industry-in-yemen-is-feeling-the-impacts-of-war-and-climate-change.
2. Jerome Socolovsky, "Portuguese Fado and a Few Dirty Tables," *Morning Edition*, NPR, June 23, 2008, https://www.npr.org/transcripts/90961630.

3. "Clint Smith Reflects On This Moment," *TED Radio Hour*, NPR, June 5, 2020, https://www.npr.org/2020/06/05/869765344/clint-smith-reflects-on-this-moment?ft=nprml&f=869765344.
4. Karen Grigsby Bates and Kamna Shastri, "A Whiteness That's Only Skin Deep," *Code Switch*, NPR, Jan. 19, 2022, https://www.npr.org/transcripts/1073804771.
5. "Excellence," *NPR Ethics Handbook*, https://www.npr.org/templates/story/story.php?storyId=688414581, quoting Jonathan Kern, *Sound Reporting: The NPR Guide to Audio Journalism and Production* (Chicago: University of Chicago Press, 2008), 242.

CHAPTER THIRTEEN

1. Marc Brysbaert, "How Many Words Do We Read per Minute? A Review and Meta-analysis of Reading Rate," Journal of Memory and Language 109 (December 2019), article 104047, https://doi.org/10.1016/j.jml.2019.104047.
2. Rebecca Hersher, Ryan Kellman, Margaret Cirino and Gabriel Spitzer, "One Park. 24 Hours," *Short Wave*, NPR, Sept. 27, 2022, https://www.npr.org/transcripts/1124118206.
3. Frank Langfitt (@franklangfitt), "London lights up for the holidays," Instagram video, Dec. 24, 2022, https://www.instagram.com/reel/Cmjn1-BMdDX/?igsh=MTR0YzIwaGhvZG1haQ%3D%3D.
4. "Climate, Migration and the Far Right: How the Ripples of Climate Change Are Radiating Outward," NPR special series, https://www.npr.org/series/1131399786/climate-change-migration-far-right-africa-europe.
5. Emma Peaslee, "Longtime Anti-nuke Activists Face Prison, Again, After Breaking Into Naval Base," *Morning Edition*, NPR, Dec. 28, 2020, https://www.npr.org/transcripts/948116757.
6. Emma Peaslee, "Longtime Anti-nuke Activists Face Prison, Again, After Breaking Into Naval Base," *Morning Edition*, NPR, Dec. 28, 2020, https://www.npr.org/2020/12/28/948116757/longtime-anti-nuclear-activists-face-prison-again-after-breaking-into-naval-base.
7. Jeff Brady, "We Need To Talk About Your Gas Stove, Your Health and Climate Change," *Morning Edition*, NPR, Oct. 7, 2021, https://www.npr.org/transcripts/1015460605.
8. Rebecca Hersher, "Meteorologists Can't Keep Up With Climate Change in Mozambique," *Morning Edition*, NPR, Dec. 11, 2019, https://www.npr.org/transcripts/782918005.
9. Steven Mufson, "The Battle Over Climate Change Is Boiling Over on the Home Front," *Washington Post*, Feb. 23, 2021, https://www.washingtonpost.com/climate-environment/2021/02/23/climate-change-natural-gas/.

10. Brad Plumer and Hiroko Tabuchi, "How Politics Are Determining What Stove You Use," *New York Times*, Dec. 16, 2021, https://www.nytimes.com/2021/12/16/climate/gas-stoves-climate-change.html.
11. Jonathan Mingle, "Why Gas Stoves Are More Hazardous Than We've Been Led To Believe," *Slate*, Dec. 3, 2020, https://slate.com/technology/2020/12/gas-stoves-hazardous-asthma.html.

CHAPTER FOURTEEN

1. Leila Fadel, "Southern Baptists Expel Five Churches Because They Have Female Pastors," interview of Linda Barnes Popham, *Morning Edition*, NPR, March 1, 2023, https://www.npr.org/2023/03/01/1160297895/southern-baptists-expel-5-churches-because-they-have-female-pastors.
2. Andrea Seabrook, "U.N. Science Panel Sees Faster Warming of Earth," interview of Jerome Socolovsky, *All Things Considered*, NPR, Nov. 17, 2007, https://www.npr.org/2007/11/17/16393441/u-n-science-panel-sees-faster-warming-of-earth.
3. Ayesha Rascoe, "Tim Nelson of Cub Sport on Their New Album 'Jesus at the Gay Bar,'" interview of Tim Nelson, *Weekend Edition Sunday*, NPR, April 16, 2023, https://www.npr.org/2023/04/16/1170293948/tim-nelson-of-cub-sport-on-their-new-album-jesus-at-the-gay-bar.
4. Melissa Block, "Schubert's 'Winterreise' Paints Bleak Landscape for Bill T. Jones," interview of Bill T. Jones, *All Things Considered*, NPR, Jan. 21, 2014, https://www.npr.org/transcripts/143579090.
5. Leila Fadel, "Biden and Lawmakers Postpone Debt Ceiling Meeting as Their Staffs Keep Negotiating," interview of Nancy Mace, *Morning Edition*, NPR, May 12, 2023, https://www.npr.org/2023/05/12/1175711799/biden-and-lawmakers-postpone-debt-ceiling-meeting-as-their-staffs-keep-negotiati.
6. Leila Fadel, "The Debt Ceiling Compromise Isn't Sitting Well With Some Conservative Republicans," interview of Ralph Norman, *Morning Edition*, NPR, May 30, 2023, https://www.npr.org/2023/05/30/1178773376/the-debt-ceiling-compromise-isnt-sitting-well-with-some-conservative-republicans.
7. Asma Khalid, "Bozoma Saint John Writes of Love, Loss and Survival in 'The Urgent Life,'" interview of Bozoma Saint John, *Morning Edition*, NPR, Feb. 16, 2023, https://www.npr.org/2023/02/16/1157398933/bozoma-saint-john-writes-of-love-loss-and-survival-in-the-urgent-life.
8. Asma Khalid, "A Debt Ceiling Clash Isn't New: Ex-treasury Secretary Lew Looks Back on Negotiations," interview of Jack Lew, *Morning Edition*, NPR, Feb. 15, 2023, https://www.npr.org/2023/02/15/1157114156/a-debt-ceiling-clash-isnt-new-ex-treasury-secretary-lew-looks-back-on-negotiatio.

9. Rhitu Chatterjee, Emily Kwong and Rebecca Ramirez, "Can the Next School Shooting Be Prevented With Compassion?" *Short Wave*, NPR, June 16, 2022, https://www.npr.org/transcripts/1105082501.
10. Steve Inskeep and Daniel Estrin, "It's a Day of Grieving and More Protests in Gaza," *Morning Edition*, NPR, May 15, 2018, https://www.npr.org/2018/05/15/611213687/its-a-day-of-grieving-and-more-protests-in-gaza.

CHAPTER FIFTEEN

1. David Greene, "A Museum With Nearly 300 Brass Horns: You've Gotta See It Tuba-lieve It," *Morning Edition*, NPR, Aug. 30, 2016, https://www.npr.org/transcripts/491068442; Wynne Davis and Barry Gordemer, "Here's How a Videographer Reimagines Hollywood Blockbusters as Cat Videos," *Morning Edition*, NPR, March 16, 2022, https://www.npr.org/2022/03/15/1086605649/cat-videos-titanic-owlkitty.
2. Alexi Horowitz-Ghazi, Sarah Gonzalez, Willa Rubin and Keith Romer, "How the Cookie Became a Monster," *Planet Money*, NPR, Nov. 18, 2022, https://www.npr.org/2022/11/18/1137657496/third-party-cookie-data-tracking-internet-user-privacy.
3. Adam Bearne, "More Small Airports Are Being Cut Off From the Air Travel Network. This Is Why," *Morning Edition*, NPR, Sept. 4, 2023, https://www.npr.org/2023/09/04/1197337454/more-small-airports-are-being-cut-off-from-the-air-travel-network-this-is-why.
4. Alexi Horowitz-Ghazi, Sarah Gonzalez and Darian Woods, "Joke Theft," *Planet Money*, NPR, April 6, 2019, https://www.npr.org/transcripts/710404327.
5. Amy Mitchell, Galen Stocking and Katerina Eva Matsa, *Long-Form Reading Shows Signs of Life in Our Mobile News World*, Pew Research Center, May 4, 2016, https://www.pewresearch.org/journalism/2016/05/05/long-form-reading-shows-signs-of-life-in-our-mobile-news-world/.

INDEX

F

N

O

P

Q

R

S

ABOUT THE AUTHOR

Jerome Socolovsky reported for National Public Radio from Spain and Portugal and served as editor on *Morning Edition* and the national, international and culture desks. Since 2018, he has served as NPR's audio journalism trainer. During an award-winning career of more than three decades in digital, audio and video journalism, he has also been an international correspondent for the Associated Press, religion reporter for Voice of America and editor-in-chief of Religion News Service. He lives in Washington, D.C.

Printed and bound by CPI Group (UK) Ltd, Croydon, CR0 4YY

25/03/2025

14647343-0002